Encyclopedia of Electronic Components Volume 1

Charles Platt

Encyclopedia of Electronic Components Volume 1

by Charles Platt

Printed in Canada.

Published by Maker Media, Inc., 1005 Gravenstein Highway North, Sebastopol, CA 95472.

Maker Media books may be purchased for educational, business, or sales promotional use. Online editions are also available for most titles (*http://safaribooksonline.com*). For more information, contact O'Reilly Media's corporate/institutional sales department: 800-998-9938 or *corporate@oreilly.com*.

Editor: Brian Jepson
Production Editor: Melanie Yarbrough
Proofreader: Melanie Yarbrough
Indexer: Judy McConville
Cover Designer: Mark Paglietti
Interior Designer: Edie Freedman and Nellie McKesson
Illustrator: Charles Platt
Photographer: Charles Platt
Cover Production: Randy Comer

October 2012: First Edition

Revision History for the First Edition:

2012-10-03: First release

2012-12-04: Second release

2013-05-03: Third release

2013-12-20: Fourth release

2014-07-25: Fifth release

2015-01-09: Sixth release

See *http://oreilly.com/catalog/errata.csp?isbn=9781449333898* for release details.

ISBN: 978-1-449-33389-8

[TI]

To Mark Frauenfelder, who reacquainted me with the pleasures of Making.

Table of Contents

> > MODERATION

> DISCRETE SEMICONDUCTOR

> > SINGLE JUNCTION

Preface

At a time when information is widely and freely available in greater quantities than ever before, the reader may wonder whether *The Encyclopedia of Electronic Components* is really necessary. Surely, anything you want to know can be found online?

Well, yes and no. Let's consider the available resources.

1. Datasheets

Datasheets are indispensable, but they have limitations. Some are detailed; others are skimpy. Some show you sample schematics as a guide to using a component; many don't. None of them tells you much about how a component works, because that's not their purpose. Often they don't mention other components that must be added. Some datasheets for DC-DC converters, for instance, say nothing at all about bypass capacitors, even though the capacitors may be essential. A datasheet for an optocoupler says nothing about the pullup resistor required by the open-collector output.

Datasheets don't facilitate comparison shopping. A datasheet from one manufacturer will not compare its products with those from another manufacturer, and may not even provide much guidance about alternatives that are available from the same manufacturer. For example, a datasheet for a linear voltage regulator won't suggest that you might do better to use a DC-DC converter in an application where high efficiency is important.

Most of all, datasheets don't tell you how to avoid common mistakes. What actually happens if you connect that tantalum capacitor the wrong way around? A datasheet gives you the customary list of absolute maximum values, and after that, you are on your own, burning things out, encountering mysterious electronic behavior, and discovering limitations that are so well known, the datasheet didn't bother to mention them. In my experience, relying on datasheets creates a significant risk of reinventing the wheel.

2. Wikipedia

Wikipedia's coverage of electronics is impressive but inconsistent. Some entries are elementary, while others are extremely technical. Some are shallow, while others are deep. Some are well organized, while others run off into obscure topics that may have interested one of the contributors but are of little practical value to most readers. Many topics are distributed over multiple entries, forcing you to hunt through several URLs. Overall, Wikipedia tends to be good if you want theory, but not-so-good if you want hands-on practicality.

3. Manufacturers' Tutorials

A few helpful and enlightened manufacturers have compiled highly authoritative, instructional overviews of the components that they sell. Littelfuse, for instance, publishes an excellent series of documents telling you everything you could possibly want to know about fuses. But now you encounter a different problem: There is so much information, you'll need a couple of hours to dig through it all. Also, because the tutorials tend not to receive high page rankings on Google, they can be hard to find. And if a manufacturer has gaps in its product line, its tutorial is unlikely to mention them. Consequently, you won't know what's missing.

4. Personal Guides

It is a well-known attribute of the Web that many individuals feel the impulse to share everything they know (or think they know) about a particular topic. These personal guides can present surprisingly thorough online coverage of relatively obscure issues, such as the types of capacitors most suitable for loudspeaker crossover circuits, or the correct derivation of amp-hour numbers for lead-acid batteries. Unfortunately, on some sites you can also find errors, unsubstantiated opinions, plagiarism, and eccentricity. My general rule is that three or more guides generally have to agree with each other before their statements can be trusted—and even then, I have a small residue of doubt. The search-inspect-and-verify process can take a while.

So—yes, the information that you want usually does exist somewhere online, but no, it may not be easy to find. The vastness of the Web is not organized like an encyclopedia.

What about books? Generally speaking, they tend to be entry-level, or they specialize in narrow areas. A few broad-ranging books are truly excellent, but they are primarily educational, organized in an instructional sequence. They are not reference books.

The Encyclopedic Solution

Scarcity or inaccessibility of information ceased to be a problem many years ago. Its vast quantity, inconsistency, and dispersal have become the new barriers to acquiring knowledge. If you have to go hunting among datasheets, Wikipedia, manufacturers' tutorials (which may or may not exist), personal guides (which may have unrevealed bias), and multiple educational books, the process will be inconvenient and time-consuming. If you plan to revisit the topic in the future, you'll have to remember which URLs were useful and which ones weren't—and you may find that many of them are not even there anymore.

When I considered these issues during my own work as an electronics columnist for *Make* magazine, I saw a real need for a fact-checked, cross-referenced encyclopedia that would compile the basic information about components concisely, in an organized, consistent format, with informative photographs, schematics, and diagrams. It might save many people a lot of search time if it could summarize how components work, how to use them, what the alternatives are, and what the common errors and problems may be.

That is the modest ambition of *The Encyclopedia of Electronic Components*.

The Audience

Like any reference work, this one hopes to serve two categories of readers: The informed and the not-yet-informed.

Perhaps you are learning electronics, and you see a part listed in a catalog. It looks interesting, but the catalog doesn't tell you exactly what the part does or how it is commonly used. You need to look it up either by function or by name, but you're not sure where to start. An encyclopedic reference can simplify the fact-finding process, can save you from ordering a part that may be inappropriate, and can tell you how it should be used.

Perhaps, instead, you are an electronics engineer or hobbyist, thinking about a new circuit. You remember using a component three or four years ago, but your recollection may not be reliable. You need to refresh your memory with a quick summary—and so, you open the encyclopedia, just to make sure.

Completeness

Obviously, this book cannot include every component that exists. Mouser Electronics claims to have more than 2 million products listed in its online database. *The Encyclopedia of Electronic Components* only has room for a fraction of that number—but still, it can refer you to the primary types. The electronic edition of this book should allow easy insertions and updates. My hope is that it can become an ever-expanding resource.

Acknowledgments

Any reference work draws inspiration from many sources, and this one is no exception. Three were of special importance:

Practical Electronics for Inventors by Paul Scherz (second edition) McGraw-Hill, 2007

Electronic Devices and Circuit Theory by Robert L. Boylestad and Louis Nashelsky (ninth edition) Pearson Education Inc., 2006

The Art of Electronics by Paul Horowitz and Winfield Hill (second edition) Cambridge University Press, 2006

I also made extensive use of information gleaned through Mouser Electronics and Jameco Electronics. And where would any of us be without *Getting Started in Electronics* by Forrest M. Mims III, or *The TTL Cookbook* by Don Lancaster?

In addition, there were individuals who provided special assistance. My editor, Brian Jepson, was immensely helpful in the development of the project. Michael Butler contributed greatly to the early concept and its structure. Josh Gates did resourceful research. My publishers, Maker Media, demonstrated their faith in my work. Kevin Kelly unwittingly influenced me with his legendary interest in "access to tools."

Primary fact checkers were Eric Moberg, Chris Lirakis, Jason George, Roy Rabey, Emre Tuncer, and Patrick Fagg. I am indebted to them for their help. Any remaining errors are, of course, my responsibility.

Lastly I should mention my school friends from decades ago: Hugh Levinson, Patrick Fagg, Graham Rogers, William Edmondson, and John Witty, who helped me to feel that it was okay to be a nerdy kid building my own audio equipment, long before the word "nerd" existed.

—Charles Platt, 2012

1 How to Use This Book

To avoid misunderstandings regarding the purpose and method of this book, here is a quick guide regarding the way in which it has been conceived and organized.

Reference vs. Tutorial

As its title suggests, this is a reference book, not a tutorial. In other words, it does not begin with elementary concepts and build sequentially toward concepts that are more advanced.

You should be able to dip into the text at any point, locate the topic that interests you, learn what you need to know, and then put the book aside. If you choose to read it straight through from beginning to end, you will not find concepts being introduced in a sequential, cumulative manner.

My book *Make:Electronics* follows the tutorial approach. Its range, however, is more circumscribed than that of this encyclopedia, because a tutorial inevitably allocates a lot of space to step-by-step explanations and instructions.

Theory and Practice

This book is oriented toward practicality rather than theory. I am assuming that the reader mostly wants to know how to use electronic components, rather than why they work the way they do. Consequently I have not included any proofs of formulae, any definitions rooted in electrical theory, or any historical background. Units are defined only to the extent that is necessary to avoid confusion.

Many books on electronics theory already exist, if theory is of interest to you.

Organization

The encyclopedia is divided into entries, each entry being devoted to one broad type of component. Two rules determine whether a component has an entry all to itself, or is subsumed into another entry:

1. A component merits its own entry if it is (a) widely used or (b) not-so-widely used but has a unique identity and maybe some historical status. A widely used component would be a **bipolar transistor**, while a not-so-widely-used component with a unique identity would be a **unijunction transistor**.
2. A component does not merit its own entry if it is (a) seldom used or (b) very similar in function to another component that is more widely used. For example, the *rheostat* is subsumed into the **potentiometer** section, while *silicon diode*, *Zener diode*, and *germa-*

nium diode are combined together in the **diode** entry.

Inevitably, these guidelines required judgment calls that in some cases may seem arbitrary. My ultimate decision was based on where I would expect to find a component if I was looking for it myself.

Subject Paths

Entries are not organized alphabetically. Instead they are grouped by subject, in much the same way that books in the nonfiction section of a library are organized by the Dewey Decimal System. This is convenient if you don't know exactly what you are looking for, or if you don't know all the options that may be available to perform a task that you have in mind.

Each primary category is divided into subcategories, and the subcategories are divided into component types. This hierarchy is shown in Figure 1-1. It is also apparent when you look at the top of the first page of each entry, where you will find the path that leads to it. The **capacitor** entry, for instance, is headed with this path:

power > moderation > capacitor

Any classification scheme tends to run into exceptions. You can buy a chip containing a *resistor array*, for instance. Technically, this is an *analog integrated circuit*, but should it really be included with solid-state relays and comparators? A decision was made to put it in the **resistor** section, because this seemed more useful.

Some components have hybrid functions. In Volume 2, in the *integrated circuit* subcategory, we will distinguish between those that are *analog* and those that are *digital*. So where should an **analog-digital converter** be listed? It will be found under *analog*, because that category seems better associated with its primary function, and people may be more likely to look for it there.

Primary Category	Secondary Category	Component Type
power	source	battery
	connection	jumper
		fuse
		pushbutton
		switch
		rotary switch
		rotational encoder
	moderation	relay
		resistor
		potentiometer
		capacitor
		variable capacitor
	conversion	inductor
		AC-AC transformer
		AC-DC power supply
		DC-DC converter
		DC-AC inverter
	regulation	voltage regulator
electro-magnetism	linear output	electromagnet
		solenoid
	rotational output	DC motor
		AC motor
		servo motor
		stepper motor
discrete semi-conductor	single junction	diode
		unijunction transistor
	multi-junction	bipolar transistor
		field-effect transistor

Figure 1-1. *The subject-oriented organization of categories and entries in this encyclopedia.*

Inclusions and Exclusions

There is also the question of what is, and what is not, a component. Is wire a component? Not for the purposes of this encyclopedia. How about a **DC-DC converter**? Because converters are now sold in small packages by component suppliers, they have been included as components.

Many similar decisions had to be made on a case-by-case basis. Undoubtedly, some readers will disagree with the outcome, but reconciling all the disagreements would have been impossible. Speaking personally, the best I could do was create a book that is organized in the way that would suit me best if I were using it myself.

Typographical Conventions

Throughout this encyclopedia, the names of components that have their own entries are presented in **bold type.** Other important electronics terms or component names are presented in *italics* where they first appear in any one section.

The names of components, and the categories to which they belong, are all set in lower-case type, except where a term is normally capitalized because it is an acronym or a trademark. *Trimpot*, for instance, is trademarked by Bourns, but *trimmer* is not. **LED** is an acronym, but *cap* (abbreviation for **capacitor**) is not.

Where formulae are used, they are expressed in a format that will be familiar to computer programmers but may be unfamiliar to others. The * (asterisk) symbol is used in place of a multiplication sign, while the / (slash symbol) is used to indicate division. Where pairs of parentheses are nested, the most deeply nested pair identifies the operations that should be performed first.

Volume Contents

Practical considerations relating to book length influenced the decision to divide *The Encyclopedia of Electronic Components* into three volumes. Each volume deals with broad subject areas as follows.

Volume 1

Power, electromagnetism, and discrete semiconductors.

The *power* category includes sources of power and methods to distribute, store, interrupt, and modify power. The *electromagnetism* category includes devices that exert force linearly, and others that create a turning force. *Discrete semiconductors* include the main types of diodes and transistors.

Volume 2

Integrated circuits, light sources, sound sources, heat sources, and high-frequency sources.

Integrated circuits are divided into analog and digital components. *Light sources* range from incandescent bulbs to LEDs and small display screens; some reflective components, such as liquid-crystal displays and e-ink, are also included. *Sound sources* are primarily electromagnetic.

Volume 3

Sensing devices.

The field of sensors has become so extensive, they easily merit a volume to themselves. *Sensing devices* include those that detect light, sound, heat, motion, pressure, gas, humidity, orientation, electricity, proximity, force, and radiation.

At the time of writing, volumes 2 and 3 are still in preparation, but their contents are expected to be as described above.

Safari® Books Online

Safari Books Online is an on-demand digital library that lets you easily search over 7,500 technology and creative reference books and videos to find the answers you need quickly.

With a subscription, you can read any page and watch any video from our library online. Read books on your cell phone and mobile devices. Access new titles before they are available for print, and get exclusive access to manuscripts in development and post feedback for the authors.

Copy and paste code samples, organize your favorites, download chapters, bookmark key sections, create notes, print out pages, and benefit from tons of other time-saving features.

Maker Media has uploaded this book to the Safari Books Online service. To have full digital access to this book and others on similar topics from MAKE and other publishers, sign up for free at *http://safaribooksonline.com*.

How to Contact Us

Please address comments and questions concerning this book to the publisher:

MAKE
1005 Gravenstein Highway North
Sebastopol, CA 95472
800-998-9938 (in the United States or Canada)
707-829-0515 (international or local)
707-829-0104 (fax)

MAKE unites, inspires, informs, and entertains a growing community of resourceful people who undertake amazing projects in their backyards, basements, and garages. MAKE celebrates your right to tweak, hack, and bend any technology to your will. The MAKE audience continues to be a growing culture and community that believes in bettering ourselves, our environment, our educational system—our entire world. This is much more than an audience, it's a worldwide movement that Make is leading—we call it the Maker Movement.

For more information about MAKE, visit us online:

MAKE magazine: *http://makezine.com/magazine/*
Maker Faire: *http://makerfaire.com*
Makezine.com: *http://makezine.com*
Maker Shed: *http://makershed.com/*

We have a web page for this book, where we list errata, examples, and any additional information. You can access this page at:

http://oreil.ly/encyc_electronic_comp_v1

To comment or ask technical questions about this book, send email to:

bookquestions@oreilly.com

For more information about our publications, events, and products, see our website at *http://makermedia.com*.

Find us on Facebook: *https://www.facebook.com/makemagazine*

Follow us on Twitter: *https://twitter.com/make*

Watch us on YouTube: *http://www.youtube.com/makemagazine*

battery 2

This entry covers electrochemical power sources. Electricity is most often generated electromagnetically, but since these sources cannot be classified as components, they are outside the scope of the encyclopedia. Electrostatic sources are excluded for similar reasons.

A battery is sometimes referred to as a *cell* or *power cell*, but can actually contain multiple cells, as defined in this entry. It used to be called an *accumulator* or a *pile*, but those terms are now archaic.

OTHER RELATED COMPONENTS

- **capacitor** (see Chapter 12)

What It Does

A battery contains one or more *electrochemical cells* in which chemical reactions create an electrical potential between two immersed terminals. This potential can be discharged as *current* passing through a *load*.

An electrochemical cell should not be confused with an *electrolytic cell*, which is powered by an external source of electricity to promote *electrolysis*, whereby chemical compounds are broken down to their constituent elements. An electrolytic cell thus consumes electricity, while an electrochemical cell produces electricity.

Batteries range in size from *button cells* to large *lead-acid* units that store power generated by solar panels or windmills in locations that can be off the grid. Arrays of large batteries can provide bridging power for businesses or even small communities where conventional power is unreliable. Figure 2-1 shows a 60KW, 480VDC self-watering battery array installed in a corporate data center, supplementing wind and solar sources and providing time-of-day peak shaving of energy usage. Each lead-acid battery in this array measures approximately 28″ × 24″ × 12″ and weighs about 1,000 lb.

Figure 2-1. *A battery array providing 60KW at 480VDC as backup for a corporate data center. (Photo by permission of Hybridyne Power Systems, Canada, Inc., and the Hybridyne group of companies. Copyright by Hybridyne, an internationally registered trademark of Hybridyne Power Systems Canada Inc. No right of further reproduction unless specifically granted by Hybridyne.)*

Schematic symbols for a battery are shown in Figure 2-2. The longer of the two lines represents the positive side of the battery, in each case. One way to remember this is by imagining that the longer line can be snipped in half so that the two segments can combine to form a + sign. Traditionally, multiple connected battery symbols indicate multiple cells inside a battery; thus the center symbols in the figure could indicate a 3V battery, while those on the right would indicate a voltage greater than 3V. In practice, this convention is not followed conscientiously.

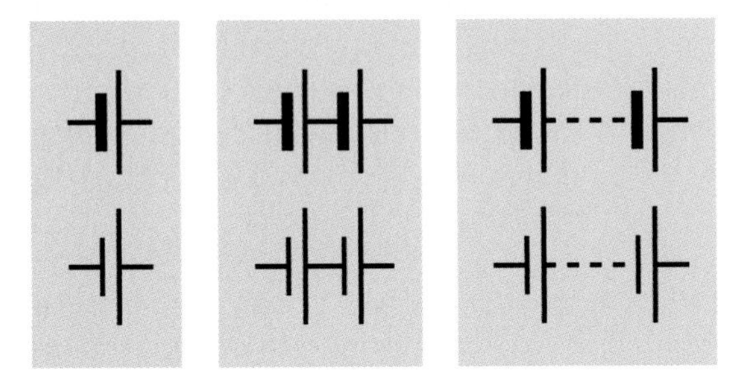

Figure 2-2. *Schematic symbols for a battery. Each pair of symbols within a blue rectangle is functionally identical.*

How It Works

In a basic battery design often used for demonstration purposes, a piece of copper serves as an *electrode*, partially immersed in a solution of copper sulfate, while a piece of zinc forms a second electrode, partially immersed in a solution of zinc sulfate. Each sulfate solution is known as an *electrolyte*, the complete battery may be referred to as a *cell*, and each half of it may be termed a *half-cell*.

A simplified cross-section view is shown in Figure 2-3. Blue arrows show the movement of electrons from the zinc terminal (the *anode*), through an external load, and into a copper terminal (the *cathode*). A *membrane separator* allows the electrons to circulate back through the battery, while preventing electrolyte mixing.

Orange arrows represent positive copper *ions*. White arrows represent positive zinc ions. (An ion is an atom with an excess or deficit of electrons.) The zinc ions are attracted into the zinc sulfate electrolyte, resulting in a net loss of mass from the zinc electrode.

Meanwhile, electrons passing into the copper electrode tend to attract positive copper ions, shown as orange arrows in the diagram. The copper ions are drawn out of the copper sulfate electrolyte, and result in a net accumulation of copper atoms on the copper electrode.

This process is energized partially by the fact that zinc tends to lose electrons more easily than copper.

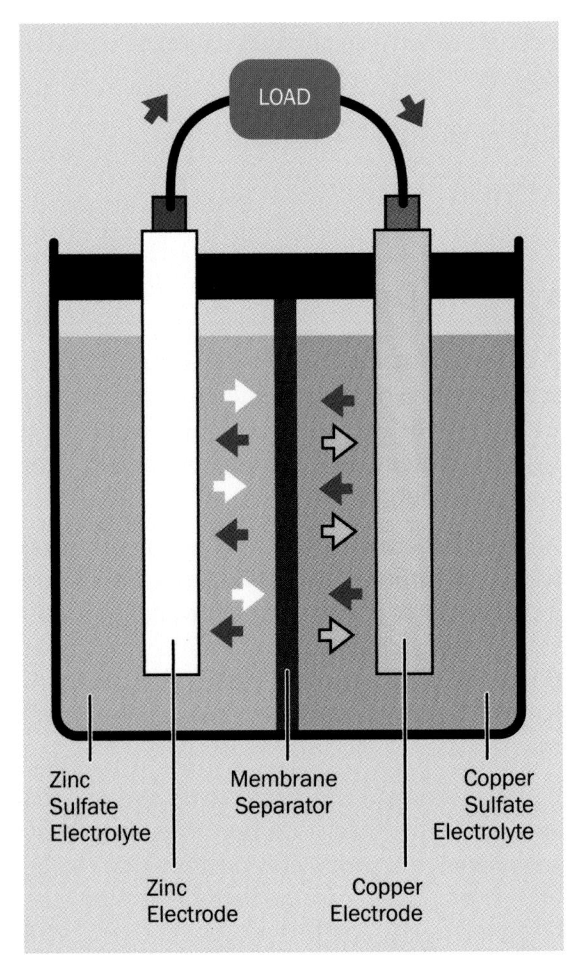

Figure 2-3. *A classically simple electrochemical cell. See text for additional details.*

Batteries for use in consumer electronics typically use a paste instead of a liquid as an electrolyte, and have been referred to as *dry cells*, although this term is becoming obsolete. The two half-cells may be combined concentrically, as in a typical 1.5-volt C, D, AA, or AAA alkaline battery (see Figure 2-4).

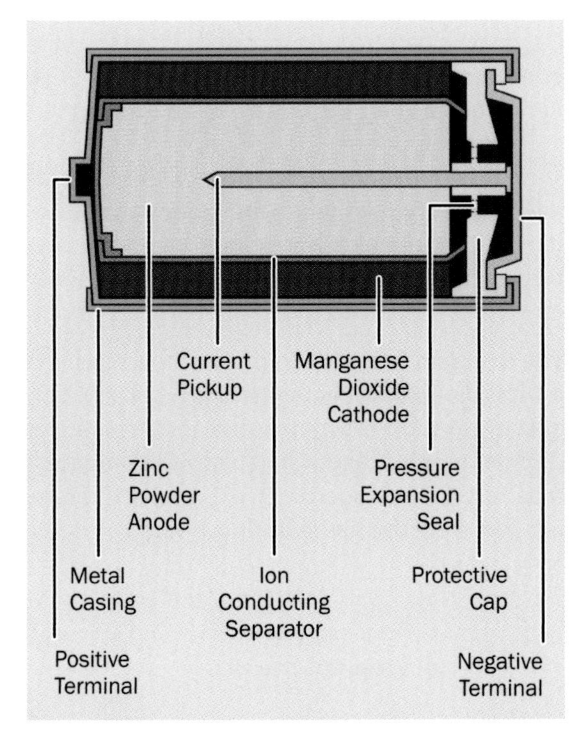

Figure 2-4. *Cross-section view of a typical 1.5-volt alkaline battery.*

A 1.5V battery contains one cell, while a 6V or 9V battery will contain multiple cells connected in series. The total voltage of the battery is the sum of the voltages of its cells.

Electrode Terminology

The electrodes of a cell are often referred to as the *anode* and the *cathode*. These terms are confusing because the electrons enter the anode inside the cell and leave it outside the cell, while electrons enter the cathode from outside the cell and leave it inside the cell. Thus, the anode is an electron emitter if you look at it externally, but the cathode is an electron emitter if you look at it internally.

Conventional current is imagined to flow in the opposite direction to electrons, and therefore, outside the cell, this current flows from the cathode to the anode, and from this perspective, the cathode can be thought of as being "more positive" than the anode. To remember this, think of the letter t in "cathode" as being a + sign, thus: ca+hode. In larger batteries, the cathode is often painted or tagged red, while the anode may be painted or tagged black or blue.

When a reusable battery is recharged, the flow of electrons reverses and the anode and the cathode effectively trade places. Recognizing this, the manufacturers of rechargeable batteries may refer to the more-positive terminal as the anode. This creates additional confusion, exacerbated further still by electronics manufacturers using the term "cathode" to identify the end of a **diode** which must be "more negative"(i.e., at a lower potential) than the opposite end.

To minimize the risk of errors, it is easiest to avoid the terms "anode" and "cathode" when referring to batteries, and speak instead of the negative and positive terminals. This encyclopedia uses the common convention of reserving the term "cathode" to identify the "more negative" end of any type of diode.

Variants

Three types of batteries exist.

1. *Disposable batteries*, properly (but infrequently) referred to as *primary cells*. They are not reliably rechargeable because their chemical reactions are not easily reversible.
2. *Rechargeable batteries*, properly (but infrequently) known as *secondary cells*. They can be recharged by applying a voltage between the terminals from an external source such as a *battery charger*. The materials used in the battery, and the care with which the battery is maintained, will affect the rate at which chemical degradation of the electrodes gradually occurs as it is recharged repeated-

ly. Either way, the number of charge/ discharge cycles is limited.

3. *Fuel Cells* require an inflow of a reactive gas such as hydrogen to maintain an electrochemical reaction over a long period. They are beyond the scope of this encyclopedia.

A large **capacitor** may be substituted for a battery for some applications, although it has a lower energy density and will be more expensive to manufacture than a battery of equivalent power storage. A capacitor charges and discharges much more rapidly than a battery because no chemical reactions are involved, but a battery sustains its voltage much more successfully during the discharge cycle. See Figure 2-5.

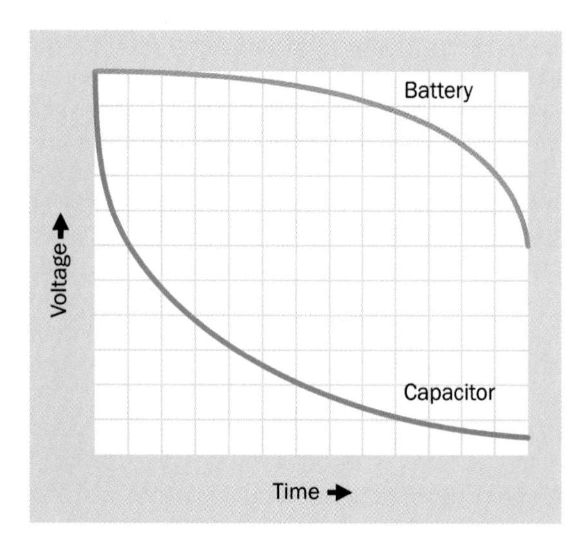

Figure 2-5. *The voltage drop of a discharging capacitor is much steeper initially than that of a battery, making capacitors unsuitable as a battery substitute in many applications. However, the ability of a capacitor to discharge very rapidly at high amperage can sometimes be a significant advantage.*

Capacitors that can store a very large amount of energy are often referred to as *supercapacitors.*

Disposable Batteries

The energy density of any disposable battery is higher than that of any type of rechargeable battery, and it will have a much longer shelf life because it loses its charge more slowly during storage (this is known as the *self-discharge rate*). Disposable batteries may have a useful life of five years or more, making them ideal for applications such as smoke detectors, handheld remotes for consumer electronics, or emergency flashlights.

Disposable batteries are not well suited to delivering high currents through loads below 75Ω. Rechargeable batteries are preferable for higher-current applications. The bar chart in Figure 2-6 shows the rated and actual capabilities of an alkaline battery relative to the three most commonly used rechargeable types, when the battery is connected with a resistance that is low enough to assure complete discharge in 1 hour.

The manufacturer's rating of watt hours per kilo is typically established by testing a battery with a relatively high-resistance load and slow rate of discharge. This rating will not apply in practice if a battery is discharged with a C-rate of 1, meaning complete discharge during 1 hour.

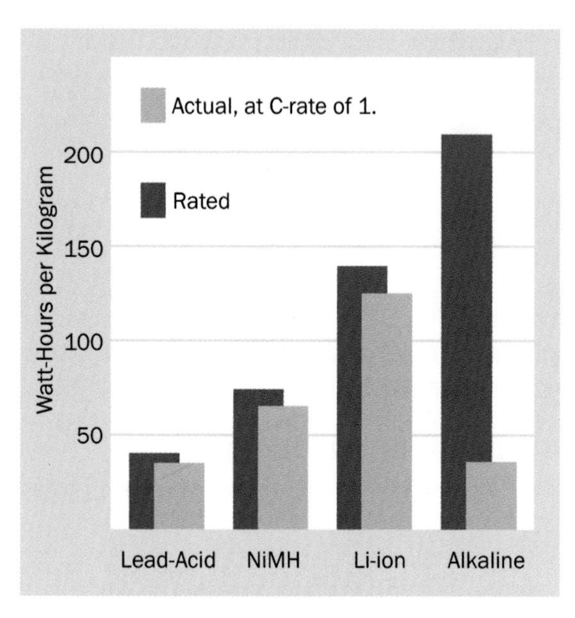

Figure 2-6. *Because of their relatively high internal resistance, alkaline batteries are especially unsuited to high discharge rates, and should be reserved for applications where a small current is required over a long period. (Chart derived from http://batteryuniversity.com.)*

Common types of disposable batteries are *zinc-carbon cells* and *alkaline cells*. In a zinc-carbon cell, the negative electrode is made of zinc while the positive electrode is made of carbon. The limited power capacity of this type of battery has reduced its popularity, but because it is the cheapest to manufacture, it may still be found where a company sells a product with "batteries included." The electrolyte is usually ammonium chloride or zinc chloride. The 9V battery in Figure 2-7 is actually a zinc-carbon battery according to its supplier, while the smaller one beside it is a 12V alkaline battery designed for use in burglar alarms. These examples show that batteries cannot always be identified correctly by a casual assessment of their appearance.

Figure 2-7. *At left, a cheap carbon-zinc battery; at right, a 12V alkaline burglar-alarm battery. See text for additional details.*

In an alkaline cell, the negative electrode is made of zinc powder, the positive electrode is manganese dioxide, and the electrolyte is potassium hydroxide. An alkaline cell may provide between three to five times the power capacity of an equal size of zinc-carbon cell and is less susceptible to voltage drop during the discharge cycle.

Extremely long shelf life is necessary in some military applications. This may be achieved by using a *reserve battery*, in which the internal chemical compounds are separated from each other but can be recombined prior to use.

Rechargeable Batteries

Commonly used types are *lead-acid*, *nickel cadmium* (abbreviated *NiCad* or *NiCd*), *nickel-metal hydride* (abbreviated *NiMH*), *lithium-ion* (abbreviated *Li-ion*), and *lithium-ion polymer*.

Lead-acid batteries have existed for more than a century and are still widely used in vehicles, burglar alarms, emergency lighting, and large power backup systems. The early design was described as *flooded*; it used a solution of sulfuric acid (generically referred to as *battery acid*) as its electrolyte, required the addition of distilled water periodically, and was vented to allow gas to escape. The venting also allowed acid to spill if the battery was tipped over.

The *valve-regulated lead-acid* battery (*VRLA*) has become widely used, requiring no addition of water to the cells. A pressure relief valve is included, but will not leak electrolyte, regardless of the position of the battery. VRLA batteries are preferred for *uninterruptible power supplies* for data-processing equipment, and are found in automobiles and in electric wheelchairs, as their low gas output and security from spillage increases their safety factor.

VRLA batteries can be divided into two types: absorbed glass mat (AGM) and gel batteries. The electrolyte in an AGM is absorbed in a fiber-glass mat separator. In a gel cell, the electrolyte is mixed with silica dust to form an immobilized gel.

The term *deep cycle battery* may be applied to a lead-acid battery and indicates that it should be more tolerant of discharge to a low level—perhaps 20 percent of its full charge (although manufacturers may claim a lower number). The plates in a standard lead-acid battery are composed of a lead *sponge*, which maximizes the surface area available to acid in the battery but can be physically abraded by deep discharge. In a deep cycle battery, the plates are solid. This means they are more robust, but are less able to supply high amperage. If a deep-discharge battery is used to start an internal combustion engine, the battery should be larger than a regular lead-acid battery used for this purpose.

A sealed lead-acid battery intended to power an external light activated by a motion detector is shown in Figure 2-8. This unit weighs several pounds and is trickle-charged during the day-time by a 6″ × 6″ solar panel.

Figure 2-8. *A lead-acid battery from an external light activated by a motion sensor.*

Nickel-cadmium (*NiCad*) batteries can withstand extremely high currents, but have been banned in Europe because of the toxicity of metallic cadmium. They are being replaced in the United States by *nickel-metal hydride* (*NiMH*) types, which are free from the *memory effect* that can prevent a NiCad cell from fully recharging if it has been left for weeks or months in a partially discharged state.

Lithium-ion and lithium-ion polymer batteries have a better energy-to-mass ratio than NiMH batteries, and are widely used with electronic devices such as laptop computers, media players, digital cameras, and cellular phones. Large arrays of lithium batteries have also been used in some electric vehicles.

Various small rechargeable batteries are shown in Figure 2-9. The NiCad pack at top-left was manufactured for a cordless phone and is rapidly becoming obsolete. The 3V lithium battery at top-right was intended for a digital camera. The three batteries in the lower half of the photograph are all rechargeable NiMH substitutes for 9V, AA, and AAA batteries. The NiMH chemistry results in the AA and AAA single-cell batteries being rated for 1.2V rather than 1.5V, but the manufacturer claims they can be substituted for 1.5V alkaline cells because NiMH units sustain their rated voltage more consistently over time. Thus, the output from a fresh NiMH battery may be comparable to that of an alkaline battery that is part-way through its discharge cycle.

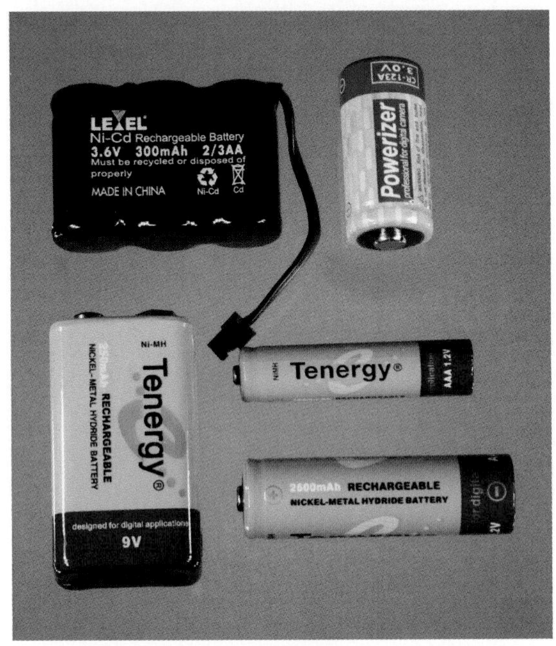

Figure 2-9. *Top left: NiCad battery pack for a cordless phone. Top right: Lithium battery for a digital camera. The other batteries are rechargeable NiMH substitutes for everyday alkaline cells.*

NiMH battery packs are available to deliver substantial power while being smaller and lighter than lead-acid equivalents. The NiMH package in Figure 2-10 is rated for 10Ah, and consists of ten D-size NiMH batteries wired in series to deliver 12VDC. This type of battery pack is useful in robotics and other applications where a small motor-driven device must have free mobility.

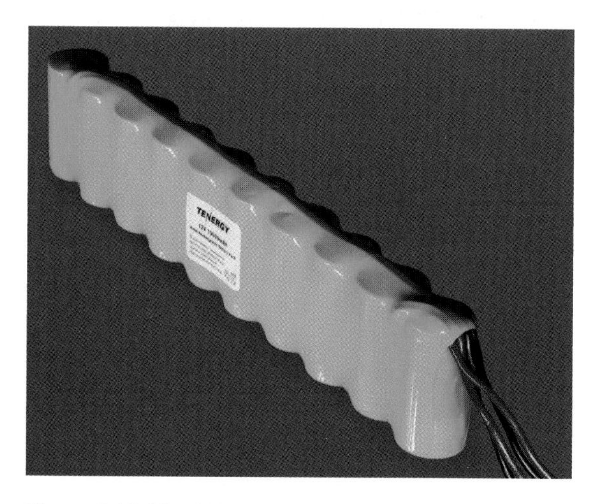

Figure 2-10. *This NiMH battery pack is rated at 10Ah and delivers 12 volts from ten D-size cells wired in series.*

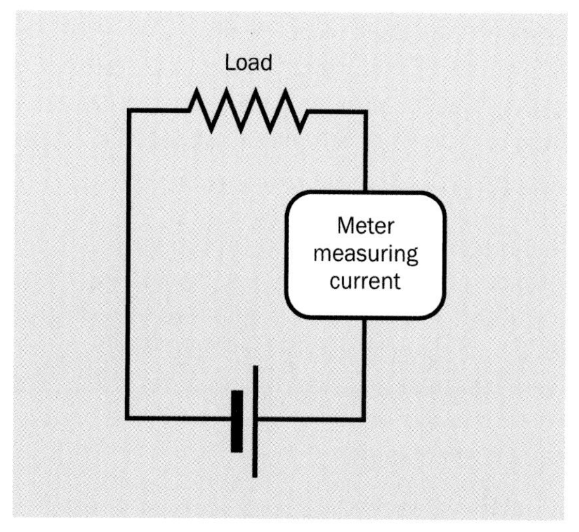

Figure 2-11. *When measuring current using an ammeter (or a multimeter configured to measure amps), the meter must be placed in series with the battery and a load. To avoid damaging the meter, it must never be applied directly across the terminals of the battery, or in parallel with a load. Be careful to observe the polarity of the meter.*

Values

Amperage

The current delivered by a battery will be largely determined by the resistance of the external load placed between its terminals. However, because ion transfer must occur inside the battery to complete the circuit, the current will also be limited by the *internal resistance* of the battery. This should be thought of as an active part of the circuit.

Since a battery will deliver no current if there is no load, current must be measured while a load is attached, and cannot be measured by a meter alone. The meter will be immediately overloaded, with destructive results, if it is connected directly between the terminals of a battery, or in parallel with the load. Current must always be measured with the meter in series with the load, and the polarity of the meter must correspond with the polarity of the battery. See Figure 2-11.

Capacity

The *electrical capacity* of a battery is measured in *amp-hours*, abbreviated *Ah*, *AH*, or (rarely) *A/H*. Smaller values are measured in *milliamp-hours*, usually abbreviated *mAh*. If I is the current being drawn from a battery (in amps) and T is the time for which the battery can deliver that current (in hours), the amp-hour capacity is given by the formula:

```
Ah = I * T
```

By turning the formula around, if we know the amp-hour rating that a manufacturer has determined for a battery, we can calculate the time in hours for which a battery can deliver a particular current:

```
T = Ah / I
```

Theoretically, Ah is a constant value for any given battery. Thus a battery rated for 4Ah should provide 1 amp for 4 hours, 4 amps for 1 hour, 5 amps for 0.8 hours (48 minutes), and so on.

In reality, this conveniently linear relationship does not exist. It quickly breaks down as the current rises, especially when using lead-acid batteries, which do not perform well when required to deliver high current. Some of the current is lost as heat, and the battery may be electrochemically incapable of keeping up with demand.

The *Peukert number* (named after its German originator in 1897) is a fudge factor to obtain a more realistic value for T at higher currents. If n is the Peukert number for a particular battery, then the previous formula can be modified thus:

$$T = Ah / I^n$$

Manufacturers usually (but not always) supply Peukert's number in their specification for a battery. So, if a battery has been rated at 4Ah, and its Peukert number is 1.2 (which is typical for lead-acid batteries), and I=5 (in other words, we want to know for how long a time, T, the battery can deliver 5 amps):

$$T = 4 / 5^{1.2} = \text{approximately } 4 / 6.9$$

This is about 0.58 hours, or 35 minutes—much less than the 48 minutes that the original formula suggested.

Unfortunately, there is a major problem with this calculation. In Peukert's era, the amp-hour rating for a battery was established by a manufacturer by drawing 1A and measuring the time during which the battery was capable of delivering that current. If it took 4 hours, the battery was rated at 4Ah.

Today, this measurement process is reversed. Instead of specifying the current to be drawn from the battery, a manufacturer specifies the time for which the test will run, then finds the maximum current the battery can deliver for that time. Often, the time period is 20 hours. Therefore, if a battery has a modern 4Ah rating, testing has probably determined that it delivered 0.2A for 20 hours, not 1A for 4 hours, which would have been the case in Peukert's era.

This is a significant distinction, because the same battery that can deliver 0.2A for 20 hours will not be able to satisfy the greater demand of 1A for 4 hours. Therefore the old amp-hour rating and the modern amp-hour rating mean different things and are incompatible. If the modern Ah rating is inserted into the old Peukert formula (as it was above), the answer will be misleadingly optimistic. Unfortunately, this fact is widely disregarded. Peukert's formula is still being used, and the performance of many batteries is being evaluated incorrectly.

The formula has been revised (initially by Chris Gibson of SmartGauge Electronics) to take into account the way in which Ah ratings are established today. Suppose that AhM is the modern rating for the battery's capacity in amp-hours, H is the duration in hours for which the battery was tested when the manufacturer calibrated it, n is Peukert's number (supplied by the manufacturer) as before, and I is the current you hope to draw from the battery. This is the revised formula to determine T:

$$T = H * (AhM / (I * H)^n)$$

How do we know the value for H? Most (not all) manufacturers will supply this number in their battery specification. Alternatively, and confusingly, they may use the term *C-rate*, which can be defined as 1/H. This means you can easily get the value for H if you know the C-rate:

$$H = 1 / \text{C-rate}$$

We can now use the revised formula to rework the original calculation. Going back to the example, if the battery was rated for 4Ah using the modern system, in a discharge test that lasted 20 hours (which is the same as a C-rate of 0.05), and the manufacturer still states that it has a Peukert number of 1.2, and we want to know for how long we can draw 5A from it:

$$T = 20 * (4/(5 * 20)^{1.2}) = \text{approximately } 20 * 0.021$$

This is about 0.42 hours, or 25 minutes—quite different from the 35 minutes obtained with the old version of the formula, which should never be used when calculating the probable discharge time based on a modern Ah rating. These issues may seem arcane, but they are of great importance when assessing the likely performance of battery-powered equipment such as electric vehicles.

Figure 2-12 shows the probable actual performance of batteries with Peukert numbers of 1.1, 1.2, and 1.3. The curves were derived from the revised version of Peukert's formula and show how the number of amp-hours that you can expect diminishes for each battery as the current increases. For example, if a battery that the manufacturer has assigned a Peukert number of 1.2 is rated at 100Ah using the modern 20-hour test, but we draw 30A from it, the battery can actually deliver only 70Ah.

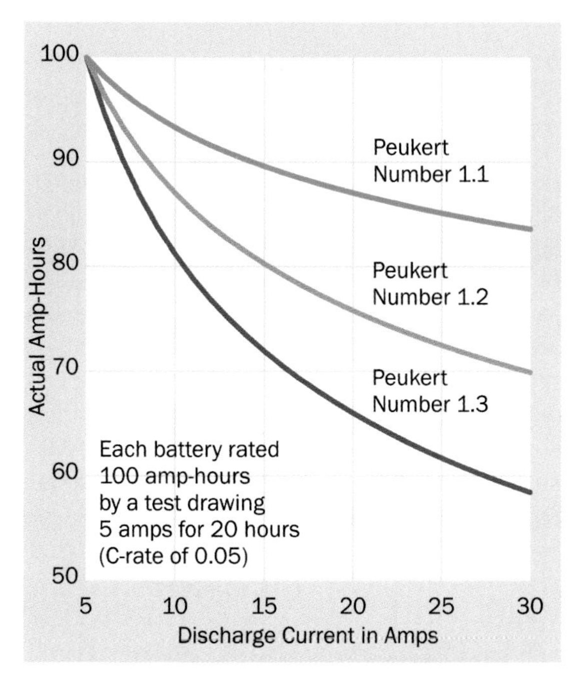

Figure 2-12. *Actual amp-hour performance that should be expected from three batteries of Peukert numbers 1.1, 1.2, and 1.3 when they discharge currents ranging from 5 to 30 amps, assuming that the manufacturer has rated each battery at 100Ah using the modern system, which usually entails a 20-hour test (a C-rate of 0.05).*

One additional factor: For any rechargeable battery, the Peukert number gradually increases with age, as the battery deteriorates chemically.

Voltage

The rated voltage of a fully charged battery is known as the *open circuit voltage* (abbreviated *OCV* or V_{oc}), defined as the potential that exists when no load is imposed between the terminals. Because the internal resistance of a volt meter (or a multimeter, when it is used to measure DC volts) is very high, it can be connected directly between the battery terminals with no other load present, and will show the OCV quite accurately, without risk of damage to the meter. A fully charged 12-volt car battery may have an OCV of about 12.6 volts, while a fresh 9-volt alkaline battery typically has an OCV of about 9.5 volts. Be extremely careful to set a multimeter to measure DC volts before connecting it across the battery. Usually this entails plugging the wire from the red probe into a socket separately reserved for measuring voltage, not amperage.

The voltage delivered by a battery will be pulled down significantly when a load is applied to it, and will decrease further as time passes during a discharge cycle. For these reasons, a **voltage regulator** is required when a battery powers components such as digital integrated circuit chips, which do not tolerate a wide variation in voltage.

To measure voltage while a load is applied to the battery, the meter must be connected in parallel with the load. See Figure 2-13. This type of measurement will give a reasonably accurate reading for the potential applied to the load, so long as the resistance of the load is relatively low compared with the internal resistance of the meter.

Figure 2-14 shows the performance of five commonly used sizes of alkaline batteries. The ratings in this chart were derived for alkaline batteries under favorable conditions, passing a small current through a relatively high-ohm load for long periods (40 to 400 hours, depending on battery type). The test continued until the final voltage for each 1.5V battery was 0.8V, and the final voltage for the 9V battery was a mere 4.8V. These voltages were considered acceptable when the Ah ratings for the batteries were calculated by the manufacturer, but in real-world situations, a final voltage of 4.8V from a 9V battery is likely to be unacceptable in many electronics applications.

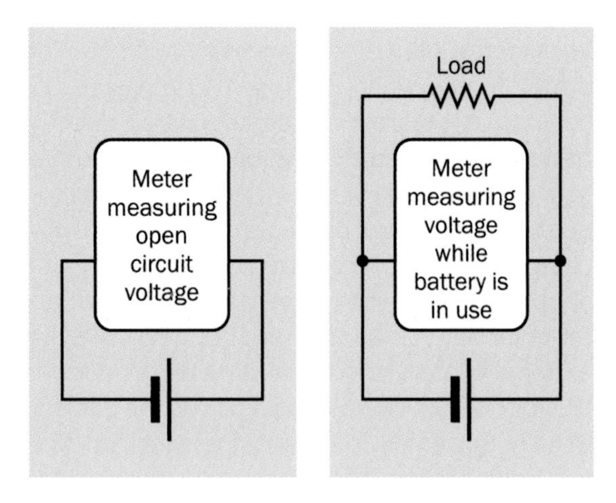

Figure 2-13. *When using a volt meter (or a multimeter configured to measure voltage), the meter can be applied directly between the battery terminals to determine the open-circuit voltage (OCV), or in parallel with a load to determine the voltage actually supplied during use. A multimeter must be set to measure DC volts before connecting it across a battery. Any other setting may damage the meter.*

Battery type	Rating (Ah)	Final voltage	Load (ohms)	Current (mA)
AAA	1.15	0.8	75	20
AA	2.87	0.8	75	20
C	7.8	0.8	39	40
D	17	0.8	39	40
9V	0.57	4.8	620	14

Figure 2-14. *The voltage delivered by a battery may drop to a low level while a manufacturer is establishing an amp-hour rating. Values for current, shown in the chart, were calculated subsequently as estimated averages, and should be considered approximate. (Derived from a chart published by Panasonic.)*

As a general rule of thumb, if an application does not tolerate a significant voltage drop, the manufacturer's amp-hour rating for a small battery may be divided by 2 to obtain a realistic number.

How to Use it

When choosing a battery to power a circuit, considerations will include the intended shelf life, maximum and typical current drain, and battery weight. The amp-hour rating of a battery can be used as a very approximate guide to determine its suitability. For 5V circuits that impose a drain of 100mA or less, it is common to use a 9V battery, or six 1.5V batteries in series, passing current through a **voltage regulator** such as the LM7805. Note that the voltage regulator requires energy to function, and thus it imposes a voltage drop that will be dissipated as heat. The minimum drop will vary depending on the type of regulator used.

Batteries or cells may be used in series or in parallel. In series, the total voltage of the chain of cells is found by summing their individual voltages, while their amp-hour rating remains the same as for a single cell, assuming that all the cells are identical. Wired in parallel, the total voltage of the cells remains the same as for a single cell, while the combined amp-hour value is found by summing their individual amp-hour ratings, assuming that all the batteries are identical. See Figure 2-15.

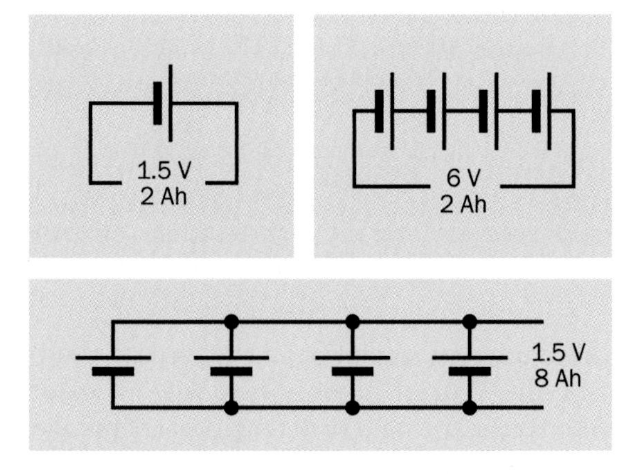

Figure 2-15. *Theoretical results of using 1.5V cells in series or in parallel, assuming a 2Ah rating for one cell.*

In addition to their obvious advantage of portability, batteries have an additional advantage of being generally free from power spikes and noise that can cause sensitive components to misbehave. Consequently, the need for smoothing will depend only on possible noise created by other components in the circuit.

Motors or other inductive loads draw an initial surge that can be many times the current that they use after they start running. A battery must be chosen that will tolerate this surge without damage.

Because of the risk of fire, United States airline regulations limit the amp-hour capacity of lithium-ion batteries in any electronic device in carry-on or checked passenger baggage. If a device may be carried frequently as passenger baggage (for example, emergency medical equipment), NiMH batteries are preferred.

What Can Go Wrong

Short Circuits: Overheating and Fire

A battery capable of delivering significant current can overheat, catch fire, or even explode if it is short-circuited. Dropping a wrench across the terminals of a car battery will result in a bright flash, a loud noise, and some molten metal. Even a 1.5-volt alkaline AA battery can become too hot to touch if its terminals are shorted together. (Never try this with a rechargeable battery, which has a much lower internal resistance, allowing much higher flow of current.) Lithium-ion batteries are particularly dangerous, and almost always are packaged with a current-limiting component that should not be disabled. A short-circuited lithium battery can explode.

If a battery pack is used as a cheap and simple workbench DC power supply, a **fuse** or circuit breaker should be included. Any device that uses significant battery power should be fused.

Diminished Performance Caused by Improper Recharging

Many types of batteries require a precisely measured charging voltage and a cycle that ends automatically when the battery is fully charged. Failure to observe this protocol can result in chemical damage that may not be reversible. A charger should be used that is specifically intended for the type of battery. A detailed comparison of chargers and batteries is outside the scope of this encyclopedia.

Complete Discharge of Lead-Acid Battery

Complete or near-complete discharge of a lead-acid battery will significantly shorten its life (unless it is specifically designed for deep-cycle use —although even then, more than an 80% discharge is not generally recommended).

Inadequate Current

Chemical reactions inside a battery occur more slowly at low temperatures. Consequently, a cold battery cannot deliver as much current as a warm battery. For this reason, in winter weather, a car battery is less able to deliver high current. At the same time, because engine oil becomes more viscous as the temperature falls, the starter motor will demand more current to turn the engine. This combination of factors explains the tendency of car batteries to fail on cold winter mornings.

Incorrect Polarity

If a battery charger or generator is connected with a battery with incorrect polarity, the battery may experience permanent damage. The **fuse** or *circuit breaker* in a charger may prevent this from occurring and may also prevent damage to the charger, but this cannot be guaranteed.

If two high-capacity batteries are connected with opposite polarity (as may happen when a clumsy attempt is made to start a stalled car with jumper cables), the results may be explosive. Never lean over a car battery when attaching cables to it, and ideally, wear eye protection.

Reverse Charging

Reverse charging can occur when a battery becomes completely discharged while it is wired (correctly) in series with other batteries that are still delivering current. In the upper section of the schematic at Figure 2-16 two healthy 6V batteries, in series, are powering a resistive load. The battery on the left applies a potential of 6 volts to the battery on the right, which adds its own 6 volts to create a full 12 volts across the load. The red and blue lines indicate volt meter leads, and the numbers show the reading that should be observed on the meter.

In the second schematic, the battery on the left has become exhausted and is now a "dead weight" in the circuit, indicated by its gray color. The battery on the right still sustains a 6-volt potential. If the internal resistance of the dead battery is approximately 1 ohm and the resistance of the load is approximately 20 ohms, the potential across the dead battery will be about 0.3 volts, in the opposite direction to its normal charged voltage. Reverse charging will result and can damage the battery. To avoid this problem, a battery pack containing multiple cells should never be fully discharged.

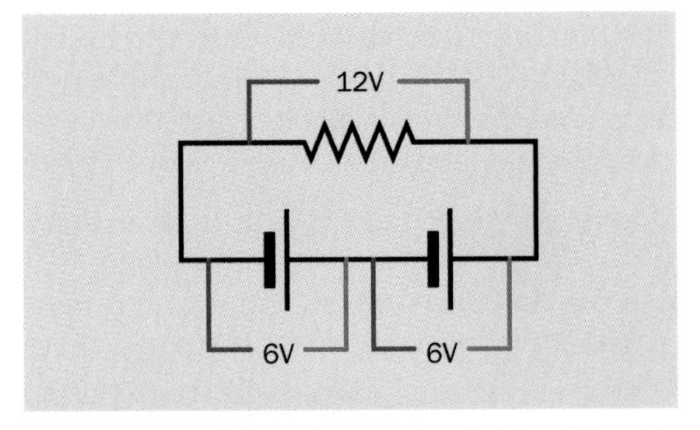

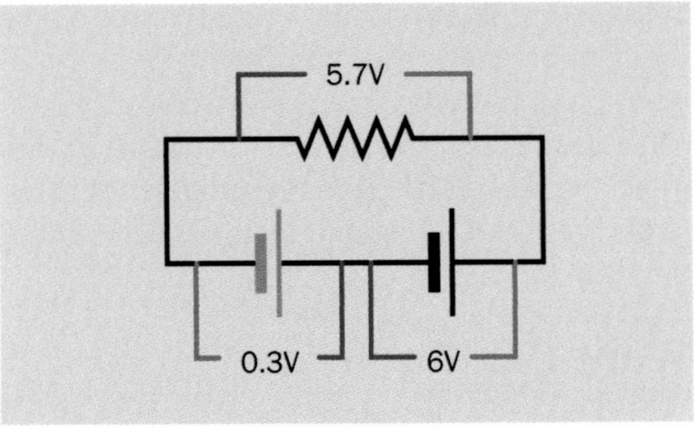

Figure 2-16. *When a pair of 6V batteries is placed in series to power a resistive load, if one of the batteries discharges completely, it becomes a load instead of a power source, and will be subjected to reverse charging, which may cause permanent damage.*

Sulfurization

When a lead-acid battery is partially or completely discharged and is allowed to remain in that state, sulfur tends to build up on its metal plates. The sulfur gradually tends to harden, forming a barrier against the electrochemical reactions that are necessary to recharge the battery. For this reason, lead-acid batteries should not be allowed to sit for long periods in a discharged condition. Anecdotal evidence suggests that even a very small trickle-charging current can prevent sulfurization, which is why some people recommend attaching a small solar panel to a battery that is seldom used—for example, on a sail boat, where the sole function of the battery is to start an auxiliary engine when there is insufficient wind.

High Current Flow Between Parallel Batteries

If two batteries are connected in parallel, with correct polarity, but one of them is fully charged while the other is not, the charged battery will attempt to recharge its neighbor. Because the batteries are wired directly together, the current will be limited only by their internal resistance and the resistance of the cables connecting them. This may lead to overheating and possible damage. The risk becomes more significant when linking batteries that have high Ah ratings. Ideally they should be protected from one another by high-current **fuses.**

jumper

3

A jumper may also be referred to as a *jumper socket* or a *shunt*. A jumper should not be confused with *jumper wires*, which are not considered components for the purposes of this encyclopedia.

OTHER RELATED COMPONENTS

- **switch** (See Chapter 6)

What It Does

A jumper is a low-cost substitute for a **switch,** where a connection has to be made (or unmade) only a few times during the lifetime of a product. Typically it allows a function or feature on a circuit board to be set on a semipermanent basis, often at the time of manufacture. A *DIP switch* performs the same function more conveniently. See "DIP" on page 43.

There is no standardized schematic symbol to represent a jumper.

How It Works

A jumper is a very small rectangular plastic tab containing two (or sometimes more) metal sockets usually spaced either 0.1" or 2mm apart. The sockets are connected electrically inside the tab, so that when they are pushed over two (or more) pins that have been installed on a circuit board for this purpose, the jumper shorts the pins together. The pins are usually 0.025" square and are often part of a *header* that is soldered into the board. In a parts catalogue, jumpers may be found in a section titled "Headers and Wire Housings" or similar.

Three jumpers are shown in Figure 3-1. The blue one contains two sockets spaced 0.1" and is deep enough to enclose the pins completely. The red one contains two sockets spaced 2mm and may allow the tips of the pins to emerge from its opposite end. The black one contains four sockets, each pair spaced 0.1" apart.

Figure 3-1. *Three jumpers containing two sockets spaced 2mm (left), two sockets spaced 0.1" (top right), and four sockets, each pair spaced 0.1" (bottom right).*

The set of pins with which a jumper is used is often referred to as a *header*. Headers are available with pins in single or dual rows. Some headers are designed to be snapped off to provide the desired number of pins. A dual 28-pin header is shown in Figure 3-2 with a black jumper pushed onto a pair of pins near the midpoint.

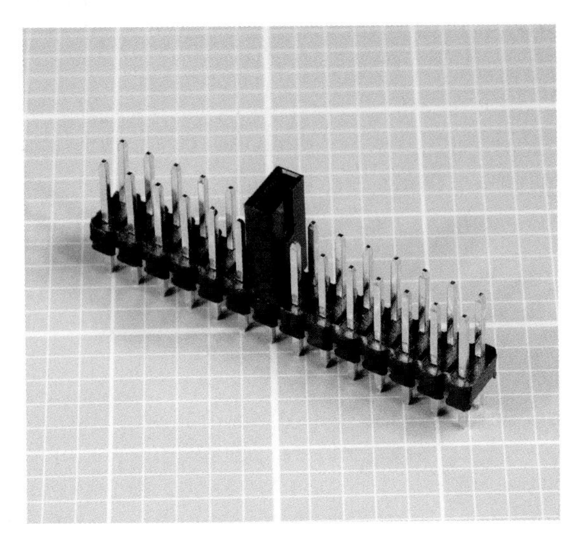

Figure 3-2. *A jumper pushed onto a pair of pins midway along a dual 28-pin header.*

Variants

A *jumper assembly* may be a kit containing not only the jumper but also the array of pins with which it is intended to be used. Check the manufacturer's datasheet to find out exactly what is included.

The most common types of jumpers have two sockets only, but variants are available with as many as 12 sockets, which may be arranged in one or two rows. *Header sockets* may be used as a substitute for purpose-made jumpers, with the advantage that they are often sold in long strips that can be snapped off to provide as many sockets as needed. However, the pins attached to header sockets must be manually connected by soldering small lengths of wire between them.

In some jumpers, the plastic tab extends upward for about half an inch and functions as a finger grip, making the jumper much easier to hold during insertion and removal. This is a desirable feature if there is room to accommodate it.

The sockets inside a jumper are often made from phosphor-bronze, copper-nickel alloy, tin alloy, or brass alloy. They are usually gold-plated, but in some instances are tin-plated.

Rarely, a jumper may consist of a metal strip with U-shaped connections suitable for being used in conjunction with screw terminals. Two jumpers of this type are shown in Figure 3-3. They should not be confused with high-amperage fuses that look superficially similar.

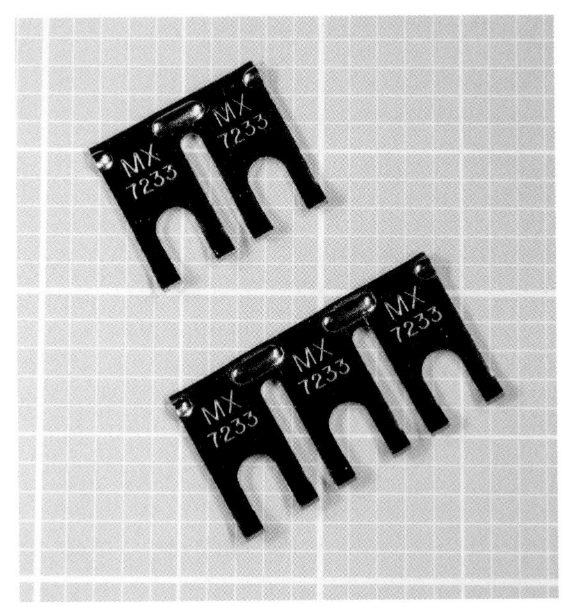

Figure 3-3. *These jumpers are designed to short together two or three screw terminals.*

Values

The spacing between the sockets in a jumper is referred to as its *pitch*. As previously noted, 0.1" and 2mm are the most popular values.

A typical maximum rating for a jumper of 0.1" pitch is 2A or 2.5A at 250V.

How to Use it

A jumper may activate a "set it and forget it" circuit function. An example would be the factory configuration of a product to work with 115VAC or 230VAC power input. End users were expected to set jumpers in some computer equipment sold during the 1980s, but this is no longer the case.

What Can Go Wrong

Jumpers are easily dropped, easily lost, and easily placed incorrectly. When purchasing jumpers, buy extras to compensate for their fragility and the ease of losing them.

Any location where a jumper may be used should be clearly labelled to define the function of each setting.

Cheap, poorly made jumpers may self-destruct from mechanical stresses when removed from their pins. The plastic casing can come away, leaving the sockets clinging naked to the pins protruding from the circuit board. This is another reason why it is a good idea to have a small stock of spare jumpers for emergencies.

Oxidation in jumpers where the contacts are not gold- or silver-plated can create electrical resistance or unreliable connections.

fuse 4

The alternate spelling *fuze* is seldom used.

OTHER RELATED COMPONENTS

- None

What It Does

A fuse protects an electrical circuit or device from excessive current when a metal element inside it melts to create an open circuit. With the exception of *resettable fuses* (discussed separately in "Resettable Fuses" on page 24), a fuse must be discarded and replaced after it has fulfilled its function.

When high current melts a fuse, it is said to *blow* or *trip* the fuse. (In the case of a resettable fuse, only the word *trip* is used.)

A fuse can work with either AC or DC voltage, and can be designed for almost any current. In residential and commercial buildings, *circuit breakers* have become common, but a large cartridge fuse may still be used to protect the whole system from short-circuits or from overcurrent caused by lightning strikes on exposed power lines.

In electronic devices, the **power supply** is almost always fused.

Schematic symbols for a fuse are shown in Figure 4-1. Those at the right and second from right are most frequently used. The one in the center is approved by ANSI, IEC, and IEEE but is seldom seen. To the left of that is the fuse symbol understood by electrical contractors in architectural plans. The symbol at far left used to be common but has fallen into disuse.

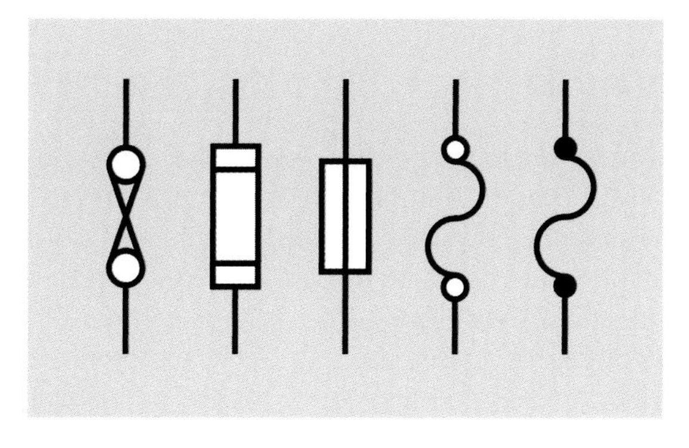

Figure 4-1. *Alternate schematic symbols for a fuse. See text for explanation.*

How It Works

The *element* in a fuse is usually a wire or thin metal strip mounted between two terminals. In a cartridge fuse, it is enclosed in a glass or ceramic cylinder with a contact at each end, or in a small metallic can. (Old-style, large, high-amperage fuses may be packaged in a paper or cardboard tube.) The traditional glass cartridge allows visual inspection to confirm that the fuse has blown.

A fuse responds only to current, not to voltage. When choosing a fuse that will be reliable in conditions of steady current consumption, a safe rule is to figure the maximum amperage when all components are functioning and add 50%. However, if current surges or spikes are likely, their duration will be relevant. If I is the current surge in amps and t is its duration in seconds, the surge sensitivity of a fuse—which is often referred to verbally or in printed format as I2t—is given by the formula:

```
I2t = I² * t
```

Some semiconductors also have an I2t rating, and should be protected with a similarly rated fuse.

Any fuse will present some resistance to the current flowing through it. Otherwise, the current would not generate the heat that blows the fuse. Manufacturer datasheets list the voltage drop that the internal resistance of a fuse is likely to introduce into a circuit.

Values

The *current rating* or *rated current* of a fuse is usually printed or stamped on its casing, and is the maximum flow that it should withstand on a continuous basis, at the ambient temperature specified by the manufacturer (usually 25 degrees Centigrade). The ambient temperature refers to the immediate environment of the fuse, not the larger area in which it may be located. Note that in an enclosure containing other components, the temperature is usually significantly higher than outside the enclosure.

Ideally a fuse should function reliably and indefinitely at its rated maximum amperage, but should blow just as reliably if the current rises by approximately 20% beyond the maximum. In reality, manufacturers recommend that continuous loading of a fuse should not exceed 75% of its rating at 25 degrees Centigrade.

The *voltage rating* or *rated voltage* of a fuse is the maximum voltage at which its element can be counted on to melt in a safe and predictable manner when it is overloaded by excess current. This is sometimes known as the *breaking capacity*. Above that rating, the remaining pieces of the fuse element may form an arc that sustains some electrical conduction.

A fuse can always be used at a lower voltage than its rating. If it has a breaking capacity of 250V, it will still provide the same protection if it is used at 5V.

Four differently rated glass cartridge fuses are shown in Figure 4-2. The one at the top is a slow-blowing type, rated at 15A. Its element is designed to absorb heat before melting. Below it is a 0.5A fuse with a correspondingly thinner element. The two smaller fuses are rated at 5A each. The center two fuses have a maximum voltage rating of 250V, while the one at the top is rated at 32V and the one at the bottom is rated at 350V. Clearly, the size of a fuse should never be used as a guide to its ratings.

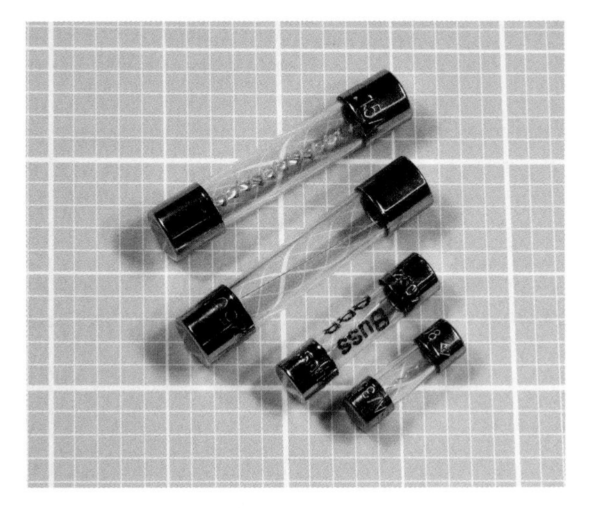

Figure 4-2. *Four glass cartridge fuses. See text for details.*

Variants

Early power fuses in residential buildings consisted of bare nichrome wire wrapped around a porcelain holder. In the 1890s, Edison developed *plug fuses* in which the fuse was contained in a porcelain module with a screw thread, compatible with the base of an incandescent bulb. This design persisted in some U.S. urban areas for more than 70 years, is still found in old buildings, and is still being manufactured.

Small Cartridge Fuses

Small cartridge fuses for appliances and electronics equipment—such as those shown in Figure 4-2--are available in sizes tabulated in Figure 4-3. With the exception of the 4.5mm diameter fuse (a European addition), these sizes were originally measured in inches; today, they are often described only with the equivalent metric measurement. Any cartridge fuse is usually available with the option of a lead attached to it at each end, so that it can be used as a through-hole component.

Fuse type	Diameter (inches)	Diameter (metric)	Length (inches)	Length (metric)
1AG	1/4"	6mm	5/8"	16mm
2AG	0.177"	4.5mm	0.588"	15mm
3AG	1/4"	6mm	1-1/4"	32mm
4AG	9/32"	7mm	1-1/4"	32mm
5AG	13/32"	10mm	1-1/2"	38mm
7AG	1/4"	6mm	7/8"	22mm
8AG	1/4"	6mm	1"	25mm

Figure 4-3. *The approximate physical sizes of commonly used small glass or ceramic cartridge fuses are shown here with the codes that are often used to identify them.*

Fuses may be fast acting, medium acting, or slow-blowing, the last of which may alternatively be referred to as *delay fuses*. Extra-fast–acting fuses are available from some manufacturers. The term *Slo-Blo* is often used but is actually a trademark of Littelfuse. None of the terms describing the speed of action of a fuse has been standardized with a specific time or time range.

Some cartridge fuses are available in a ceramic format as an alternative to the more common glass cylinder. If accidental application of extremely high current is possible (for example, in a multimeter that can be set to measure amps, and may be accidentally connected across a powerful battery), a ceramic cartridge is preferable because it contains a filler that will help to stop an arc from forming. Also, if a fuse is physically destroyed by application of very high current, ceramic fragments may be preferable to glass fragments.

Automotive Fuses

Automotive fuses are identifiable by their use of blades designed for insertion in flat sockets where the fuse is unlikely to loosen as a result of vibration or temperature changes. The fuses come in various sizes, and are uniformly color-coded for easy identification.

A selection of automotive fuses is shown in Figure 4-4. The type at the top is typically described as a "maxi-fuse" while the type at bottom-left is a "mini-fuse." Here again, size is irrelevant to function, as all three of those pictured are rated 30A at 32V.

Figure 4-4. *Three automotive fuses. All have the same rating: 30A at 32V.*

In Figure 4-5, the largest of the fuses from Figure 4-4 has been cut open to reveal its element.

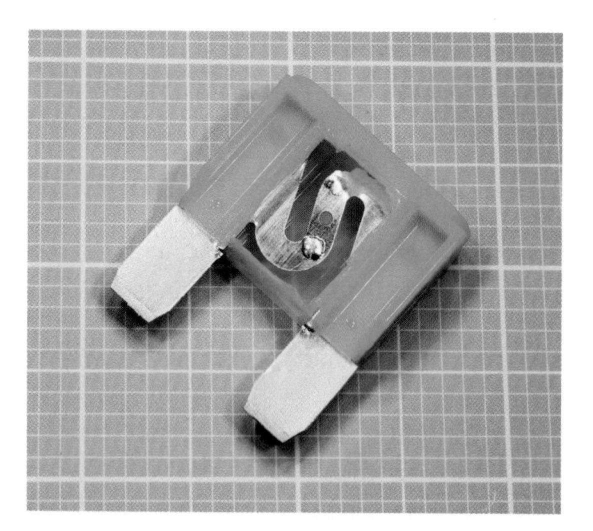

Figure 4-5. *The largest fuse from the previous figure, cut open to reveal its element.*

Usually automotive fuses are mounted together in a block, but if aftermarket accessory equipment is added, it may be protected by an inline fuse in a holder that terminates in two wires. This is shown with two sample fuses in Figure 4-6. Similar inline fuse holders are manufactured for other types of fuses.

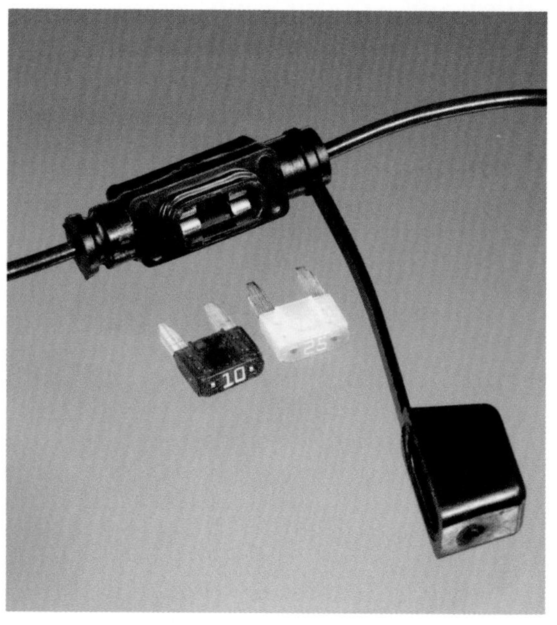

Figure 4-6. *Two blade-type fuses, commonly used for automotive applications, shown with an inline fuse holder. The plastic cap, at right, is closed over the holder when a fuse has been installed.*

Strip Fuses

High-amperage fuses for vehicles may be sold in "strip fuse" format, also known as a *fusible link*, designed to be clamped between two screw-down terminals. Since some **jumpers** may look very similar, it is important to keep them separate. A strip fuse is shown in Figure 4-7.

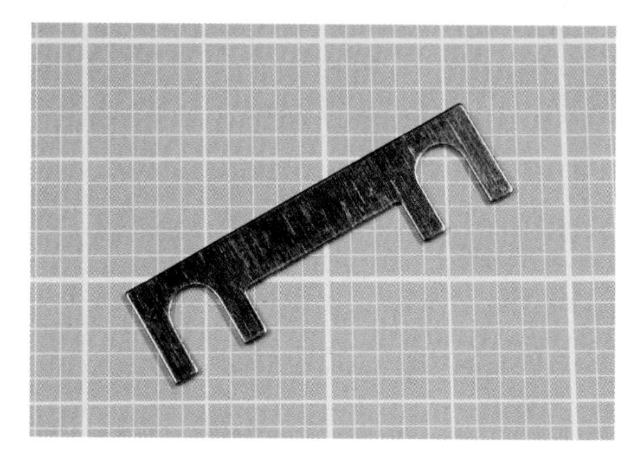

Figure 4-7. *This strip fuse is intended for use in diesel vehicles. The example shown is rated 100A at 36V.*

Through-Hole Fuses

Small fuses with radial leads, which seem appropriate for through-hole insertion in printed circuit boards, are actually often used in conjunction with appropriate sockets, so that they can be easily replaced. They are described in catalogues as "subminiature fuses" and are typically found in laptop computers and their power supplies, also televisions, battery chargers, and air conditioners. Three examples are shown in Figure 4-8. All have slow-blowing characteristics.

Resettable Fuses

Properly known as a *polymeric positive temperature coefficient* fuse (often abbreviated *PTC* or *PPTC*), a *resettable fuse* is a solid-state, encapsulated component that greatly increases its resistance in response to a current overload, but gradually returns to its original condition when the flow of current is discontinued. It can be thought

Figure 4-8. *Three subminiature fuses terminating in wire leads. From left to right: 10A at 250V, 2.5A at 250V, and 5A at 250V.*

of as a **thermistor** that has a nonlinear response. Three through-hole examples are shown in Figure 4-9. While different sizes of cartridge fuse may share the same ratings, differently rated resettable fuses may be identical in size. The one on the left is rated 40A at 30V, while the one on the right is rated 2.5A at 30V. (Note that the codes printed on the fuses are not the same as their manufacturer part numbers.) The fuse at the top is rated 1A at 135V.

When more than the maximum current passes through the fuse, its internal resistance increases suddenly from a few ohms to hundreds of thousands of ohms. This is known as *tripping* the fuse. This inevitably entails a small delay, but is comparable to the time taken for a slow-blowing fuse to respond.

A resettable fuse contains a polymer whose crystalline structure is loaded with graphite particles that conduct electricity. As current flowing through the fuse induces heat, the polymer transitions to an amorphous state, separating the graphite particles and interrupting the conductive pathways. A small current still passes through the component, sufficient to maintain its amorphous state until power is disconnected. After the resettable fuse cools, it gradually recrystallizes, although its resistance does not fall

Figure 4-9. *Some through-hole resettable fuses. See text for details.*

back completely to its original value for more than an hour.

The maximum safe level of current for a resettable fuse is known as the *hold current*, while the current that triggers its response is termed the *trip current*. Resettable fuses are available with trip-current ratings from 20mA to 100A. While conventional appliance and electronics fuses may be rated as high as 600V, resettable fuses are seldom rated above 100V.

Typical cartridge fuses are affected only to a minor extent by temperature, but the current rating of a resettable fuse may diminish to 75% of its normal value at 50 degrees Centigrade and may drop to 50% of its normal value at 80 degrees Centigrade. In other words, a fuse that is rated for 4A at 25 degrees may tolerate a maximum of only 3A when it operates at twice that temperature. See Figure 4-10.

Conventional slow-blowing fuses are temperature-sensitive, but to a lesser degree than resettable fuses.

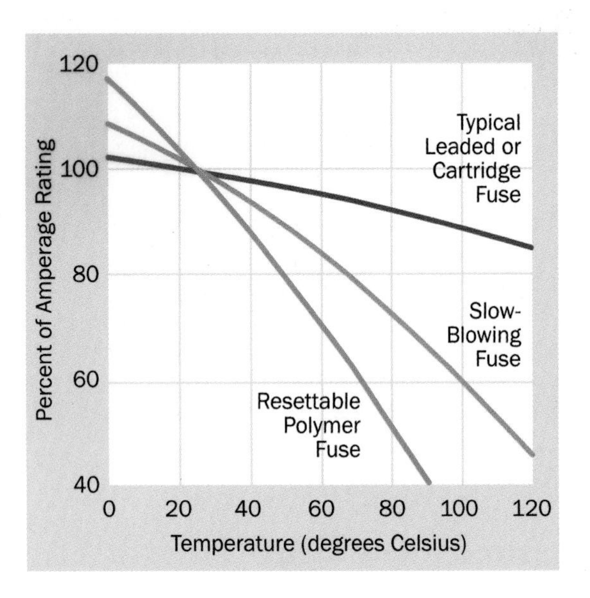

Figure 4-10. *The three curves give a very approximate idea of the temperature sensitivity of three types of fuses commonly used to protect electronic equipment. The left-hand scale provides an approximate value for the amperage which will trip the fuse.*

Resettable fuses are used in computer power supplies, USB power sources, and loudspeaker enclosures, where they protect the speaker coils from being overdriven. They are appropriate in situations where a fuse may be tripped relatively often, or where technically unsophisticated users may feel unable to replace a fuse or reset a circuit breaker.

Brand names for resettable fuses include PolySwitch, OptiReset, Everfuse, Polyfuse, and Multifuse. They are available in *surface-mount* (SMT) packages or as *through-hole* components, but not in cartridge format.

Surface Mount Fuses

Because surface-mount fuses are difficult or impossible to replace after they have been soldered onto the board, they are often resettable.

A surface-mount resettable fuse approximately 0.3" square is shown in Figure 4-11. It is rated for 230V and has an internal resistance of 50 ohms. Its hold current is 0.09A and its trip current is 0.19A.

Figure 4-11. *A surface-mount resettable fuse. See text for details.*

How to Use it

Any equipment that may be plugged into a wall outlet should be fused, not only to protect its components but also to protect users who may open the box and start investigating with a screwdriver.

Equipment that contains powerful motors, pumps, or other inductive loads should be protected with slow-blowing fuses, as the initial surge of current when the equipment is switched on is likely to rise well above the rating of the fuse. A slow-blowing fuse will tolerate a surge for a couple of seconds. Other fuses will not.

Conversely, fast-acting fuses should be used with electronic equipment, especially integrated circuits that are quickly and easily damaged.

Any device using substantial **battery** power should be fused because of the unpredictable and generally bad behavior of batteries when they are short-circuited. Parallel connections between multiple large batteries should be fused to avoid the possibility that a highly charged battery may attempt to recharge its neighbor(s). Large "J size" fuses rated from 125A to 450A have become common in the solar power community, where banks of lead-acid batteries are often used. These fuses have a thick brass tab at each end, drilled so that they can be bolted into place. Alternatively, they will push-fit into an appropriate *fuseholder*.

For cartridge fuses up to 1/4" in diameter that don't have leads attached, appropriately sized fuseholders are available in several formats:

Panel mounted fuse enclosure is probably the most common, consisting of a plastic tube with a spring-contact at the bottom, and a plastic cap with a second contact inside. The cap either screws onto the tube of the fuse, or is pushed down and turned to hold it in place. A nut is provided to secure the fuseholder after it has been inserted into a hole drilled in the panel. The fuse is dropped into the tube, and the cap is applied. This type of holder is available in full-length or shorter, "low profile" formats. A low-profile holder is shown in Figure 4-12. It is shown assembled at right, with its component parts disassembled alongside.

Figure 4-12. *A low-profile panel-mounted fuse holder shown disassembled (left) and assembled (right).*

Circuit board mounted fuse enclosure is basically the same as the panel-mounted version, but with through-hole solder pins attached.

Fuse block is a small plastic block with two clips on its upper surface for insertion of a cartridge fuse.

Fuse clips can be bought individually, with solder pins for through-hole mounting.

Inline fuse holder is designed to be inserted in a length of wire. Usually made of plastic, it will either terminate it, wires or will have metal contacts to crimp or solder at each end. See Figure 4-6.

Through-hole fuse holders are available for subminiature fuses.

What Can Go Wrong

Repeated Failure

When a fuse in a circuit blows frequently, this is known as *nuisance opening*. Often it can result from failure to take into account all the aspects of the circuit, such as a large filtering capacitor in a power supply that draws a major surge of current when the power supply is switched on. The formally correct procedure to address this problem is to measure the power surge, properly known as *peak inrush current*, with an oscilloscope, calculate the $I^2 * t$ of the wave form, and select a fuse with a rating at least 5 times that value.

A fuse should never be replaced with an equivalent length of wire or any other conductor.

Soldering Damage

When a through-hole or surface-mount fuse is soldered into place, heat from the soldering process can cause the soft metal element inside the fuse to melt partially and reflow. This is likely to change the rating of the fuse. Generally, fuses should be treated with the same caution as semiconductors when they are fixed in place with solder.

Placement

A fuse should be placed close to the power source or power input point in a circuit, so that it protects as much of the circuit as possible.

5 pushbutton

Often referred to as a *pushbutton switch* and sometimes as a *momentary switch*. In this encyclopedia, a pushbutton is considered separately from a **switch**, which generally uses a lever-shaped actuator rather than a button, and has at least one *pole* contact where a pushbutton generally has contacts that are not distinguishable from each other.

OTHER RELATED COMPONENTS

- **switch** (See Chapter 6)
- **rotary switch** (See Chapter 7)

What It Does

A pushbutton contains at least two contacts, which close or open when the button is pressed. Usually a spring restores the button to its original position when external pressure is released. Figure 5-1 shows schematic symbols for pushbuttons. The symbols that share each blue rectangle are functionally identical. At top is a normally-open single-throw pushbutton. At center is a normally-closed single-throw pushbutton. At bottom is a double-throw pushbutton.

Unlike a **switch**, a basic pushbutton does not have a primary contact that can be identified as the *pole*. However, a single pushbutton may close or open two separate pairs of contacts, in which case it can be referred to, a little misleadingly, as a double-pole pushbutton. See Figure 5-2. Different symbols are used for slider pushbuttons with multiple contact pairs; see "Slider" on page 31.

A generic full-size, two-contact pushbutton is shown in Figure 5-3.

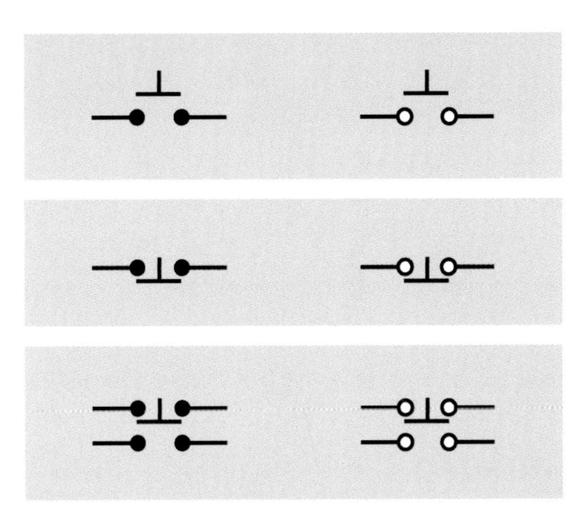

Figure 5-1. *Commonly used schematic symbols to represent a simple pushbutton. See text for details.*

How It Works

Figure 5-4 shows a cross-section of a pushbutton that has a single steel return spring, to create resistance to downward force on the button, and a pair of springs above a pair of contacts, to hold each contact in place and make a firm connection when the button is pressed. The two upper contacts are electrically linked, although this feature is not shown.

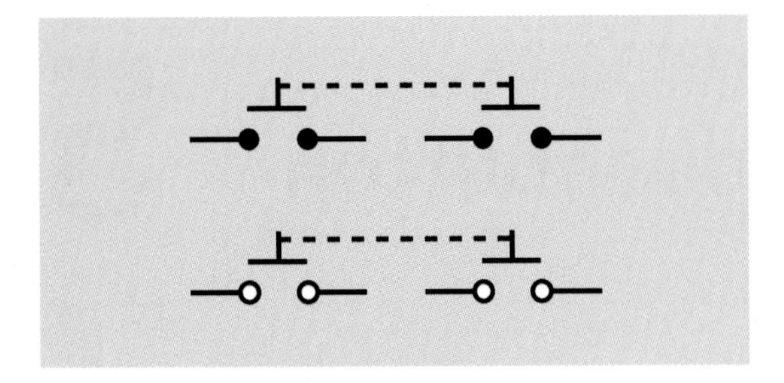

Figure 5-2. *Commonly used schematic symbols to represent a double-pole pushbutton.*

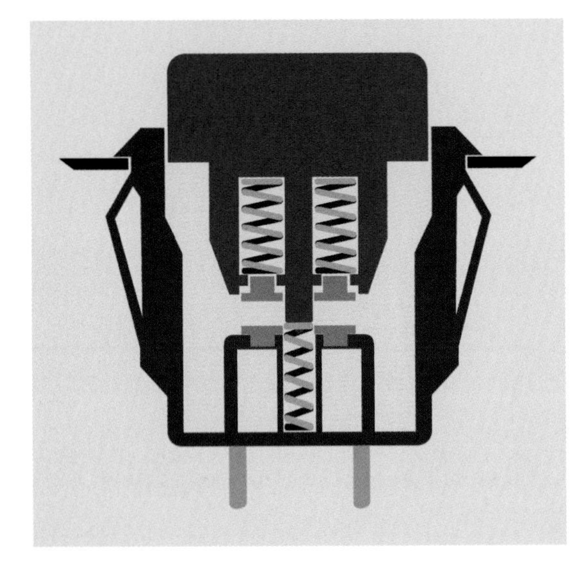

Figure 5-4. *Cross-section of a pushbutton showing two spring-loaded contacts and a single return spring.*

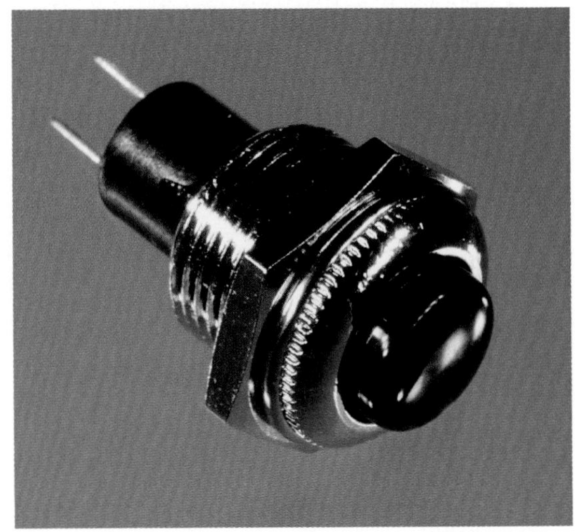

Figure 5-3. *The simplest, traditional form of pushbutton, in which pressing the button creates a connection between two contacts.*

Variants

Poles and Throws

Abbreviations that identify the number of poles and contacts inside a pushbutton are the same as the abbreviations that identify those attributes in a **switch**. A few examples will make this clear:

SPST, also known as 1P1T
: Single pole, single throw

DPST also known as 2P1T
: Double pole, single throw

SPDT also known as 1P2T
: Single pole, double throw

3PST also known as 3P1T
: Three pole, single throw

While a switch may have an additional center position, pushbuttons generally do not.

On-Off Behavior

Parentheses are used to indicate the momentary state of the pushbutton while it is pressed. It will return to the other state by default.

OFF-(ON) or (ON)-OFF
: Contacts are normally open by default, and are closed only while the button is pressed. This is sometimes described as a *make-to-make* connection, or as a *Form A* pushbutton.

ON-(OFF) or (OFF)-ON
: Contacts are normally closed by default, and are open only while the button is pressed. This is sometimes described as a *make-to-break* connection, or as a *Form B* pushbutton.

ON-(ON) or (ON)-ON
: This is a double-throw pushbutton in which one set of contacts is normally closed. When the button is pressed, the first set of contacts

is opened and the other set of contacts is closed, until the button is released. This is sometimes described as a *Form C* pushbutton.

For a single-throw pushbutton, the terms NC or NO may be used to describe it as *normally closed* or *normally open*.

Slider

This type, also known as a *slide* pushbutton, contains a thin bar or rod that slides in and out of a long, narrow enclosure. Contacts on the rod rub across secondary contacts inside the enclosure. Closely resembling a slider switch, it is cheap, compact, and well adapted for multiple connections (up to 8 separate poles in some models). However, it can only tolerate low currents, has limited durability, and is vulnerable to contamination.

A four-pole, double-throw pushbutton is shown in Figure 5-5. A variety of plastic caps can be obtained to press-fit onto the end of the white nylon actuator.

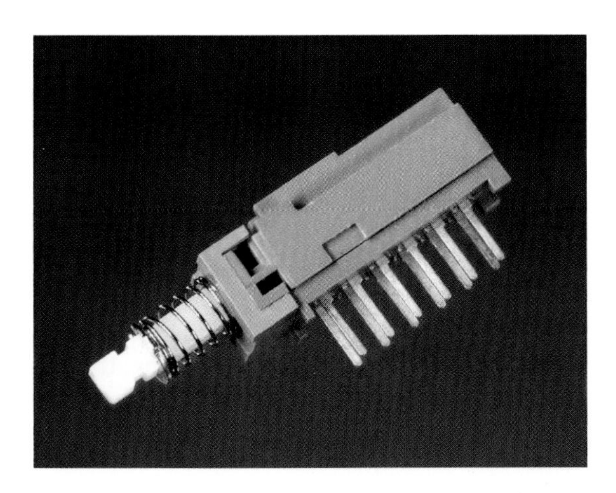

Figure 5-5. *A 4PDT slider pushbutton, shown without the cap that can be snapped onto the end of the actuator.*

Figure 5-6 shows schematic symbols for two possible slide pushbuttons, with a black rectangle indicating each sliding contact. The lead that functions as a pole is marked with a P in each case. Standardization for slide pushbutton schematic symbols does not really exist, but these examples are fairly typical. An insulating section that connects the sliding contacts internally is shown here as a gray rectangle, but in some datasheets may appear as a line or an open rectangle.

Since the symbols for a slide pushbutton may be identical to the symbols for a slide switch, care must be taken when examining a schematic, to determine which type of component is intended.

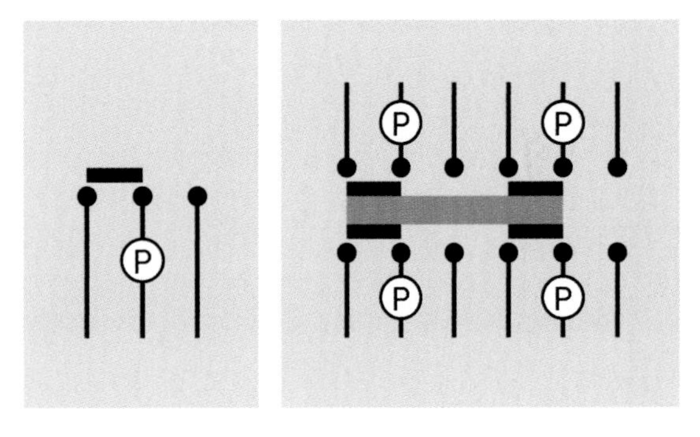

Figure 5-6. *Left: schematic symbol for a simple SPDT slide pushbutton, where a movable contact shorts together either the left pair or right pair of fixed contacts. Right: A 4PDT pushbutton in which the same principle has been extended. The movable contacts are attached to each other mechanically by an insulator. Each pole terminal is marked with a P.*

Styles

Many pushbutton switches are sold without caps attached. This allows the user to choose from a selection of styles and colors. Typically the cap is a push-fit onto the end of the rod or bar that activates the internal contacts. Some sample caps are shown in Figure 5-7, alongside a DPDT pushbutton. Any of the caps will snap-fit onto its actuator.

An *illuminated* pushbutton contains a small *incandescent bulb*, *neon bulb*, or **LED** (*light-emitting diode*). The light source almost always has its own two terminals, which are isolated from the other terminals on the button housing and can be wired to activate the light when the button is pressed, when it is released, or on some other basis. Pushbuttons containing LEDs usually

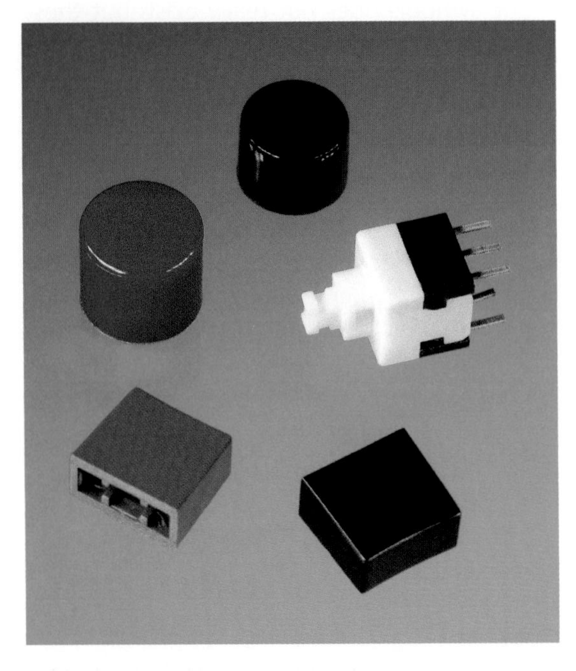

Figure 5-7. *Caps (buttons or knobs) that may be sold as separate accessories for some pushbuttons, shown here alongside a compatible pushbutton switch.*

require external series resistors, which should be chosen according to the voltage that will be used. See the **LED** entry in Volume 2 for additional commentary on appropriate series resistors. An example of an illuminated pushbutton is shown in Figure 5-8. This is a DPDT component, designed to be mounted on a printed circuit board, with an additional lead at each end connecting with an internal LED underneath the translucent white button.

Termination and Contact Plating

These options are the same as for a **switch** and are described in that entry.

Mounting Style

The traditional panel-mounted button is usually secured through a hole in the panel by tightening a nut that engages with a thread on the bushing of the pushbutton. Alternatively, a pushbutton housing can have flexible plastic protrusions on either side, allowing it to be snapped

Figure 5-8. *This pushbutton contains an LED underneath the white translucent button.*

into place in an appropriate-sized panel cutout. This style is shown in Figure 5-4.

PC pushbuttons (pushbuttons mounted in a printed circuit board, or PCB) are a common variant. After the component has been installed in the circuit board, either the button must align with a cutout in the front panel and poke through it when the device is assembled, or an external (non-electrical) button that is part of the product enclosure must press on the actuator of the pushbutton after assembly.

Surface-mount pushbuttons that allow direct fingertip access are uncommon. However, about one-quarter of tactile switches are designed for surface mount at the time of writing. They are typically found beneath membranes that the user presses to activate the switch beneath—for example, in remotes that are used to operate electronic devices.

Sealed or Unsealed

A sealed pushbutton will include protection against water, dust, dirt, and other environmental hazards, at some additional cost.

Latching

This variant, also known as a *press-twice* pushbutton, contains a mechanical ratchet, which is rotated each time the button is pressed. The first press causes contacts to latch in the closed state. The second press returns the contacts to the open state, after which, the process repeats. This press-twice design is typically found on flashlights, audio equipment, and in automotive applications. While *latching* is the most commonly used term, it is also known as *push-push*, *locking*, *push-lock push-release*, *push-on push-off*, and *alternate*.

In a latching pushbutton with *lockdown*, the button is visibly lower in the latched state than in the unlatched state. However, buttons that behave this way are not always identified as doing so on their datasheets.

A six-pole double-throw pushbutton that latches and then unlatches each time it is pressed is shown in Figure 5-9.

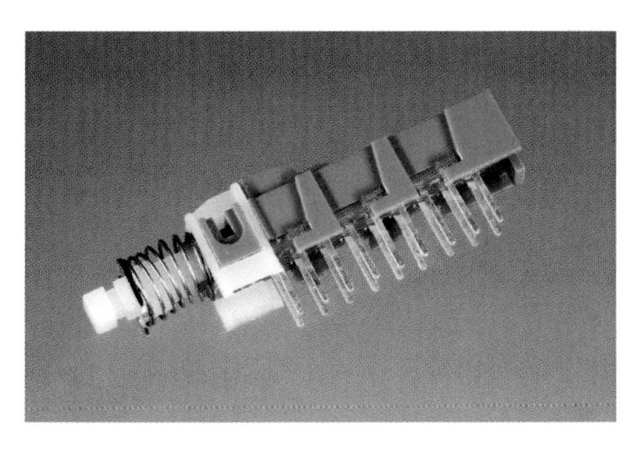

Figure 5-9. *This 6PDT pushbutton latches and then unlatches, each time it is pressed.*

Two more variants are shown in Figure 5-10. On the right is a simple DPDT latching pushbutton with lockdown. On the left is a latching pushbutton that cycles through four states, beginning with one "off" state, the remaining three connecting a different pair of its wires in turn.

A simple OFF-(ON) button may appear to have a latching output if it sends a pulse to a **microcontroller** in which software inside the microcontroller toggles an output between two states. The microcontroller can step through an unlimited number of options in response to each button press. Examples are found on cellular phones or portable media players.

Figure 5-10. *At right, a simple DPDT latching pushbutton with lockdown. At left, this pushbutton cycles through four states, one of them an "off" state, the others connecting a different pair of its wires in turn.*

A mechanically latching pushbutton has a higher failure rate than a simple OFF-(ON) button, as a result of its internal mechanism, but has the advantage of requiring no additional microcontroller to create its output. Microcontrollers are discussed in Volume 2.

Foot Pedal

Foot pedal pushbuttons generally require more actuation force than those intended for manual use. They are ruggedly built and are commonly found in vacuum cleaners, audio-transcription foot pedals, and "stomp boxes" used by musicians.

Keypad

A keypad is a rectangular array of usually 12 or 16 OFF-(ON) buttons. Their contacts are accessed via a *header* suitable for connection with a ribbon cable or insertion into a printed circuit board. In some keypads, each button connects with a separate contact in the header, while all the buttons share a common ground. More often, the buttons are *matrix encoded*, meaning that each of them bridges a unique pair of conductors in a matrix. A 16-button matrix is shown in Figure 5-11. This configuration is suitable for polling by a **microcontroller**, which can be programmed to send an output pulse to each of the four horizontal wires in turn. During each pulse, it checks the remaining four vertical wires in sequence, to determine which one, if any, is carrying a signal. Pullup or pulldown resistors should be added to the input wires to prevent the inputs of the microcontroller from behaving unpredictably when no signal is present. The external appearance of two keypads is shown in Figure 5-12.

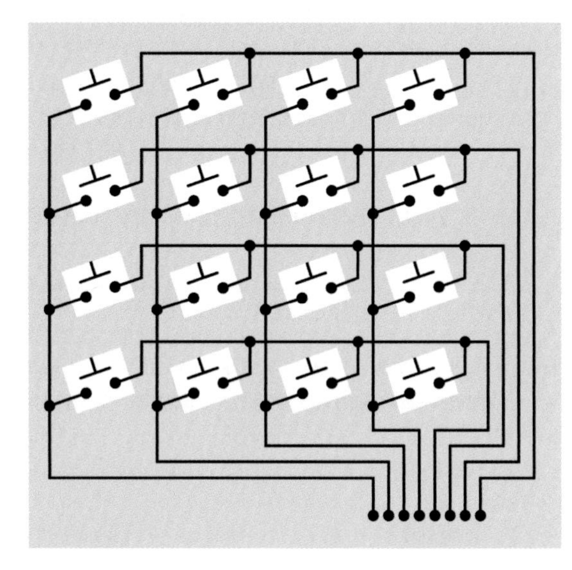

Figure 5-11. *Buttons in a numeric keypad are usually wired as a matrix, where each button makes a connection between a unique pair of wires. This system is suitable for being polled by a microcontroller.*

Figure 5-12. *The keypad on the left is matrix-encoded, and is polled via seven through-hole pins that protrude behind it. The keypad on the right assigns each button to a separate contact in its header. See the text for details about matrix encoding.*

Figure 5-13. *A typical tactile switch.*

Tactile Switch

Despite being called a switch, this is a miniature pushbutton, less than 0.4" square, designed for insertion in a printed-circuit board or in a solderless breadboard. It is almost always a SPST device but may have four pins, one pair connected to each contact. Tactile switches may be PC-mounted behind membrane pads. An example is shown in Figure 5-13.

Membrane Pad

Typically found on devices such as microwave ovens where contacts must be sealed against particles and liquids. Finger pressure on a membrane pad closes hidden or internal pushbuttons. They are usually custom-designed for specific product applications and are not generally available as generic off-the-shelf components. Some surplus pads may be found for sale on auction websites.

Radio Buttons

The term *radio buttons* is sometimes used to identify a set of pushbuttons that are mechanically interlinked so that only one of them can make an electrical connection at a time. If one button is pressed, it latches. If a second button is pressed, it latches while unlatching the first button. The buttons can be pressed in any sequence. This system is useful for applications such as component selection in a stereo system, where only one input can be permitted at a time. However, its use is becoming less common.

Snap-Action Switches

A snap-action switch (described in detail in the **switch** section of this encyclopedia) can be fitted with a pushbutton, as shown in Figure 5-14. This provides a pleasingly precise action, high reliability, and capability of switching currents of around 5A. However, snap-action switches are almost always single-pole devices.

Figure 5-14. *A pushbutton mounted on top of a SPDT snap-action switch.*

Emergency Switch

An emergency switch is a normally-closed device, usually consisting of a large pushbutton that clicks firmly into its "off" position when pressed, and does not spring back. A flange around the button allows it to be grasped and pulled outward to restore it to its "on" position.

Values

Pushbutton current ratings range from a few mA to 20A or more. Many pushbuttons have their current ratings printed on them but some do not. Current ratings are usually specified for a particular voltage, and may differ for AC versus DC.

How to Use it

Issues such as appearance, tactile feel, physical size, and ease of product assembly tend to dictate the choice of a pushbutton, after the fundamental requirements of voltage, current, and durability have been satisfied. Like any electromechanical component, a pushbutton is vulnerable to dirt and moisture. The ways in which a device may be used or abused should be taken into account when deciding whether the extra expense of a sealed component is justified.

When a pushbutton controls a device that has a high inductive load, a *snubber* can be added to minimize arcing. See "Arcing" on page 47 in the **switch** entry of this encyclopedia, for additional information.

What Can Go Wrong

No Button

When ordering a pushbutton switch, read datasheets carefully to determine whether a cap is included. Caps are often sold separately and may not be interchangeable between switches from different manufacturers.

Mounting Problems

In a panel-mount pushbutton that is secured by turning a nut, the nut may loosen with use, allowing the component to fall inside its enclosure when the button is pressed. Conversely, overtightening the nut may strip the threads on the pushbutton bushing, especially in cheaper components where the threads are molded into plastic. Consider applying a drop of Loc-Tite or similar adhesive before completely tightening the nut. Nut sizes vary widely, and finding a replacement may be time-consuming.

LED Issues

When using a pushbutton containing an LED, be careful to distinguish the LED power terminals from the switched terminals. The manufacturer's datasheet should clarify this distinction, but the polarity of the LED terminals may not be clearly indicated. If a diode-testing meter function is unavailable, a sample of the switch should be tested with a source of 3 to 5VDC and a 2K series resistor. Briefly touching the power to the LED terminals, through the resistor, should cause the LED to flash dimly if the polarity is correct, but should not be sufficient to burn out the LED if the polarity is incorrect.

Other Problems

Problems such as arcing, overload, short circuits, wrong terminal type, and contact bounce are generally the same as those associated with a **switch**, and are summarized in that entry in this encyclopedia.

switch

The term **switch** refers here to a physically operated mechanical switch, controlled by flipping a lever or sliding a knob. Although there is some overlap of function, **rotary switches** and **pushbuttons** have their own separate entries. Solid-state switching components are described in entries for **bipolar transistor**, **unijunction transistor**, and **field-effect transistor**. Integrated-circuit switching devices will be found in Volume 2. *Coaxial switches* are used for high-frequency signals, and are not included in this encyclopedia. *Multidirectional switches* differentiate up, down, left, right, diagonal, rotational, and other finger inputs, and are not included in this encyclopedia.

OTHER RELATED COMPONENTS

- **pushbutton** (See Chapter 5)
- **rotary switch** (See Chapter 7)

What It Does

A switch contains at least two contacts, which close or open when an external lever or knob is flipped or moved. Schematic symbols for the most basic type of on-off switch are shown in Figure 6-1.

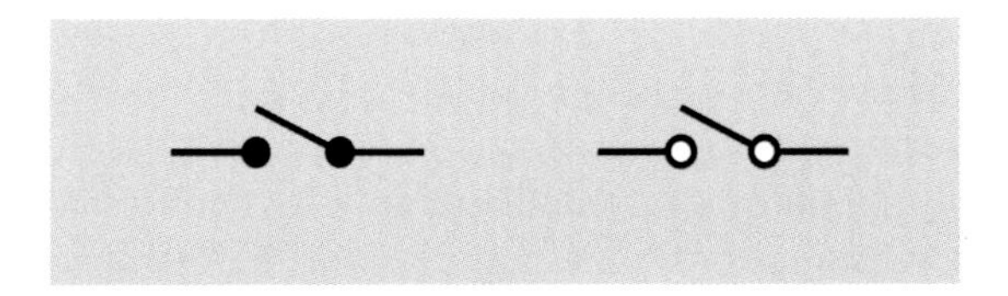

Figure 6-1. *The two most common schematic symbols for a SPST switch, also known as an on-off switch. The symbols are functionally identical.*

The most fundamental type of switch is a *knife switch*, illustrated in Figure 6-2. Although it was common in the earliest days of electrical discovery, today it is restricted to educational purposes in schools, and (in a more robust format) to AC electrical supply panels, where the large contact area makes it appropriate for conducting high amperages, and it can be used for "hot switching" a substantial load.

How It Works

The *pole* of a switch is generally connected with a movable contact that makes or breaks a connection with a secondary contact. If there is only one pole, this is a *single-pole* switch. If there is an additional pole, electrically isolated from the first, with its own contact or set of contacts, this is a *two-pole* switch, also known as a *double-pole* switch. Switches with more than 4 poles are uncommon.

If there is only one secondary contact per pole, this is a *single-throw* or *ST* switch, which may also be described as an *on-off* or *off-on* switch. If there is an additional secondary contact per pole, and the pole of the switch connects with the second contact while disconnecting from the first, this is a *double-throw* or *DT* switch, also known as a *two-way* switch.

A double-throw switch may have an additional center position. This position may have no con-

Figure 6-2. *A DPST knife switch intended for the educational environment.*

nection (it is an "off" position) or in some cases it connects with a third contact.

Where a switch is spring-loaded to return to one of its positions when manual pressure is released, it functions like a **pushbutton** even though its physical appearance may be indistinguishable from a switch.

Variants

Float switch, *mercury switch*, *reed switch*, *pressure switch*, and *Hall-effect switch* are considered as sensing devices, and will be found in Volume 3.

Terminology

Many different types of switches contain parts that serve the same common functions. The *actuator* is the lever, knob, or toggle that the user turns or pushes. A *bushing* surrounds the actuator on a toggle-type switch. The *common contact* inside a switch is connected with the *pole* of the switch. Usually a *movable contact* is attached to it internally, to touch the secondary contact, also known as a *stationary contact* when the movable contact is flipped to and fro.

Poles and Throws

Abbreviations identify the number of poles and contacts inside a switch. A few examples will make this clear:

SPST also known as 1P1T
Single pole, single throw

DPST also known as 2P1T
Double pole, single throw

SPDT also known as 1P2T
Single pole, double throw

3PST also known as 3P1T
Three pole, single throw

Other combinations are possible.

In Figure 6-3, schematic symbols are shown for double-throw switches with 1, 2, and 3 poles. The dashed lines indicate a mechanical connection, so that all sections of the switch move together when the switch is turned. No electrical connection exists between the poles.

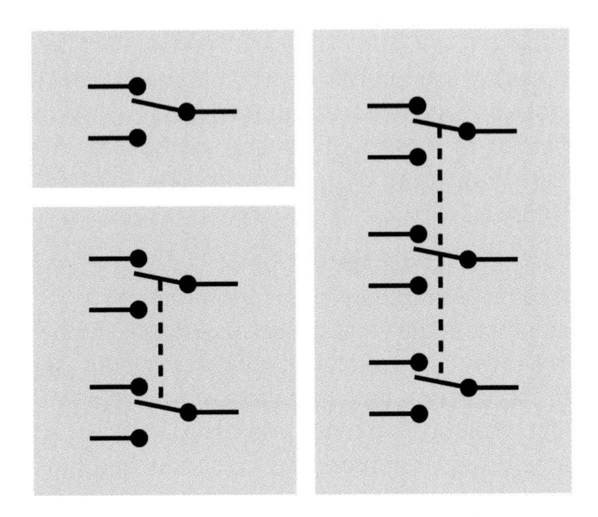

Figure 6-3. *Schematic symbols to represent three types of double-throw switch. Top left: Single-pole. Bottom left: Double-pole. Right: Triple-pole, more commonly known as 3-pole.*

On-Off Behavior

The words ON and OFF are used to indicate the possible states of a switch. The additional word NONE is used by some manufacturers to indicate that a switch does not have a center position. Some manufacturers don't bother with the word NONE, assuming that if the word is omitted, a center position does not exist.

ON-OFF or ON-NONE-OFF
: A basic on-off SPST switch with no center position.

ON-ON or ON-NONE-ON
: A basic SPDT switch with no center position.

ON-OFF-ON
: A double-throw switch with center-off position (no connection when the switch is centered).

ON-ON-ON
: A triple-throw switch where the center position connects with its own set of terminals.

Parentheses are used in descriptions of spring-loaded switches to indicate a momentary state that lasts only as long as pressure is applied to the actuator.

(ON)-OFF or OFF-(ON)
: A spring-loaded switch that is normally off and returns to that position when pressure is released. Also known as *NO* (normally open), and sometimes described as *FORM A*. Its performance is similar to that of a pushbutton and is sometimes described as a *make-to-make* connection.

ON-(OFF) or (OFF)-ON
: A spring-loaded switch that is normally on and returns to that position when pressure is released. This is sometimes described as a *make-to-break* connection. Also known as *NC* (normally closed), and sometimes described as *FORM B*.

(ON)-OFF-(ON)
: A spring-loaded double-throw switch with a no-connection center position to which it returns when pressure on its actuator is released.

Other combinations of these terms are possible.

Most double-throw switches break the connection with one contact (or set of contacts) before making the connection with the second contact (or set of contacts). This is known as a *break before make* switch. Much less common is a *make before break* switch, also known as a *shorting switch*, which establishes the second connection a moment before the first connection is broken. Use of a shorting switch may cause unforeseen consequences in electronic components attached to it, as both sides of the switch will be briefly connected when the switch is turned.

Snap-Action

Also known as a *limit switch* and sometimes as a *microswitch* or *basic switch*. This utilitarian design is often intended to be triggered mechanically rather than with finger pressure, for example in 3D printers. It is generally cheap but reliable.

Two snap-action switches are shown in Figure 6-4. A sectional view of a snap-action ON-(ON) limit switch is shown in Figure 6-5. The pole contacts are mounted on a flexible strip which can move up and down in the center of the switch. The strip has a cutout which allows an inverted U-shaped spring to flip to and fro. It keeps the contacts pressed together in either of the switch states.

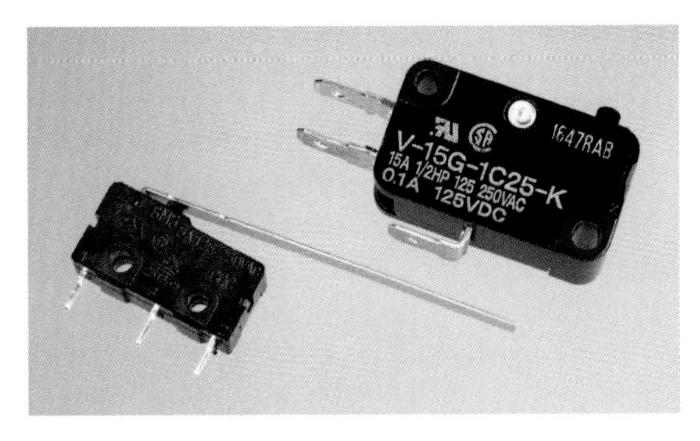

Figure 6-4. *Two SPDT snap-action switches, also known as limit switches. The one on the right is full-size. The one on the left is miniature, with an actuator arm to provide additional leverage. The arm may be trimmed to the required length.*

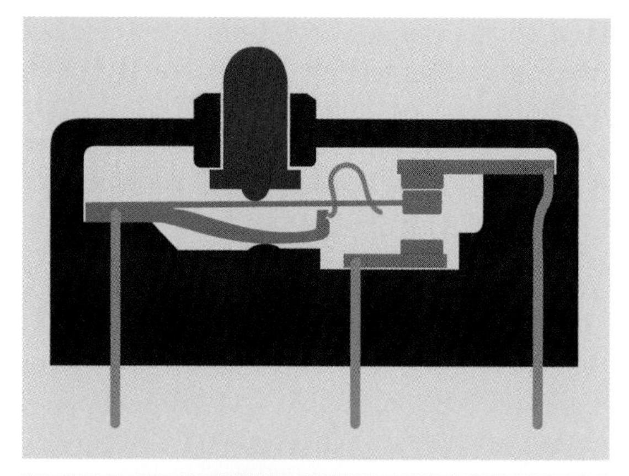

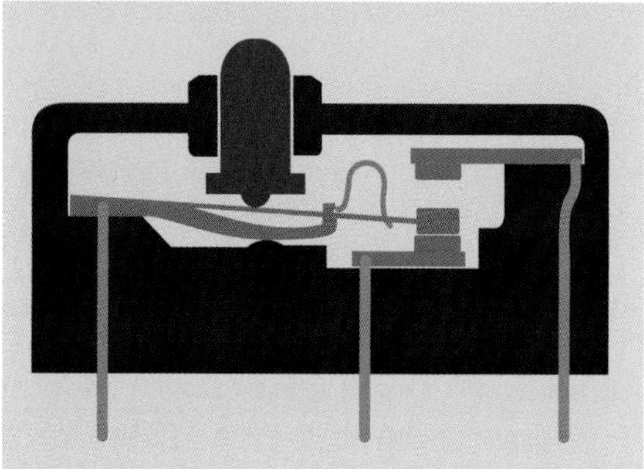

Figure 6-5. *Top: Two contacts inside this limit switch are touching by default. Bottom: When the external button is pressed, it pushes a flexible metal strip downward until it connects with the lower contact. The inverted-U–shaped component is a spring that rests inside a cutout in the flexible strip and resists motion through the central part of its travel.*

The term *snap action* refers to a spring-loaded internal mechanism which snaps to and fro between its two positions. This type of switch is usually SPDT and has a momentary action; in other words, it functions in ON-(ON) mode, although OFF-(ON) and (less often) ON-(OFF) versions are available. The body of the switch is sealed, with a small button protruding through a hole. A thin metal arm may provide additional leverage to press the button. A roller may be mounted at the end of the arm so that the switch can be activated as it slides against a moving mechanical component such as a cam or a wheel. The switch is commonly used to limit the travel or rotation of such a component. Literally thousands of variants are available, in different sizes, requiring different amounts of force for activation. Subminiature snap-action switches can often be actuated by a pressure of only a few grams.

Rocker

Three rocker switches are shown in Figure 6-6. A sectional view of a rocker switch is shown in Figure 6-7. In this design, a spring-loaded ball bearing rolls to either end of a central rocker arm when the switch is turned. Rocker switches are often used as power on-off switches.

Figure 6-6. *Three rocker switches, the upper two designed for push-insertion into a suitably sized rectangular hole in a panel. The switch at front-center is intended to be screwed in place, and is more than 20 years old, showing that while the choice of materials has changed, the basic design has not.*

Slider

Many types of slider switch (also known as *slide switch*) are widely used as a low-cost but versatile way to control small electronic devices, from clock-radios to stereos. The switch is usually mounted on a circuit board, and its knob or cap protrudes through a slot in the panel. This design is more vulnerable to dirt and moisture than other types of switch. It is usually cheaper than a toggle switch but is seldom designed for use with a high current.

Figure 6-7. *This sectional view of a rocker switch shows a spring-loaded ball-bearing that rolls to and fro along a rocker arm, connecting either pair of contacts when the switch is turned.*

Most slide switches have two positions, and function as SPDT or DPDT switches, but other configurations are less commonly available with more poles and/or positions. A subminiature slide switch is shown in Figure 6-8, while some schematic representations are shown in Figure 6-9, where a black rectangle indicates a sliding internal contact, and a terminal that functions as a pole is identified with letter P in each case. Top left: A SPDT switch using a two-position slider. Top right: A 4PDT slide switch. Bottom left: There are no poles in this switch, as such. The slider can short together any of four pairs of contacts. Bottom right: The slider shorts together three possible pairs of contacts out of four. Here again, there is no pole.

Note that the schematic representation of a slide switch may be identical to that of a slide **push-button**. A schematic should be inspected carefully to determine which type is intended.

The representation of sliders in schematics has not been standardized, but the samples shown are common.

Figure 6-8. *This subminiature slide switch is less than half an inch long, rated 0.3A at 30VDC. Larger versions look almost identical, but can handle only slightly more current.*

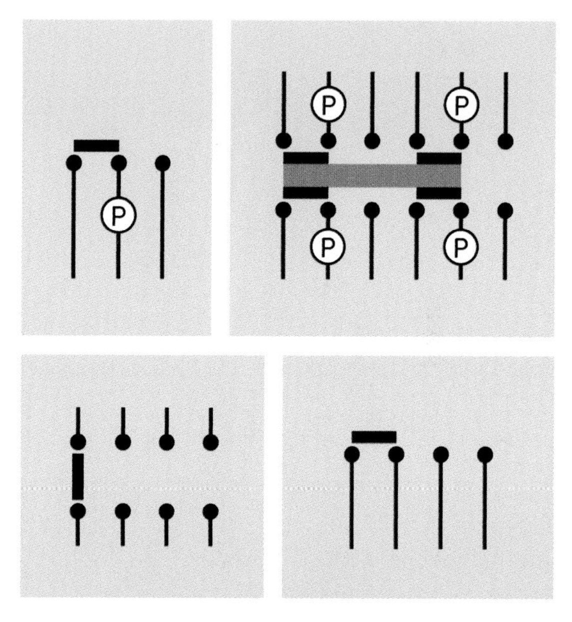

Figure 6-9. *Slide switch schematics. Each black rectangle represents a movable contact that connects two pairs of fixed contacts at a time. Detailed commentary on these variants will be found in the body of the text. Manufacturers may use variants of these symbols in their datasheets (for example, the gray rectangle indicating an insulating contact carrier, at top right, may be represented as a single line, or a black outline with a white center).*

Toggle

A *toggle switch* provides a firm and precise action via a lever (the *toggle*) that is usually tear-drop shaped and nickel plated, although plastic toggles are common in cheaper variants. Formerly used to control almost all electronic components (including early computers), the toggle has declined in popularity but is still used in applications such as automobile accessory kits, motorboat instrument panels, and industrial controls.

Three miniature DPDT toggle switches are shown in Figure 6-10. Two full-size, heavy-duty toggle switches are shown in Figure 6-11. A full-size, four-pole, double-throw heavy-duty toggle switch is shown in Figure 6-12. Toggle switches with more poles are extremely rare.

Figure 6-10. *Three miniature toggle switches with current ratings ranging from 0.3A to 6A at 125VAC. Each small square in the background grid measures 0.1" x 0.1".*

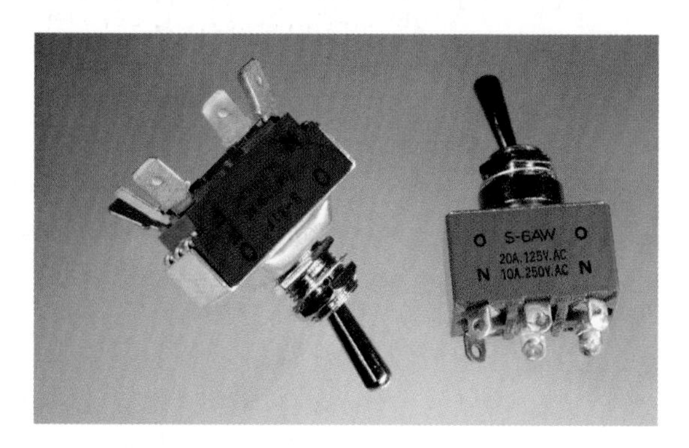

Figure 6-11. *Two full-size toggle switches capable of handling significant current. At left, the switch terminates in quick-connect terminals. At right, the switch has solder terminals (some of them containing residual traces of solder).*

Figure 6-12. *A 4PDT full-size toggle switch with solder terminals, capable of switching 25A at 125VAC. Four-pole switches are relatively unusual.*

An automotive toggle switch is shown in Figure 6-13. Its plastic toggle is extended to minimize operating error.

Figure 6-13. *A toggle switch intended for control of automotive accessories.*

High-end toggle switches are extremely durable and can be sealed from environmental contamination with a thin *boot* made from molded rubber or vinyl, which screws in place over the toggle, using the thread on the switch bushing. See Figure 6-14.

A *locking toggle switch* has a toggle that must be pulled out against the force of a retaining spring, before the toggle can be moved from one position to another. The toggle then snaps back into place, usually engaging in a small slot in the bushing of the switch.

Figure 6-14. *A rubber or vinyl boot can be used to protect a toggle switch from contamination with dirt or water. Each boot contains a nut that screws onto the threads of a toggle switch, as shown at left.*

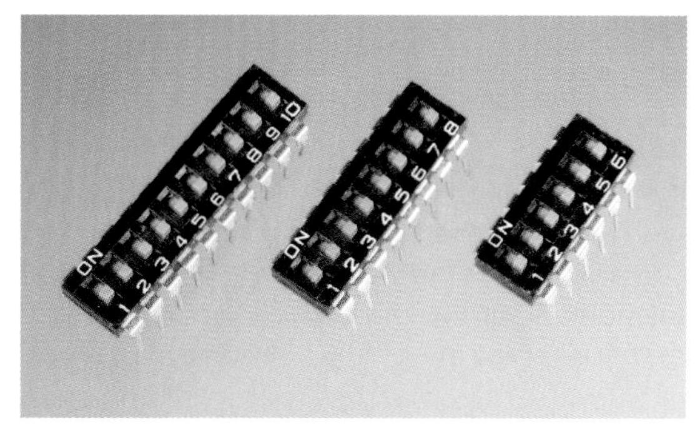

Figure 6-15. *As shown here, DIP switches are available with a variety of "positions," meaning the number of switches, not the number of switch states.*

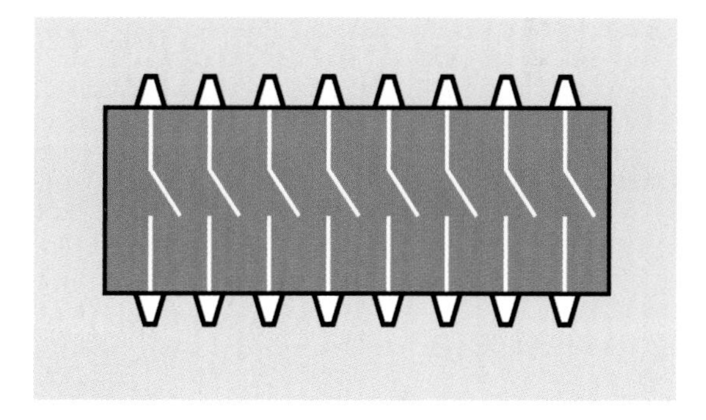

Figure 6-16. *The interior connections of a 16-pin DIP switch.*

DIP

A DIP switch is an array of very small, separate switches, designed for mounting directly on a circuit board, either in through-hole or surface-mount format. Through-hole DIP switches have two rows of pins with a 0.1" pitch, the rows being spaced 0.3" apart to fit a standard DIP (dual-inline package) socket or comparable configuration of holes in the board. Surface-mount DIP switches may have 0.1" or 0.05" pitch.

Most DIP arrays consist of SPST switches, each of which can close or open a connection between two pins on opposite sides of the switch body. The switch positions are usually labelled ON and OFF. Figure 6-15 shows a selection of DIP switches. Figure 6-16 shows the internal connections in a DIP switch.

The number of switches in a DIP array is usually referred to as its number of "positions." This should not be confused with the two positions of each physical switch lever. SPST DIP switches are made with 1, 2, 3, 4, 5, 6, 7, 8, 9, 10, 12, and 16 positions.

Early IBM-compatible desktop computers often required the user to set the position of an internal DIP switch when making routine upgrades such as installing an additional disk drive. While this feature is now obsolete, DIP switches are still used in scientific equipment where the user is expected to be sufficiently competent to open a cabinet and poke around inside it. Because of the 0.1" spacing, a small screwdriver or the tip of a pen is more appropriate than a finger to flip individual levers to and fro.

DIP switches may also be used during prototype development, as they allow a convenient way to test a circuit in numerous different modes of operation.

Most DIP switches have wire terminals which are just long enough for insertion into a standard breadboard.

DIP switch package options include standard, low-profile, right-angle (standing at 90 degrees relative to the circuit board), and piano (with switch levers designed to be pressed, like tiny rocker switches, instead of being flipped to and fro).

Some SPDT, DPST, DPDT, 3PST, and 4PST variants exist, but are uncommon. Multiple external pins connect with the additional internal switch contacts, and a manufacturer's datasheet should be consulted to confirm the pattern of internal connections. A surface-mount, 0.1″ pitch, DPST DIP switch is shown in Figure 6-17, with a plastic cover to protect the switches from contamination during wave soldering (at left), and with the cover peeled off (at right).

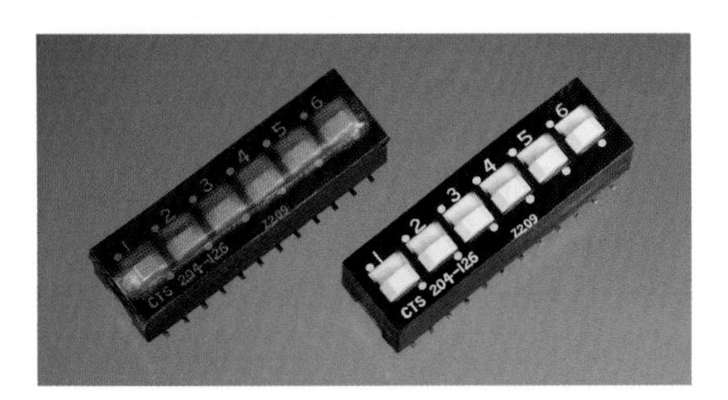

Figure 6-17. *A SPDT surface-mount double-throw DIP switch, sold with a plastic cover (shown at left) to protect it during wave soldering. The cover has been removed at right.*

SIP

A SIP switch is an array of small, separate switches, identical in concept to a DIP switch, but using only one row of pins instead of a double row. The applications for SIP switches are the same as DIP switches; the primary difference is simply that the SIP switch occupies a little less space, while being perhaps slightly less convenient to use.

One terminal of each switch usually shares a common bus. The internal connections in a typical 8-pin SIP array are shown in Figure 6-18. Pin spacing is 0.1″, as in a typical DIP switch.

Paddle

A paddle switch has a flat-sided tab-shaped plastic actuator, relatively large to allow a firm, error-free grip. Internally it is often comparable with a rocker switch, and is generally used with AC pow-

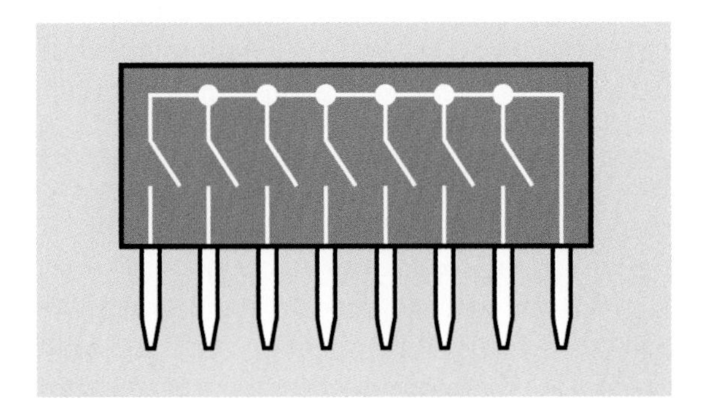

Figure 6-18. *The interior connections of an 8-pin SIP switch incorporating a common bus.*

er. Some toggle-switch bodies are also sold with paddle-shaped actuators. A subminiature paddle switch is shown in Figure 6-19.

Figure 6-19. *A subminiature paddle switch. Full-size versions are often used as power switches.*

Vandal Resistant Switch

Typically fabricated from stainless steel, this is designed to withstand most types of abuse and is also weather-proof. The pushbuttons that allow pedestrians to trigger a traffic signal are a form of vandal-resistant switch.

Tactile Switch

This is considered to be a **pushbutton**, and is described in that entry. See "Tactile Switch" on page 34.

Mounting Options

A *panel mount* switch generally has a threaded bushing that is inserted from behind the front panel of a product, through a hole of appropriate size. It is supplied with a lockwasher and a nut (often, two nuts) that fit the thread on the switch bushing.

Front panel mount usually means that screws visible from the front of the panel are attached to a bracket on the switch behind the panel. The actuator of the switch is accessible through a cutout in the panel. This mounting style is mostly used for rocker switches and sometimes for slide switches.

Subpanel mount means that the switch is attached to a separate plate or chassis behind the control panel. The actuator of the switch is accessible through a cutout.

Snap-in mount requires a switch with flexible plastic or metal tabs each side, designed to push through a cutout in the panel, at which point the tabs spring out and retain the switch.

PC mount switches have pins that are soldered into a printed circuit board. They may have additional solderable lugs to provide mechanical support.

Surface mount switches are attached to a board in the same manner as other surface-mount components.

Termination

Switches (and pushbuttons) are available with a variety of terminals.

Solder lugs are small tabs, each usually perforated with a hole through which the end of a wire can be inserted prior to soldering.

PC terminals are pins that protrude from the bottom of the switch, suitable for insertion in a printed circuit board. This style is also known as a *through-hole*. The terminals may have a right-angle bend to allow the component to be mounted flat against the board, with the switch actuator sticking out at the side. This termination style is known as *right-angle PC*. Many manufacturers offer a choice of straight or bent pin terminals, but the component may be listed in a catalog under either of those options, with no indication that other options exist. Check manufacturer datasheets carefully.

Quick connect terminals are spade-shaped to accept push-on connectors, commonly used in automotive applications. Hybrid quick-connect terminals that can also function as solder lugs are sometimes offered as an option.

Screw terminals have screws premounted in flat terminals, for solderless attachment of wires.

Wire leads are flexible insulated wires, often with stripped and tinned ends, protruding at least an inch from the body of the component. This option is becoming uncommon.

Contact Plating Options

The internal electrical contacts of a switch are usually plated with silver or gold. Nickel, tin, and silver alloys are cheaper but less common. Other types are relatively rare.

Values

Switches designed for electronic devices vary widely in power capability, depending on their purpose. Rocker switches, paddle switches, and toggle switches are often used to turn power on and off, and are typically rated for 10A at 125VAC, although some toggle switches go as high as 30A. Snap-action or limit switches may be similarly rated, although miniature versions will have reduced capability. Slide switches cannot handle significant power, and are often rated around 0.5A (or less) at 30VDC. DIP and SIP switches have a typical maximum rating of 100mA at 50V and are not designed for frequent use. Generally they are used only when the power to the device is off.

How to Use it

Power Switches

When a simple SPST switch is used to turn DC power on and off, it conventionally switches the positive side of the power supply, also sometimes known as the *high side*. The primary reason for following this convention is that it is widely used; thus, following it will reduce confusion.

More importantly, an on-off switch that controls AC power must be used on the "live" side of the supply, not the "neutral" side. If you have any doubts about these concepts (which go beyond the scope of this book), consult a reference guide on this subject. Using a DPST component to switch both sides of an AC supply may be a worthwhile additional precaution in some applications. The ground wire of an AC supply should never be switched, because the device should always be grounded when it is plugged into an electric outlet.

Limit Switches

An application for two limit switches with a DC motor and two rectifier **diodes** is shown in Figure 6-20. This diagram assumes that the motor turns clockwise when its lower terminal is positive, and counter-clockwise when its upper terminal is positive. Only two terminals are used (and shown) in each limit switch; they are chosen to be normally-closed. Other terminals inside a switch may exist, may be normally-open, and can be ignored.

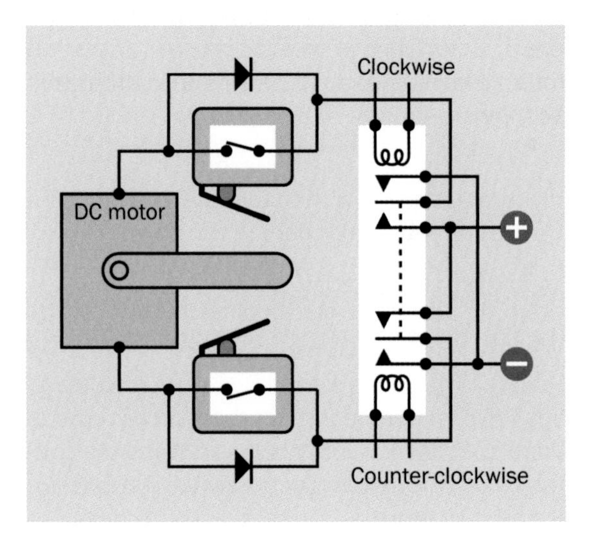

Figure 6-20. *In this schematic, normally-closed limit switches are opened by pressure from an arm attached to a motor, thus switching off its power at each end of its permitted travel and preventing overload and burnout. A two-coil latching relay activates the motor. Rectifier diodes allow power to reach the motor to reverse its rotation when a limit switch is open.*

The motor is driven through a dual-coil, DPDT latching **relay**, which will remain in either position indefinitely without drawing power. When the upper coil of the relay receives a pulse from a pushbutton or some other source, the relay flips to its upper position, which conducts positive current through the lower limit switch, to the lower terminal of the motor. The motor turns clockwise until the arm attached to its shaft hits the lower limit switch and opens it. Positive current is blocked by the lower diode, so the motor stops.

When the lower coil of the relay is activated, the relay flips to its lower position. Positive current can now reach the upper side of the motor through the upper limit switch. The motor runs counter-clockwise until its arm opens the upper limit switch, at which point the motor stops again. This simple system allows a DC motor to be run in either direction by a button-press of any duration, without risk of burnout when the motor reaches the end of its travel. It has been used for applications such as raising and lowering powered windows in an automobile.

A DPDT pushbutton could be substituted for the latching relay if manual control, only, is acceptable. However, in this scenario, sustained pressure on the pushbutton would be necessary to move the motor arm all the way to the opposite end of its travel. A DPDT switch might be more appropriate than a pushbutton.

Logic Circuits

Logic circuits that depend purely on switches can be constructed (for example, to add binary numbers) but are rare and have no practical applications. The most familiar and simplest example of manually switched logic is a pair of SPDT switches in house wiring, one positioned at the top of a flight of stairs and the other at the bottom, as shown in Figure 6-21. Either switch will turn the light on if it is currently off, or off if it is currently on. To extend this circuit by incorporating a third switch that has the same function as the other two, a DPDT switch must be inserted. See Figure 6-22.

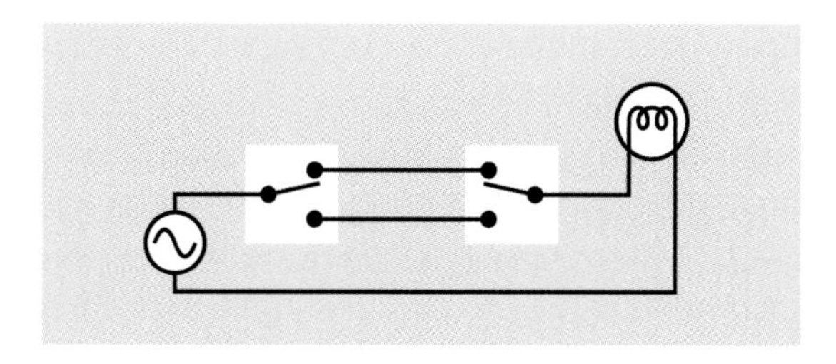

Figure 6-21. *SPDT switches are commonly used in house wiring so that either of them will turn a shared light on if it is off, or off it is on.*

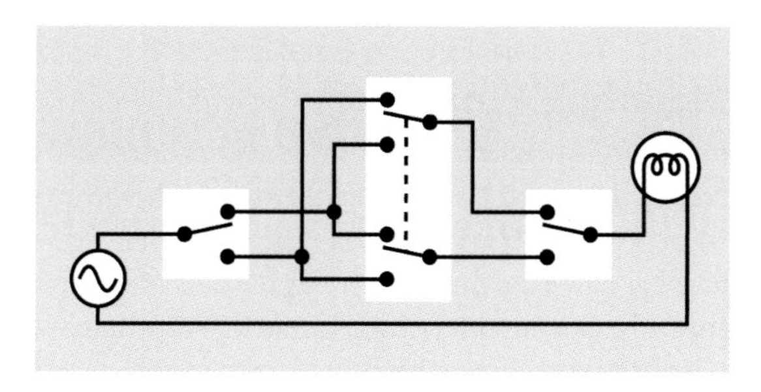

Figure 6-22. *A DPDT switch must be inserted if three switches must have identical function to control the on-off state of a single light bulb.*

Alternatives

As **microcontrollers** have become cheaper and more ubiquitous, they have taken over many functions in electronic products that used to be served by switches. A menuing system driven by a microcontroller can use one **rotational encoder** with a SPST pushbutton built into it to select and adjust numerous parameters in a device such as a car stereo, where functions were once selected and adjusted by individual switches and potentiometers. The rotational-encoder option takes up less space, is cheaper to build (assuming a microcontroller is going to be used in the device for other purposes anyway), and can be more reliable, as it reduces the number of electromechanical parts. Whether it is easier to use is a matter of taste. Cost and ergonomics may be the primary factors to consider when choosing where and how to use switches.

What Can Go Wrong

Arcing

The contacts inside a switch will be rapidly eroded if *arcing* (pronounced "arking") occurs. An electric arc is a spark that tends to form when a switch is opened while conducting a high current or high voltage (typically 10A or more and 100V or more). The most common cause is an inductive load that generates back-EMF when it is switched on and forward-EMF when it is switched off. The surge can be many times the amperage that the load draws during continuous operation. In DC circuits, arcing can be reduced by using a rectifier **diode** in parallel with the load (with its polarity blocking normal current flow). This is often referred to as a *flyback diode* or *freewheeling diode*. In AC circuits, where a diode cannot be used in this way, a *snubber* (a simple combination of capacitor and resistor) may be placed around the load. A snubber can also be used around the switch itself, in DC circuits. See "Snubber" on page 102.

When switching an inductive load, it is generally prudent to use switches rated for twice as much current as the circuit will normally draw.

Dry Joints

Switches that control significant current will have substantial terminals, and these terminals will be attached to heavy-gauge wire. When using solder to make this type of connection, the combined heat capacity of the wire and the terminal will sink a lot more heat than a small component on a circuit board. At least a 30W soldering iron should be used. Lower-wattage irons may be incapable of melting the solder completely (even though they seem to), and a "dry joint" will result, which can have a relatively high electrical resistance and will be mechanically weak, liable to break later. Any good solder joint should withstand flexing of the wire attached to it.

Short Circuits

Because many switches are still wired in with solder tabs, screw terminals, or quick-connect terminals, wires that become accidentally detached can be a significant hazard. Heat-shrink tubing should be applied to enclose wires and terminals at the rear of a power switch, as an additional precaution. Power switches should always be used in conjunction with appropriate **fuses**.

Contact Contamination

Sealed switches should be used in any environment where dirt or water may be present. Slide switches are especially vulnerable to contamination, and are difficult to seal. Switches used in audio components will create "scratchy" sounds if their contacts deteriorate.

Wrong Terminal Type

Because switches are available with a wide variety of terminal types, it's easy to order the wrong type. Switches may be supplied with pins for through-hole insertion in circuit boards; screw terminals; quick-disconnect terminals; or solder lugs. Variants may also be available for surface mount. If a project requires, for example, the insertion of pins in a printed circuit board, and a switch is supplied with solder lugs, it will be unusable.

Part numbers generally include codes to identify each terminal variant, and should be studied carefully.

Contact Bounce

Also known as *switch bounce*. When two contacts snap together, extremely rapid, microscopic vibrations occur that cause brief interruptions before the contacts settle. While this phenomenon is not perceptible to human senses, it can be perceived as a series of multiple pulses by a **logic chip**. For this reason, various strategies are used to *debounce* a switch that drives a logic input. This issue is explored in detail in the entry on logic chips in Volume 2 of the encyclopedia.

Mechanical Wear

Any toggle or rocker switch contains a mechanical pivot, which tend to deteriorate in harsh environments. Friction is also an issue inside these switches, as the design often entails the rounded tip of a lever rubbing to and fro across the center of a movable contact.

The spring inside a snap switch or limit switch may fail as a result of metal fatigue, although this is rare. A slide switch is far less durable, as its contacts rub across each other every time the switch changes position.

In any application that entails frequent switching, or where switch failure is a critical issue, the most sensible practice is to avoid using cheap switches.

Mounting Problems

In a panel-mount switch that is secured by turning a nut, the nut may loosen with use, allowing the component to fall inside its enclosure. Conversely, overtightening the nut may strip the threads on the switch body, especially in cheaper components where the threads are molded into plastic. Consider applying a drop of Loc-Tite or similar adhesive after moderately tightening the nut. Note that nut sizes vary widely, and finding a replacement may be time-consuming.

Cryptic Schematics

In some circuit schematics, the poles of a multipole switch may be visually separated from each other, even at opposite sides of the page, for convenience in drawing the schematic. Dotted lines usually, but not always, link the poles. In the absence of dotted lines, switch segments are often coded to indicate their commonality. For example, SW1(a) and SW1(b) are almost certainly different parts of the same switch, with linked poles.

rotary switch

7

Not to be confused with **rotational encoder**, which has its own entry in this encyclopedia.

OTHER RELATED COMPONENTS

- **switch** (See Chapter 6)
- **rotational encoder** (See Chapter 8)

What It Does

A rotary switch makes an electrical connection between a *rotor*, mounted on a shaft that is turned by a knob, and one of two or more stationary contacts. Traditionally, it was the component of choice to select wavebands on a radio receiver, broadcast channels on a television or inputs on a stereo preamplifier. Since the 1990s, it has been substantially superceded by the **rotational encoder**. However it still has applications in military equipment, field equipment, industrial control systems, and other applications requiring a rugged component that will withstand heavy use and a possibly harsh environment. Also, while the output from a rotational encoder must be decoded and interpreted by a device such as a microcontroller, a rotary switch is an entirely passive component that does not require any additional electronics for its functionality.

Two typical schematic symbols for a rotary switch are shown in Figure 7-1. They are functionally identical. A simplified rendering of the interior of a traditional-style rotary switch is shown in Figure 7-2. A separate contact (not shown) connects with the rotor, which connects with each of the stationary contacts in turn. The colors were chosen to differentiate the parts more clearly, and do not correspond with colors in an actual switch.

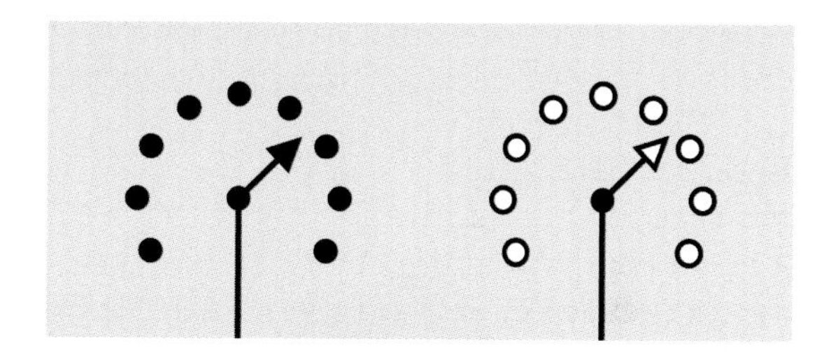

Figure 7-1. *Typical schematic symbols for a rotary switch. The two symbols are functionally identical. The number of contacts will vary depending on the switch.*

A selection of rotary switches is shown in Figure 7-3. At top-left is an *open frame* switch, providing no protection to its contacts from contaminants. This type of component is now rare. At top-right is a twelve-position, single-pole switch rated 2.5A at 125VAC. At front-left is a four-position, single-pole switch rated 0.3A at 16VDC or 100VAC. At front right is a two-position, two-pole switch with the same rating as the one beside it. All the sealed switches allow a choice of panel mounting or through-hole printed circuit board mounting.

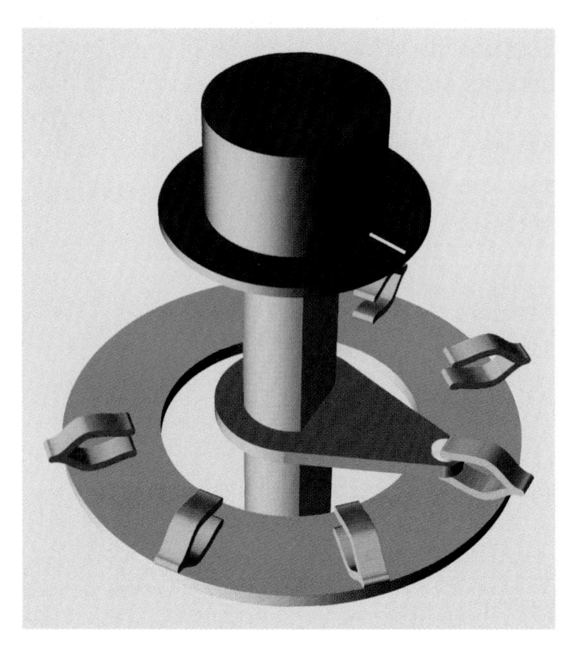

Figure 7-2. *A simplified rendering of interior parts in a basic SP6T rotary switch. Arbitrary colors have been added for clarity.*

Figure 7-3. *A selection of rotary switches. See text for details.*

How It Works

A switch may have multiple poles, each connecting with its own rotor. The rotors are likely to be on separate *decks* of the switch, but two, three, or four rotors, pointing in different directions, may be combined on a single deck if the switch has only a small number of positions.

Rotary switches are usually made with a maximum of twelve positions, but include provision for limiting the number of positions with a *stop*. This is typically a pin, which may be attached to a washer that fits around the bushing of the switch. The pin is inserted into a choice of holes to prevent the switch from turning past that point. For example, an eight-position rotary switch can be configured so that it has only seven (or as few as two) available positions.

A specification for a rotary switch usually includes the angle through which the switch turns between one position and the next. A twelve-position switch usually has a 30-degree turn angle.

Variants

Conventional

The traditional style of rotary switch is designed to be panel-mounted, with a body that ranges from 1″ to 1.5″ in diameter. If there is more than one deck, they are spaced from each other by about 0.5″. The switch makes an audible and tactile "click" as it is turned from one position to the next.

A rugged sealed five-deck rotary switch is shown in Figure 7-4. It has five poles (one per deck), and a maximum of 12 positions. The contacts are rated 0.5A at 28VDC. This type of heavy-duty component is becoming relatively rare.

If the rotor in a switch establishes a connection with the next contact a moment before breaking the connection with the previous contact, this is known as a *shorting* switch, which may also be described as a *make-before-break* switch. In a *nonshorting* or *break-before-make* switch, a tiny interval separates one connection from the next. This can be of significant importance, depending

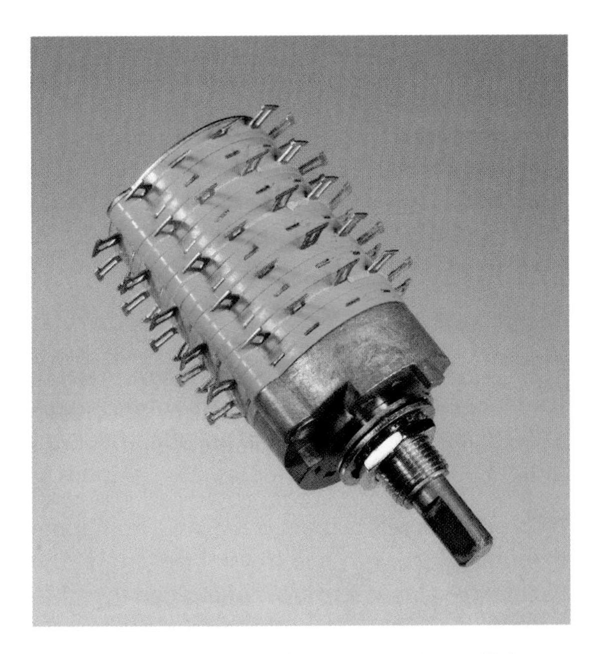

Figure 7-4. *A five-pole, twelve-position rotary switch.*

on the components that are connected with the switch.

The shaft may be round, splined, or D-shaped in section. A knob is seldom supplied with a switch and must be chosen to match the shaft. Some shaft dimensions are metric, while others are measured in inches, with 1/4" diameter being the oldest standard. Some switches with a splined shaft are supplied with an adapter for a knob of D-shaped internal section; the adapter can be slipped onto the shaft in any of 12 or more positions, to minimize the inconvenience of positioning the body of the switch itself so that the knob is correctly oriented in relation to positions printed on the face of the panel.

Miniature rotary switches may be as small as 0.5" diameter, and usually terminate in pins for through-hole mounting on a PC board. Miniature switches usually have lower current ratings than full-size switches.

Rotary switches must be securely anchored to resist the high turning forces that can be inflicted upon them by users. In a panel-mount design, a nut is tightened around a thread on the bushing of the switch. Through-hole versions can be secured to the PC board with the shaft protruding loosely through a cutout in the panel. To minimize mechanical stress on the circuit board, the detents in a PC-board switch are usually weaker than in a full-size switch, and the knob is usually smaller, allowing less leverage.

Rotary DIP

A conventional *DIP switch* is a linear array of miniature SPST switches designed to fit a standard DIP (dual-inline package) layout of holes in a circuit board. It is described in the **switch** entry of this encyclopedia. A *rotary DIP* switch (also known as an *encoded output rotary switch* or a *coded rotary switch*) does not conform with a DIP layout, despite its name. It is approximately 0.4" square and usually has five pins, one of which can be considered the input or common pin while the other four can function as outputs. The pins are spaced at 0.1" pitch from one another. Pin function and layout are not standardized.

A dial on top of the switch has either 10 positions (numbered 0 through 9) or 16 positions (0 through 9 followed by letters A through F). One switch of each type is shown in Figure 7-5.

Figure 7-5. *A rotary DIP switch, also known as an encoded output rotary switch, may be used as a substitute for a DIP switch in some applications.*

Each position of the dial closes pairs of contacts inside the component to create a unique binary-coded decimal pattern (in a 10-position switch) or binary-coded hexadecimal pattern (in the 16-position switch) on the four output pins. The pin states are shown in Figure 7-6. A rotary DIP switch is a relatively flimsy device, and is not designed for frequent or heavy use. It is more likely to be a "set it and forget it" device whose state is established when it is installed in a circuit board.

Switch Position	Pin 1	Pin 2	Pin 3	Pin 4
0	●	●	●	●
1	○	●	●	●
2	●	○	●	●
3	○	○	●	●
4	●	●	○	●
5	○	●	○	●
6	●	○	○	●
7	○	○	○	●
8	●	●	●	○
9	○	●	●	○
A	●	○	●	○
B	○	○	●	○
C	●	●	○	○
D	○	●	○	○
E	●	○	○	○
F	○	○	○	○

Figure 7-6. *Positive and negative states of the four output pins of a real-coded 16-position rotary DIP switch, assuming that the common pin of the switch is connected with a positive supply voltage. A ten-position rotary DIP switch would use only the states from 0 through 9. In a complement-coded switch, the positive and negative states would be reversed.*

Because each position of the switch is identified with a unique binary pattern, this is an example of *absolute encoding*. By contrast, a typical **rotational encoder** uses *relative encoding*, as it merely generates a series of undifferentiated pulses when the shaft is turned.

A *real-coded* rotary DIP makes a connection between input and output pins wherever a binary 1 would exist. In the *complement-coded* version, the output is inverted. The switch is primarily intended for use with a **microcontroller**, enabling only four binary input pins on the microcontroller to sense up to sixteen different switch positions.

A six-pin rotary DIP variant is available from some manufacturers, with two rows of three pins, the two center pins in each row being tied together internally, and serving as the pole of the switch.

Rotary DIPs are available with a screw slot, small knurled knob, or larger knob. The screw-slot version minimizes the height of the component, which can be relevant where circuit boards will be stacked close together. A *right-angle PC* variant stands at 90 degrees to the circuit board, with pins occupying a narrower footprint. The switch on the left in Figure 7-5 is of this type.

While most rotary DIPs are through-hole components, surface-mount versions are available.

Most rotary DIPs are sealed to protect their internal components during wave-soldering of circuit boards.

Gray Code

A Gray code (named after its originator, Frank Gray) is a system of absolute encoding of a switch output, using a series of nonsequential binary numbers that are chosen in such a way that each number differs by only one digit from the preceding number. Such a series is useful because it eliminates the risk that when a switch turns, some bits in the output will change before others, creating the risk of erroneous interpretation. A minority of rotary switches or rotational encoders are available with Gray-coded outputs.

Typically, a microcontroller must use a lookup table to convert each binary output to an angular switch position.

PC Board Rotary Switch

Miniature switches with a conventional, non-encoded output are available for printed-circuit board mounting, sometimes requiring a screwdriver or hex wrench to select a position. A single-pole eight-position switch of this type is shown in Figure 7-7. Its contacts are rated to carry 0.5A at 30VDC, but it is not designed to switch this current actively.

Figure 7-7. *This miniature switch is designed for insertion on a printed circuit board. It can be used to make a setting before a device is shipped to the end user.*

Mechanical Encoder

A *mechanical encoder* functions similarly to a rotary DIP switch but is intended for much heavier use. It outputs a binary-coded-decimal value corresponding with its shaft position, is typically the size of a miniature rotary switch, and is designed for panel mounting. The Grayhill Series 51 allows 12 positions, each generating a code among four terminals. The Bourns EAW provides 128 positions, each generating a code among 8 terminals.

Pushwheel and Thumbwheel

A *pushwheel switch* is a simple electromechanical device that enables an operator to provide a code number as input to data processing equipment, often in industrial process control. The decimal version contains a wheel on which numbers are printed, usually in white on black, from 0 through 9, visible one at a time through a window in the face of the switch. A button above the wheel, marked with a minus sign, rotates it to the next lower number, while a button below the wheel, marked with a plus sign, rotates it to the next higher number. A connector at the rear of the unit includes a common (input) pin and four output pins with values 1, 2, 4, and 8. An additional set of pins with values 1, 2, 3, and 4 is often provided. The states of the output pins sum to the value that is currently being displayed by the wheel. Often two, three, or four pushwheels (each with an independent set of connector pins) are combined in one unit, although individual pushwheels are available and can be stacked in a row.

A *thumbwheel switch* operates like a pushwheel switch, except that it uses a thumbwheel instead of two buttons. Miniaturized thumbwheel switches are available for through-hole mounting on PC boards.

Hexadecimal versions are also available, with numbers from 0 through 9 followed by letters A through F, although they are less common than decimal versions.

Keylock

A *keylock switch* is generally a two-position rotary switch that can be turned only after insertion of a key in a lock attached to the top of the shaft. This type of switch almost always has an OFF-(ON) configuration and is used to control power. Keylock switches are found in locations such as elevators, for fire-department access; in cash registers; or on data-processing equipment where switching power on or off is reserved for a system administrator.

Values

A full-size rotary switch may be rated from 0.5A at 30VDC to 5A at 125VAC, depending on its purpose. A very few switches are rated 30A at 125VAC; these are high-quality, durable, expensive items.

A typical rotary DIP switch is rated 30mA at 30VDC and has a carrying current rating (continuous current when no switching occurs) of no more than 100mA at 50VDC.

How to Use it

In addition to its traditional purpose as a mode or option selector, a rotary switch provides a user-friendly way to input data values. Three ten-position switches, for instance, can allow user input of a decimal number ranging from 000 to 999.

When used with a **microcontroller**, a rotary switch can have a *resistor ladder* mounted around its contacts, like a multi-point *voltage divider*, so that each position of the rotor provides a unique potential ranging between the positive supply voltage and negative ground. This concept is illustrated in Figure 7-8, where all the resistors have the same value. The voltage can be used as an input to the microcontroller, so long as the microcontroller shares a common ground with the switch. An analog-digital converter inside the microcontroller translates the voltage into a digital value. The advantage of this scheme is that it allows very rapid control by the user, while requiring only one pin on the microcontroller to sense as many as twelve input states.

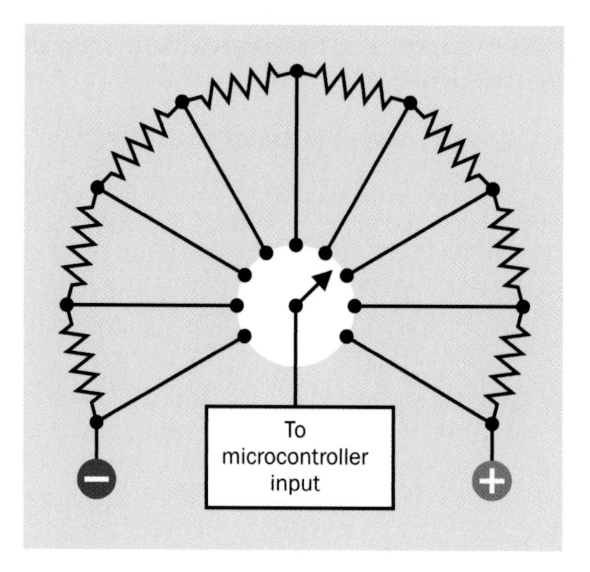

Figure 7-8. *A resistor ladder can be formed around the contacts of a rotary switch, with the pole of the switch connected to a microcontroller that has an analog-digital converter built in. The microcontroller converts the voltage input to an internal digital value. Thus, one pin can sense as many as twelve input states.*

For a ladder consisting of 8 resistors, as shown, each resistor could have a value of 250Ω. (The specifications for a particular microcontroller might require other values.) To avoid ambiguous inputs, a nonshorting rotary switch should be used in this scheme. A *pullup resistor* of perhaps 10K should be added to the microcontroller input, so that there is no risk of it "floating" when the switch rotor is moving from one contact to the next. The code that controls the microcontroller can also include a blanking interval during which the microcontroller is instructed to ignore the switch.

Because the rotary switch is an electromechanical device, it has typical vulnerabilities to dirt and moisture, in addition to being bulkier, heavier, and more expensive than a **rotational encoder**. Rotary switches have also been partially replaced by pushbuttons wired to a microcontroller. This option is found on devices ranging from digital alarm clocks to cellular phones. In addition to being cheaper, the pushbutton alternative is preferable where space on a control panel, and behind it, is limited.

What Can Go Wrong

Vulnerable Contacts

Most modern rotary switches are sealed, but some are not. Any switch with exposed contacts will be especially vulnerable to dirt and moisture, leading to unreliable connections. This was an issue in old-fashioned TV sets, where periodic contact cleaning of the channel selector switch was needed.

Exposed contacts are also more vulnerable to side-effects from temperature cycling (when a device warms up and then cools down).

Contact Overload

The contacts on a cheap rotary switch are especially vulnerable to arcing, as the user may turn the switch slowly, causing gradual engagement and disengagement of contacts instead of the snap-action that is characteristic of a well-made toggle switch. If a rotary switch may control significant currents or current surges, it must be appropriately rated, regardless of the extra expense. For more information on arc suppression in switches, see "Arcing" on page 47.

Misalignment

Most knobs for rotary switches consist of a pointer, or have a white line engraved to provide clear visual indication of the position of the switch. If this does not align precisely with indications printed on the panel, confusion will result. For hand-built equipment, the switch can be installed first, after which the control-panel indications can be glued or riveted in place on a separate piece of laminated card, plastic, or metal for precise alignment. If the switch is not secured tightly, its body may turn slightly under repeated stress, leading to erroneous interpretation of the knob position.

Misidentified Shorting Switch

If a shorting switch is used where a nonshorting switch was intended, the results can be disconcerting or even destructive, as one terminal will be briefly connected with the adjacent terminal while the switch is being turned. Multiple functions of a circuit may be activated simultaneously, and in a worst case scenario, adjacent terminals may be connected to opposite sides of the same power supply.

User Abuse

The turning force that must be applied to a full-size conventional rotary switch is significantly greater than the force that is applied to most other types of panel-mounted switches. This encourages aggressive treatment, and the turning motion is especially likely to loosen a nut holding the switch in place. The lighter action characteristic of miniature rotary switches does not necessarily solve this problem, as users who are accustomed to older-style switches may still apply the same force anyway.

Rotary switches should be mounted in expectation of rough use. It is prudent to use Loc-Tite or a similar compound to prevent nuts from loosening, and a switch should not be mounted in a thin or flimsy panel. When using a miniature rotary switch that has through-hole mounting in a circuit board, the board must be sufficiently robust and properly secured.

Wrong Shaft, Wrong Knobs, Nuts That Get Lost, Too Big to Fit

These problems are identical to those that can be encountered with a **potentiometer**, which are discussed in that entry in this encyclopedia.

rotational encoder

8

The term *rotational encoder* used to be reserved for high-quality components, often using optical methods to measure rotation with precision (more than 100 intervals in 360 degrees). Cheaper, simpler, electromechanical devices were properly referred to as *control shaft encoders*. However, the term *rotational encoder* is now applied to almost any device capable of converting rotational position to a digital output via opening and closing internal mechanical contacts; this is the sense in which the term is used here. It is sometimes distinguished from other types of encoder with the term *mechanical rotary encoder*. Magnetic and optical rotary encoders do not contain mechanical switches, are classified as *sensors* by this encyclopedia, and will appear in Volume 3. They are found in a device such as an *optical mouse*.

OTHER RELATED COMPONENTS

- **rotary switch** (See Chapter 7)

What It Does

A rotational encoder has a knob that a user can turn to display a series of prompts on an LCD screen, or to adjust the input or output on a product such as a stereo receiver. The component is almost always connected to inputs on a **microcontroller** and is usually fitted with *detents* that provide tactile feedback suggesting many closely spaced positions. The encoder often allows the user to make a selection by pushing the knob in, which closes an internal momentary switch. Thus, this type of encoder functions as a pushbutton as well as a switch.

A rotational encoder is an *incremental* or *relative* device, meaning that it merely creates and breaks internal switch connections when rotation occurs, without providing a unique code to identify each absolute rotational position. An *absolute* encoder is discussed in the **rotary switch** entry of this encyclopedia.

No schematic symbol exists to represent a rotational encoder.

How It Works

An encoder contains two pairs of contacts, which open and close out of phase with each other when the shaft rotates. In a clockwise direction, the A pair of contacts may be activated momentarily before the B pair; in a counter-clockwise direction, the B pair may be activated before the A pair. (Some encoders reverse this phase difference.) Thus if one contact from each pair is connected with two inputs of an appropriately programmed microcontroller, and if the other contact of each pair is connected with negative ground, the microcontroller can deduce which way the knob is turning by sensing which pair of contacts closes first. The microcontroller can then count the number of pulses from the contacts and interpret this to adjust an output or update a display.

A simplified schematic is shown in Figure 8-1. The two buttons inside the dashed line represent the two pairs of contacts inside the encoder, while the chip is a microcontroller. The knob and shaft that activate the internal switches are not shown. The schematic assumes that when a contact closes, it pulls the chip input to a low state. A pullup resistor is added to each input of the chip to prevent the pins from "floating" when either pair of contacts is open.

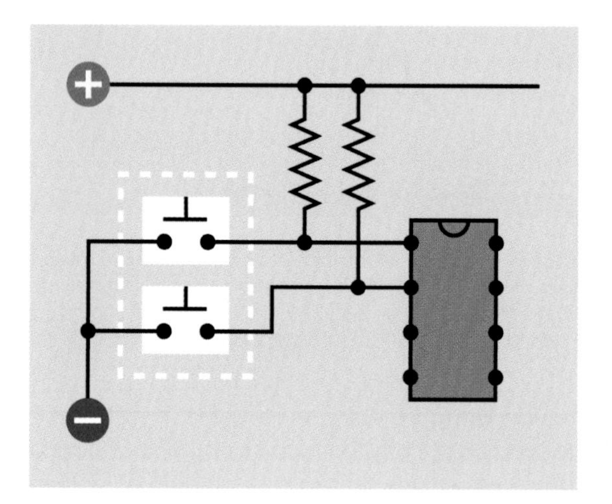

Figure 8-1. *Simplified schematic showing the typical set-up for a rotational encoder. The pushbuttons inside the dashed line represent the contacts inside the encoder. The chip is a microcontroller.*

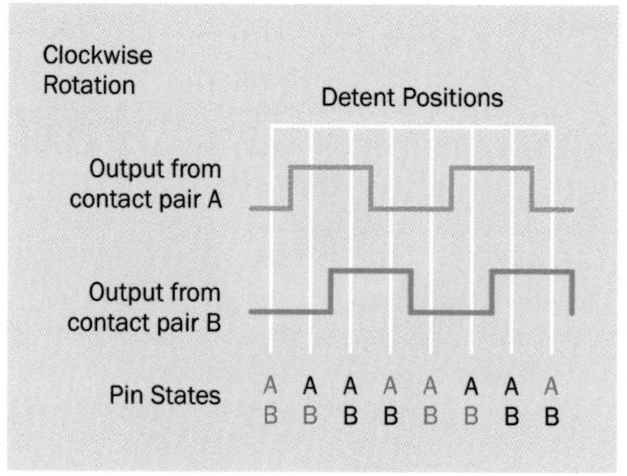

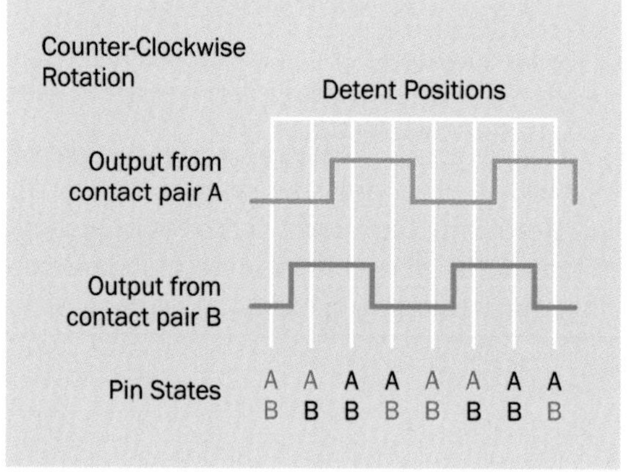

Figure 8-2. *Hypothetical outputs from a rotational encoder, assuming that the common terminals of the contact pairs are connected to negative ground. A high pulse in the graphical representation therefore indicates that the contact pair is grounded. The number of detents relative to the number of pulses per rotation varies from one type of encoder to another.*

Figure 8-2 gives a conceptual view of the outputs of an encoder that is turned clockwise (top) and then counter-clockwise (bottom). Some encoders may reverse this phase sequence. Red and black colors have been assigned to the pin states on the assumption that the terminals that are common to both pairs of contacts are connected with negative ground. Thus a "high" pulse in the graphical representation actually indicates that the encoder is grounding its output.

Microcontrollers have become so ubiquitous, and rotational encoders are so cheap, they have displaced rotary switches in many applications where a low current is being switched. The combination of a rotational encoder and a microcontroller is very versatile, allowing display and control of an almost unlimited number of menus and options.

Variants

There are two types of rotational encoders containing mechanical contacts: absolute and relative. An absolute encoder generates a code corresponding with each specific rotational position. The code is usually a binary output among four or more pins. It is discussed under *mechanical encoder* in the **rotary switch** section of this encyclopedia. The variants listed here are all relative encoders.

Pulses and Detents

Rotational encoders from different manufacturers may have as few as 4 or as many as 24 pulses per rotation (PPR), with 12 to 36 detents (or no detents at all, in a few models.) The relationship between pulses and detents shown in Figure 8-2 is typical but is far from being universal. The number of detents may be equal to, greater than, or less than the number of pulses per rotation.

Format

Rotational encoders are generally panel-mounted or through-hole devices. In the latter category, most are horizontally mounted, with a minority being at 90 degrees to the board.

Output

In an encoder containing two switches, four switch-state combinations are possible: OFF-OFF, ON-OFF, OFF-ON, and ON-ON. This is known as a *quadrature* output. All of the rotational encoders discussed here conform with that system.

Rotational Resistance

Rotational encoders vary widely in the resistance that they offer when the user turns the knob. This is largely a function of the detents, if they are included. Still, all rotational encoders generally offer less rotational resistance than a rotary switch, and do not have the kind of heavy-duty knobs that are typically used with rotary switches. Since an encoder creates only a stream of pulses without any absolute positional information, a knob with any kind of pointer on it is inappropriate.

Values

Virtually all rotational encoders are designed to work with a low-voltage supply, 12VDC or less. All of them are intended for low currents, reflecting their purpose to drive microcontroller inputs. Some sample rotational encoders are pictured in Figure 8-3. At rear: nine pulses per rotation (PPR), 36 detents, 10mA at 10VDC. Far left: 20PPR, 20 detents, with switch. Far right: 24PPR, no detents, 1mA at 5VDC. Center (blue): 16PPR, no detents, 1mA at 5VDC. Front: 12PPR, 24 detents, 1mA at 10VDC, requires Allen wrench or similar hexagonal shaft to engage with the rotor.

Figure 8-3. *Rotational encoders with a variety of specifications. See text for details.*

Contact Bounce

Any mechanical switch will suffer some degree of *contact bounce* when its contacts close. Datasheets for rotational encoders may include a specification for bounce duration ranging from around 2ms to 5ms, which is sometimes known as the *settling time*. Naturally, a lower value is preferred. The microcontroller that interprets the positional information from the encoder can include a debouncing routine that simply disregards any signals during the bounce period following switch closure.

Sliding Noise

Sliding noise is the opposite of contact bounce. When two contacts have made a connection and then rub across each other (as occurs inside a rotational encoder while the knob is being turned), the connection may suffer momentary lapses.

Datasheets for rotational encoders generally do not supply ratings for this.

How to Use it

As noted above, a rotational encoder can only be used in conjunction with a microcontroller or similar device that is capable of interpreting the phase difference between the pairs of contacts, and is capable of counting the number of opening/closing events while the knob is being turned. (Some dedicated chips are designed for this specific purpose.)

It can be adapted to be driven by a **stepper motor**, to provide feedback regarding the rotation of the motor shaft, and its output can also be interpreted to calculate angular acceleration.

Programming the microcontroller is the most significant obstacle. Generally the program should follow a sequence suggested by this pseudocode:

Check:

- If the encoder contains a pushbutton switch, check it. If the pushbutton is being pressed, go to an appropriate subroutine.
- The status of contacts A.
- The status of contacts B.

Compare their status with previously saved states for A and B. If the status has not changed, repeat from Check.

Debounce:

- Recheck the contacts status rapidly and repeatedly for 50ms, and count the states for contacts A and B. (The 50ms duration may be adjusted for different encoders, as an encoder with a higher number of pulses per rotation will tend to create shorter pulses.)
- Compare the total number of changed states with unchanged states.

If the changed states are in a small minority, probably the signal was erroneous, caused by bounce or sliding noise. Go back to Check and start over.

Interpret:

- Deduce the rotational direction from these four possibilities:
  - — Contacts A were open and have closed.
  - — Contacts A were closed and have opened.
  - — Contacts B were open and have closed.
  - — Contacts B were closed and have opened. (The specific type of encoder will determine how these transitions are interpreted.)
- Revise the variable storing the direction of rotation if necessary.
- Depending on the direction of rotation, increment or decrement a variable that counts pulses.
- Take action that is appropriate to the direction of rotation and the cumulative number of pulses.
- Go back to Check again.

What Can Go Wrong

Switch Bounce

In addition to a debouncing algorithm in the microcontroller, a 0.1µF bypass capacitor can be used with each of the output terminals from the encoder, to help reduce the problem of switch bounce.

Contact Burnout

Rotational encoders are TTL-compatible. They are not generally designed to drive even a small output device, such as an LED. The contacts are extremely delicate and will be easily damaged by any attempt to switch a significant current.

relay

Properly known as an *electromagnetic armature relay* to distinguish it from a *solid-state relay*. However, the full term is very rarely used. It may also be described as an *electromechanical relay*, but the term **relay** is normally understood to mean a device that is not solid state.

OTHER RELATED COMPONENTS

- **solid state relay** (Volume 2)
- **switch** (See Chapter 6)

What It Does

A relay enables a signal or pulse of electricity to switch on (or switch off) a separate flow of electricity. Often, a relay uses a low voltage or low current to control a higher voltage and/or higher current. The low voltage/low current signal can be initiated by a relatively small, economical switch, and can be carried to the relay by relatively cheap, small-gauge wire, at which point the relay controls a larger current near to the load. In a car, for example, turning the ignition switch sends a signal to a relay positioned close to the starter motor.

While solid-state switching devices are faster and more reliable, relays retain some advantages. They can handle double-throw and/or multiple-pole switching and can be cheaper when high voltages or currents are involved. A comparison of their advantages relative to **solid state relays** and **transistors** is tabulated in the entry on **bipolar transistor** in Figure 28-15.

Common schematic symbols for single-throw relays are shown in Figure 9-1 and for double-throw relays in Figure 9-2. The appearance and orientation of the coil and contacts in the symbols may vary significantly, but the functionality remains the same.

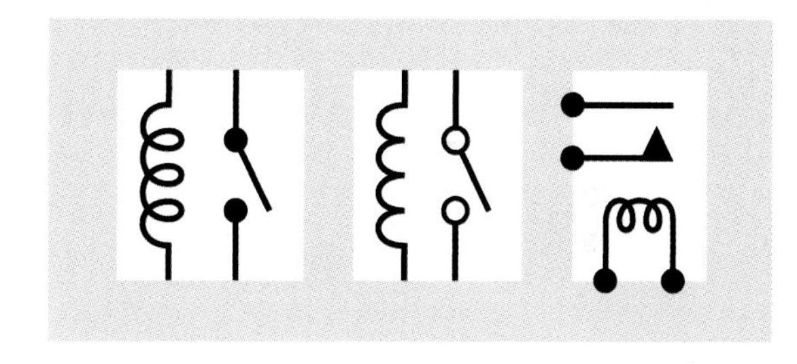

Figure 9-1. *Commonly used schematic symbols for a SPST relay. The symbols are functionally identical.*

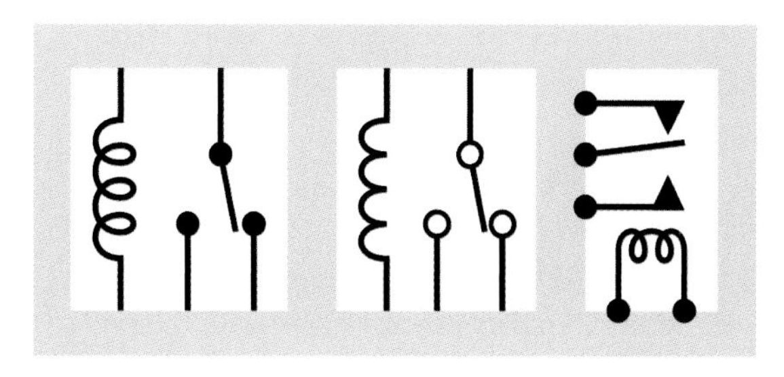

Figure 9-2. *Commonly used schematic symbols for a SPDT relay. The symbols are functionally identical.*

How It Works

A relay contains a *coil*, an *armature*, and at least one pair of *contacts*. Current flows through the coil, which functions as an **electromagnet** and generates a magnetic field. This pulls the armature, which is often shaped as a pivoting bracket that closes (or opens) the contacts. These parts are visible in the simplified rendering of a DPST relay in Figure 9-3. For purposes of identification, the armature is colored green, while the coil is red and the contacts are orange. The two blue blocks are made of an insulating material, the one on the left supporting the contact strips, the one on the right pressing the contacts together when the armature pivots in response to a magnetic field from the coil. Electrical connections to the contacts and the coil have been omitted for simplicity.

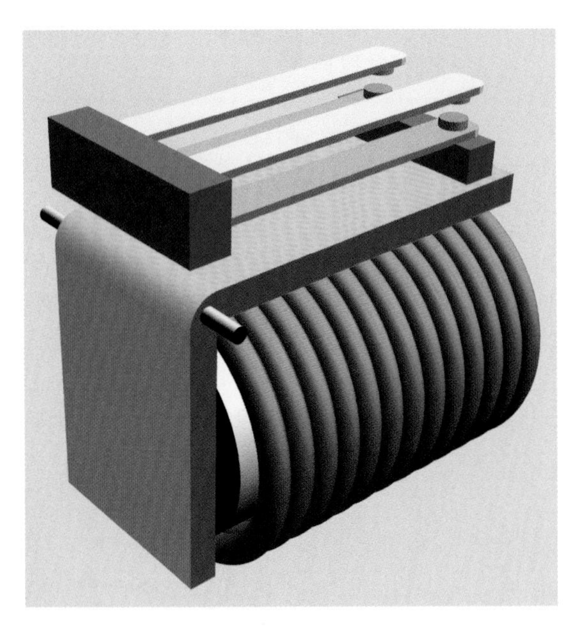

Figure 9-3. *This simplified rendering shows the primary parts of a DPST relay. See text for details.*

Various small relays, capable of handling a variety of voltages and currents, are pictured in Figure 9-4. At top-left is a 12VDC automotive relay, which plugs into a suitable socket shown immediately below it. At top-right is a 24VDC SPDT relay with exposed coil and contacts, making it suitable only for use in a very clean, dry environment. Continuing downward, the four sealed relays in colored plastic cases are designed to switch currents of 5A at 250VAC, 10A at 120VAC, 0.6A at 125VAC, and 2A at 30VDC, respectively. The two blue relays have 12VDC coils, while the red and yellow relays have 5V coils. All are nonlatching, except for the yellow relay, which is a latching type with two coils. At bottom-left is a 12VDC relay in a transparent case, rated to switch up to 5A at 240VAC or 30VDC.

Figure 9-4. *An assortment of small DC-powered relays. See text for details.*

The configuration of a relay is specified using the same abbreviations that apply to a **switch**. SP, DP, 3P, and 4P indicate 1, 2, 3, or 4 poles (relays with more than 4 poles are rare). ST and DT indicate single-throw or double-throw switching. These abbreviations are usually concatenated, as in 3PST or SPDT. In addition, the terminology Form A (meaning normally open), Form B (normally closed), and Form C (double-throw) may be used, preceded by a number that indicates the number of poles. Thus "2 Form C" means a DPDT relay.

Variants

Latching

There are two basic types of relay: *latching* and *nonlatching*. A nonlatching relay, also known as a *single side stable* type, is the most common, and resembles a *momentary switch* or **pushbutton** in that its contacts spring back to their default state when power to the relay is interrupted. This can be important in an application where the relay should return to a known state if power is lost. By contrast, a latching relay has no default state. Latching relays almost always have double-throw contacts, which remain in either position without drawing power. The relay only requires a short pulse to change its status. In semiconductor terms, its behavior is similar to that of a *flip-flop*.

In a *single-coil* latching relay, the polarity of voltage applied to the coil determines which pair of contacts will close. In a *dual-coil* latching relay, a second coil moves the armature between each of its two states.

Schematic symbols for a dual-coil latching relay are shown in Figure 9-5. Some symbol styles do not make it clear which switch position each coil induces. It may be necessary to read the manufacturer's datasheet or test the relay by applying its rated voltage to randomly selected terminal pairs while testing for continuity between other terminal pairs.

Polarity

There are three types of DC relay. In a *neutral relay*, polarity of DC current through the coil is irrelevant. The relay functions equally well either way. A *polarized relay* contains a diode in series with the coil to block current in one direction. A *biased relay* contains a permanent magnet near the armature, which boosts performance when current flows through the coil in one direction,

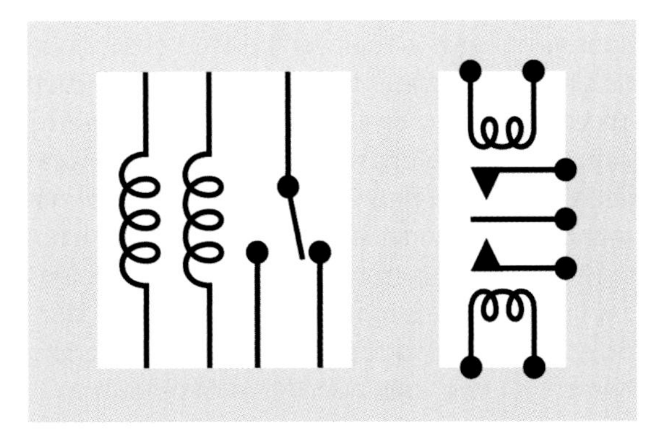

Figure 9-5. *Schematic symbols for a two-coil latching relay. The symbols are functionally identical.*

but blocks a response when the current flows through the coil in the opposite direction. Manufacturers' datasheets may not use this terminology, but will state whether the relay coil is sensitive to the polarity of a DC voltage.

All relays can switch AC current, but only an *AC relay* is designed to use AC as its coil current.

Pinout Variations

The layout and function of relay pins or quick connects is not standardized among manufacturers. Often the component will have some indication of pin functions printed on it, but should always be checked against the manufacturer's datasheet and/or tested for continuity with a meter.

Figure 9-6 shows four sample pin configurations, adapted from a manufacturer's datasheet. These configurations are functionally quite different, although all of them happen to be for DPDT relays. In each schematic, the coil of the relay is shown as a rectangle, while the pins are circles, black indicating an energized state and white indicating a non-energized state. The bent lines show the possible connections between the poles and other contacts inside the relay. The contacts are shown as arrows. Thus, pole 4 can connect with either contact 3 or contact 5, while pole 9 can connect with either contact 8 or contact 10.

Top-left: Polarized nonlatching relay in its resting condition, with no power applied. Top right: Single-coil latching relay showing energized contacts (black circles) when the coil is powered with the polarity indicated. If the polarity is reversed, the relay flips to its opposite state. Some manufacturers indicate the option to reverse polarity by placing a minus sign alongside a plus sign, and a plus sign alongside a minus sign. Bottom-left and bottom-right: Polarized latching relays with two coils, with different pinouts.

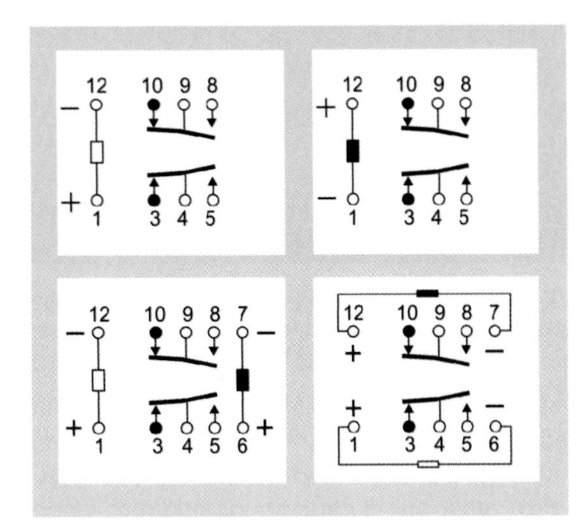

Figure 9-6. *Relay pinouts depicted in the style commonly found in manufacturers' datasheets, showing different relay types. Top-left: Single coil, nonlatching. Top-right: Single coil, latching. Bottom left: Two-coil, latching. Bottom right: Two-coil, latching, alternate pinouts. (Adapted from a Panasonic datasheet.)*

In these diagrams, the relay is seen from above. Some datasheets show the relay seen from below, and some show both views. Some manufacturers use slightly different symbols to indicate interior functions and features. When in doubt, use a meter for verification.

Reed Relay

A *reed relay* is the smallest type of electromechanical relay with applications primarily in test equipment and telecommunications. With a coil resistance ranging from 500 to 2000 ohms, these relays consume very little power. The design consists of a *reed switch* with a coil wrapped around it. Figure 9-7 shows a simplified rendering. The two black contacts are enclosed in a glass or plastic envelope and magnetized in such a way that a magnetic field from the surrounding coil bends them together, creating a connection. When power to the coil is disconnected, the magnetic field collapses and the contacts spring apart.

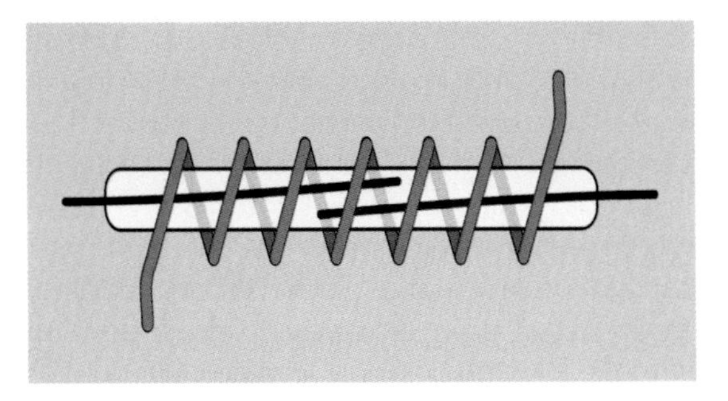

Figure 9-7. *This simplified rendering shows a reed relay, consisting of a magnetized reed switch inside a glass or plastic pod, activated by a coil wrapped around it.*

In Figure 9-8, two reed relays are shown, at top-left and center-right. At bottom-left, the type of relay on the right has been opened by a belt sander to reveal its copper coil and inside that, a capsule in which the relay contacts are visible.

Surface-mount reed relays can be smaller than 0.5″ × 0.2″. Through-hole versions are often around 0.7″ × 0.3″ with pins in two rows, though some are available in SIP packages.

Reed relays have limited current switching capacity and are not suitable to switch inductive loads.

Small Signal Relay

A *small signal relay* is also known as a *low signal relay*. This type may have a footprint as small as a reed relay but generally stands slightly taller, requires slightly more coil current, and is available in versions that can switch slightly higher voltages and currents. There are usually two rows of pins, spaced either 0.2″ or 0.3″ apart. The red

Figure 9-8. *Three reed relays, one of which has had its packaging partially removed by a belt sander to reveal its copper coil and internal contacts.*

and orange relays in Figure 9-4 are small signal relays.

Automotive Relays

An *automotive relay* is typically packaged in a cube-shaped black plastic case with quick-connect terminals at the bottom, typically plugged into a socket. Naturally they are designed to switch, and be switched by, a 12VDC supply.

General Purpose/Industrial

These relays cover a very wide range and are usually built without significant concern for size. They may be capable of switching high currents at high voltages. Typically they are designed to plug into a socket such as an *octal base* of the type that was once used for **vacuum tubes**. The base, in turn, terminates in solder tabs, screws, or quick connects and is designed to be screwed to a chassis. It allows the relay to be unplugged and swapped without resoldering.

Two industrial relays are shown in Figure 9-9. Both are DPDT type with 12VDC coils and rated to switch up to 10A at 240VAC. The one on the left has an octal base. An octal socket that fits an octal base is shown in Figure 9-10.

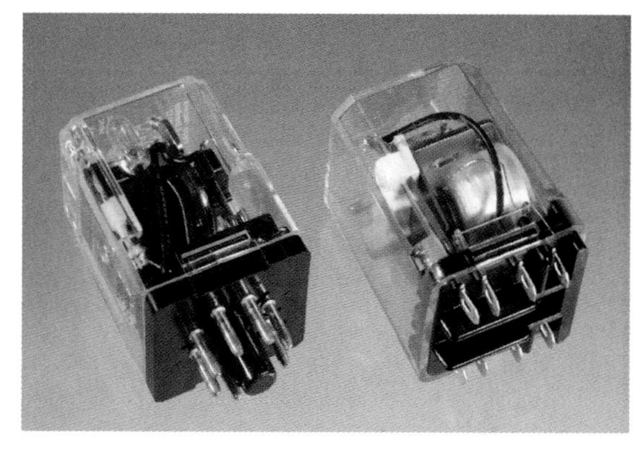

Figure 9-9. *Two relays powered by 12VDC, capable of switching up to 10A at 240VAC.*

Figure 9-10. *An octal socket with screw terminals, designed to accept a relay with an octal base.*

Time Delay Relay

Generally used to control industrial processes, a *time delay relay* switches an output on and off at preset time intervals that can be programmed to repeat. The example in Figure 9-11 has a 12VDC coil and is rated to switch up at 10A at 240VAC. It has an octal base.

Figure 9-11. *The control switches on a time-delay relay, allowing separately configured "on" and "off" intervals.*

Contactor

A contactor functions just like a relay but is designed to switch higher currents (up to thousands of amperes) at higher voltages (up to many kilovolts). It may range from being palm-sized to measuring more than one foot in diameter, and may be used to control heavy loads such as very large motors, banks of high-wattage lights, and heavy-duty power supplies.

Values

Datasheets usually specify maximum voltage and current for the contacts, and nominal voltage and current for the coil, although in some cases the coil resistance is stated instead of nominal coil current. The approximate current consumption can be estimated, if necessary, by using Ohm's Law. The minimum voltage that the relay needs for activation is sometimes described as the *Must Operate By* voltage, while the *Must Release By* voltage is the maximum coil voltage that the relay will ignore. Relays are rated on the assumption that the coil may remain energized for long periods, unless otherwise stated.

While the contact rating may suggest that a relay can switch a large load, this is not necessarily true if the load has significant inductance.

Reed relays
: Usually use a coil voltage of 5VDC and have a contact rating of up to 0.25A at 100V. Through-hole (PCB) versions may have a coil voltage of 5VDC, 6VDC, 12VDC, or 24VDC and in some cases claim to switch 0.5A to 1A at up to 100V, although this rating is strictly for a noninductive load.

Small signal/low signal relays
: Usually use a coil voltage ranging from 5VDC to 24VDC, drawing about 20mA. Maximum switching current for noninductive loads ranges from 1A to 3A.

Industrial/general purpose relays
: A very wide range of possible values, with coil voltages ranging up to 48VDC or 125VAC to 250VAC. Contact rating is typically 5A to 30A.

Automotive relays
: Coil voltage of 12VDC, and contact rating often 5A at up to 24VDC.

Timer relays
: Usually these specify a coil voltage of 12VDC, 24VDC, 24VAC, 125VAC, or 230VAC. The timed interval can range from 0.1 sec to 9999 hours in some cases. Common values for contact ratings are 5A up to 20A, with a voltage of 125V to 250V, AC or DC.

How to Use it

Relays are found in home appliances such as dishwashers, washing machines, refrigerators, air conditioners, photocopy machines, and other products where a substantial load (such as a motor or compressor) has to be switched on and off by a control switch, a thermostat, or an electronic circuit.

Figure 9-12 shows a common small-scale application in which a signal from a microcontroller (a few mA at 5VDC) is applied to the base of a transistor, which controls the relay. In this way, a logic output can switch 10A at 125VAC. Note the rectifier diode wired in parallel with the relay coil.

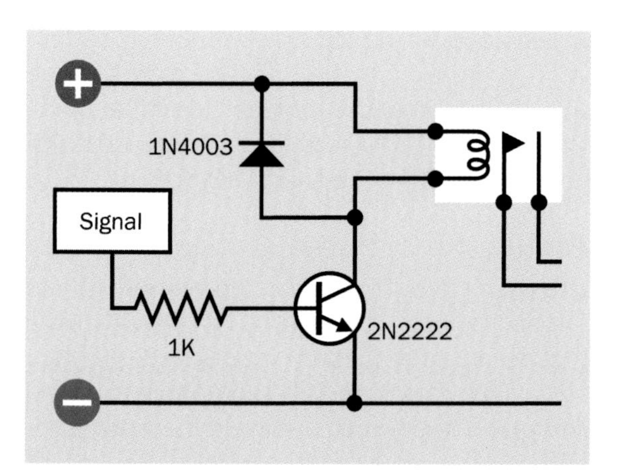

Figure 9-12. *A signal from a digital source such as a microcontroller can switch substantial voltage and current if it is applied to the base of a transistor that activates a relay.*

A latching relay is useful wherever a connection should persist when power is switched off or interrupted, or if power consumption must be minimized. Security devices are one common application. However, the circuit may require a "power reset" function to restore known default settings of latched relays.

A circuit including every possible protection against voltage spikes is shown in Figure 9-13, including a snubber to protect the relay contacts, a rectifier diode to suppress back-EMF generated by the relay coil, and another rectifier diode to protect the relay from EMF generated by a motor when the relay switches it on and off. The snubber can be omitted if the motor draws a relatively low current (below 5A) or if the relay is switching a noninductive load. The diode around the relay coil can be omitted if there are no semiconductors or other components in the circuit that are vulnerable to voltage spikes. However, a spike can affect components in adjacent circuits that appear to be electrically isolated. A severe spike can even be transmitted back into 125VAC house wiring. For information on using a resistor-capacitor combination to form a snubber, see "Snubber" on page 102.

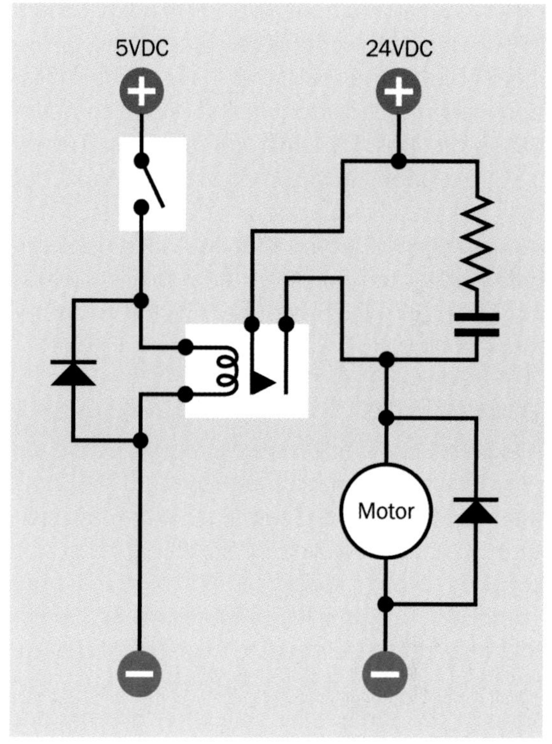

Figure 9-13. *This hypothetical schematic shows three types of protection against voltage spikes induced by an inductive load (a motor, in this instance) and the coil of the relay.*

What Can Go Wrong

Wrong Pinouts

The lack of standardization of relay pinouts can cause errors if one relay is replaced with another that appears to be the same, but isn't. In particular, the pins that connect with normally-closed contacts may trade places with pins connected with normally-open contacts, in relays from different manufacturers.

Pinouts are also confusing in that some datasheets depict them from above, some from below, and some from both perspectives.

Wrong Orientation

Small relays of through-hole type usually have pins spaced in multiples of 0.1". This allows them to be inserted the wrong way around in a perforated board. Almost all relays have an identification mark molded into one end or one corner of the plastic shell. Manufacturers do not standardize the position and meaning of these marks, but they are usually replicated in datasheets. When using a relay of a type that you have not used before, it is a sensible precaution to test it with a meter to verify the functions of its terminals before installing it.

Wrong Type

A latching relay may have exactly the same appearance as a nonlatching relay from the same manufacturer, and the same two pins may energize the coil. However, in a latching relay, the contacts won't spring back to their nonenergized position, causing functional errors that may be difficult to diagnose. The part numbers printed on latching and nonlatching versions of the same relay may differ by only one letter or numeral and should be checked carefully.

Wrong Polarity

A relay with a DC-energized coil may require power to be applied with correct polarity and may malfunction otherwise.

AC and DC

A relay coil designed to be powered by DC will not work from AC and vice-versa. The contact rating of a relay is likely to be different depending whether it is switching AC or DC.

Chatter

This is the noise created by relay contacts when they make rapid intermittent connection. Chatter is potentially damaging to relay contacts and should be avoided. It can also create electrical noise that interferes with other components. Likely correctible causes of chatter include insufficient voltage or power fluctuations.

Relay Coil Voltage Spike

A relay coil is an inductive device. Merely switching a large relay on and off can create voltage spikes. To address this problem, a rectifier diode should be placed across the coil terminals with polarity opposing the energizing voltage.

Arcing

This problem is discussed in the **switch** entry of this encyclopedia. See "Arcing" on page 47. Note that because the contacts inside a *reed relay* are so tiny, they are especially susceptible to arcing and may actually melt and weld themselves together if they are used to control excessive current or an inductive load.

Magnetic Fields

Relays generate magnetic fields during operation and should not be placed near components that are susceptible.

The reed switch inside a reed relay can be unexpectedly activated by an external magnetic field. This type of relay may be enclosed in a metal shell to provide some protection. The adequacy of this protection should be verified by testing the relay under real-world conditions.

Environmental Hazards

Dirt, oxidation, or moisture on relay contacts is a significant problem. Most relays are sealed and should remain sealed.

Relays are susceptible to vibration, which can affect the contacts and can accelerate wear on moving parts. Severe vibration can even damage a relay permanently. **Solid-state relays** (discussed in Volume 2) should be used in harsh environments.

resistor

10

OTHER RELATED COMPONENTS

- **potentiometer** (See Chapter 11)

What It Does

A resistor is one of the most fundamental components in electronics. Its purpose is to impede a flow of current and impose a voltage reduction. It consists of two wires or conductors attached at opposite ends or sides of a relatively poor electrical conductor, the resistance of which is measured in ohms, universally represented by the Greek omega symbol, Ω.

Schematic symbols that represent a resistor are shown in Figure 10-1 (Left: The traditional schematic symbol. Right: The more recent European equivalent). The US symbol is still sometimes used in European schematics, and the European symbol is sometimes used in US schematics. Letters K or M indicate that the value shown for the resistor is in thousands of ohms or millions of ohms, respectively. Where these letters are used in Europe, and sometimes in the US, they are substituted for a decimal point. Thus, a 4.7K resistor may be identified as 4K7, a 3.3M resistor may be identified as 3M3, and so on. (The numeric value in Figure 10-1 was chosen arbitrarily.)

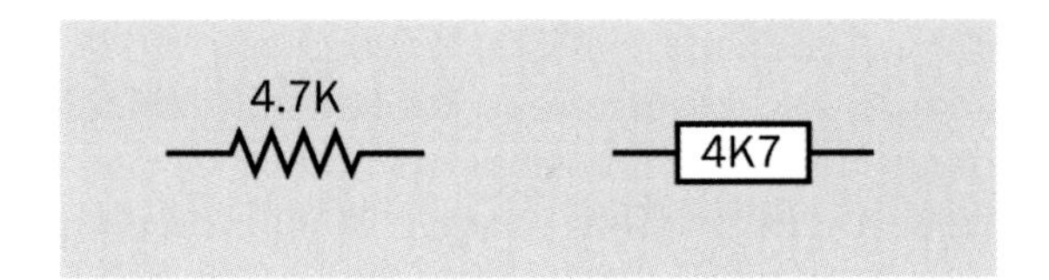

Figure 10-1. *Resistor symbols. The left one is more common in the United States, while the right one is widely used in Europe. The 4.7K value was chosen arbitrarily.*

A resistor is commonly used for purposes such as limiting the charging rate of a **capacitor**; providing appropriate control voltage to semiconductors such as **bipolar transistors**; protecting **LEDs** or other semiconductors from excessive current; adjusting or limiting the frequency response in an audio circuit (in conjunction with other components); pulling up or pulling down the voltage at the input pin of a digital logic chip; or controlling a voltage at a point in a circuit. In this last application, two resistors may be placed in series to create a *voltage divider*.

A **potentiometer** may be used instead of a resistor where variable resistance is required.

Sample resistors of various values are shown in Figure 10-2. From top to bottom, their power dissipation ratings are 3W, 1W, 1/2W, 1/4W, 1/4W, 1/4W, and 1/8W. The accuracy (tolerance) of each resistor, from top to bottom, is plus-or-minus 5%, 5%, 5%, 1%, 1%, 5%, and 1%. The beige-colored body of a resistor is often an indication that its tolerance is 5%, while a blue-colored body often indicates a tolerance of 1% or 2%. The blue-bodied resistors and the dark brown resistor contain metal-oxide film elements, while the beige-bodied resistors and the green resistor contain carbon film. For more information on resistor values, see the upcoming *Values* section.

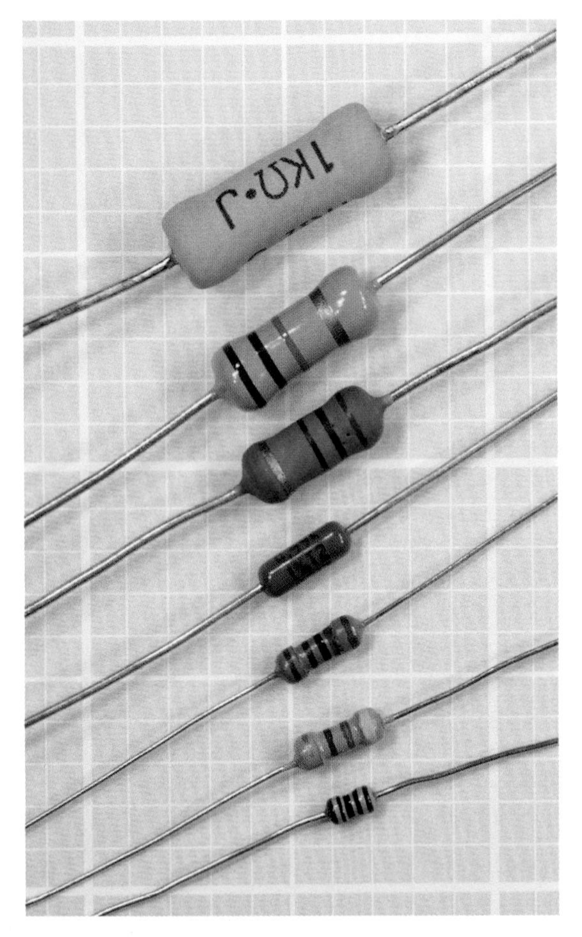

Figure 10-2. *A range of typical resistors. See text for details.*

How It Works

In the process of impeding the flow of current and reducing voltage, a resistor absorbs electrical energy, which it must dissipate as heat. In most modern electronic circuits, the heat dissipation is typically a fraction of a watt.

If R is the resistance in ohms, I is the current flowing through the resistor in amperes, and V is the voltage drop imposed by the resistor (the difference in electrical potential between the two contacts that are attached to it), Ohm's law states:

$$V = I * R$$

This is another way of saying that a resistor of 1Ω will allow a current of 1 amp when the potential difference between the ends of the resistor is 1 volt.

If W is the power in watts dissipated by the resistor, in a DC circuit:

$$W = V * I$$

By substitution in Ohm's law, we can express watts in terms of current and resistance:

$$W = I^2 * R$$

We can also express watts in terms of voltage and resistance:

$$W = V^2 / R$$

These alternates may be useful in situations where you do not know the voltage drop or the current, respectively.

Approximately similar relationships exist when using alternating current, although the power will be a more complex function.

Variants

Axial resistors have two leads that emerge from opposite ends of a usually cylindrical body. *Radial resistors* have parallel leads emerging from one side of the body and are unusual.

Precision resistors are generally defined as having a tolerance of no more than plus-or-minus 1%.

General-purpose resistors are less stable, and their value is less precise.

Power resistors are generally defined as dissipating 1 or 2 watts or more, particularly in power supplies or power amplifiers. They are physically larger and may require *heat sinks* or fan cooling.

Wire-wound resistors are used where the component must withstand substantial heat. A wire-wound resistor often consists of an insulating tube or core that is flat or cylindrical, with multiple turns of resistive wire wrapped around it. The wire is usually a nickel-chromium alloy known as *nichrome* (sometimes written as *Ni-chrome*) and is dipped in a protecting coating.

The heat created by current passing through resistive wire is a potential problem in electronic circuits where temperature must be limited. However, in household appliances such as hair dryers, toaster ovens, and fan heaters, a nichrome element is used specifically to generate heat. Wire-wound resistors are also used in 3D printers to melt plastic (or some other compound) that forms the solid output of the device.

Thick film resistors are sometimes manufactured in a flat, square format. A sample is shown in Figure 10-3, rated to dissipate 10W from its flat surface. The resistance of this component is 1K.

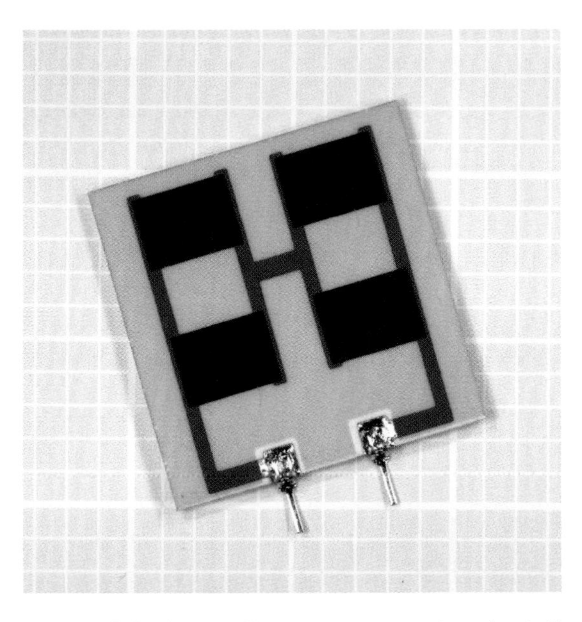

Figure 10-3. *A thick-film resistor measuring about 1" square and 0.03" thick.*

Surface-mount resistors generally consist of a resistive ink film printed on top of a tablet of aluminum oxide ceramic compound, often approximately 6mm long, known as a 2512 form factor. Each surface-mount resistor has two nickel-plated terminations coated in solder, which melts when the resistor is attached to the circuit board. The upper surface is coated, usually with black epoxy, to protect the resistive element.

Resistor Array

This is also known as a *resistor network* or *resistor ladder*, and consists of a chip containing multiple equal-valued resistors.

A resistor array in a *single-inline package* (or *SIP*) may have three possible internal configurations: isolated, common bus, and dual terminator. These options are shown at top, center, and bottom, respectively, in Figure 10-4. The isolated variant is commonly available in SIPs with 6, 8, or 10 pins. The common-bus and dual-terminator configurations generally have 8, 9, 10, or 11 pins.

In the isolated configuration, each resistor is electrically independent of the others and is accessed via its own pair of pins. On a common bus, one end of each resistor shares a bus accessed by a single pin, while the other ends of the resistors are accessed by their own separate pins. A dual-terminator configuration is more complex, consisting of pairs of resistors connected between ground and an internal bus, with the midpoint of each resistor pair accessible via a separate pin. The resistor pairs this function as voltage dividers and are commonly used in emitter-coupled logic circuits that require termination with -2 volts.

A *dual-inline package* (*DIP*) allows a similar range of internal configurations, as shown in Figure 10-5. At top, isolated resistors are commonly available in DIPs with 4, 7, 8, 9, or 10 pins. At center, the common bus configuration is available in DIPs with 8, 14, 16, 18, or 20 pins. At bottom, the dual-terminator configuration usually has 8, 14, 16, 18, or 20 pins.

The external appearance of SIP and DIP resistor arrays is shown in Figure 10-6. From left to right, the packages contain seven 120Ω resistors in isolated configuration; thirteen 120Ω resistors in bussed configuration; seven 5.6K resistors in bussed configuration; and six 1K resistors in bussed configuration.

Resistor arrays with isolated or common-bus configurations are a convenient way to reduce the component count in circuits where pullup, pulldown, or terminating resistors are required

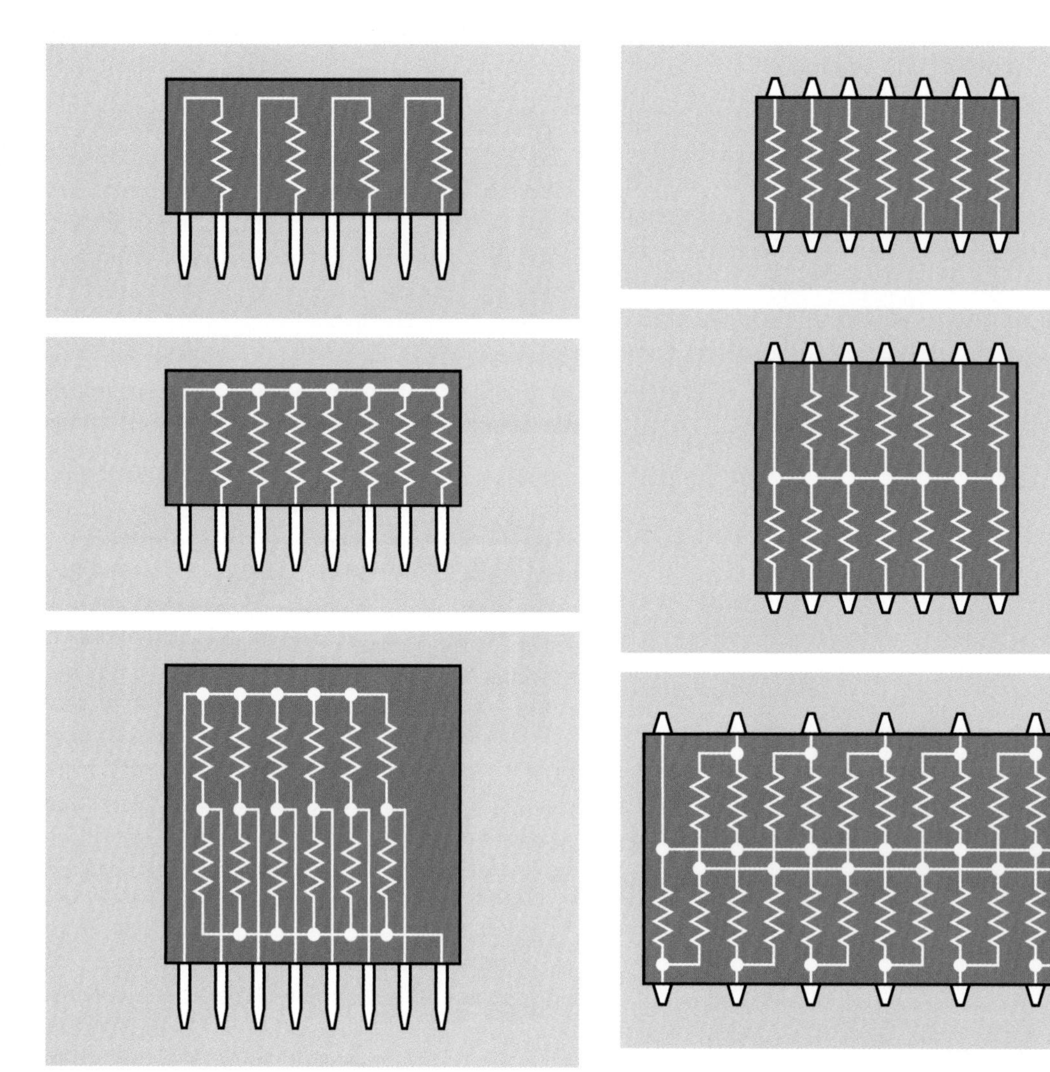

Figure 10-4. *Multiple resistors can be embedded in a single-inline package (SIP) in a variety of formats. See text for additional details.*

Figure 10-5. *Multiple resistors can be obtained embedded in a dual-inline package (DIP). See text for additional details.*

for multiple chips. The common-bus configuration is also useful in conjunction with a 7-segment **LED display**, where each segment must be terminated by a series resistor and all the resistors share a common ground or common voltage source.

Surface-mount chips are available containing a pair of resistors configured as a single voltage divider.

Chips containing multiple *RC circuits* (each consisting of a capacitor and a resistor in series) are available, although uncommon. A package containing a single RC circuit may be sold as a *snubber* to protect contacts in a **switch** or **relay** that switches a large inductive load. More information on snubber circuits is in the **capacitor** entry of this encyclopedia; see "Snubber" on page 102.

Figure 10-6. *Resistor arrays in DIP and SIP packages. See text for values.*

Values

1 *kilohm,* usually written as 1K, is 1,000Ω. 1 *megohm,* usually written as 1M or 1 meg, is 1,000K. 1 *gigaohm* is 1,000 megs, although the unit is rarely used. Resistances of less than 1Ω are uncommon and are usually expressed as a decimal number followed by the Ω symbol. The term *milliohms* (thousandths of an ohm) is used in special applications. Equivalent resistor values are shown in Figure 10-7.

Ohms	Kilohms	Megohms
1	0.001	0.000001
10	0.01	0.00001
100	0.1	0.0001
1,000	1	0.001
10,000	10	0.01
100,000	100	0.1
1,000,000	1,000	1

Figure 10-7. *Equivalent values in ohms, kilohms, and megohms.*

A resistance value remains unchanged in DC and AC circuits, except where the AC reaches an extremely high frequency.

In common electronics applications, resistances usually range from 100Ω to 10M. Power ratings may vary from 1/16 watt to 1000 watts, but usually range from 1/8 watt to 1/2 watt in most electronic circuits (less in surface-mount applications).

Tolerance

The tolerance, or precision, of a resistor may range from plus-or-minus 0.001% up to plus-or-minus 20%, but is most commonly plus-or-minus 1%, 2%, 5%, or 10%.

The traditional range of resistor values was established when a tolerance of 20% was the norm. The values were spaced to allow minimum risk of a resistor at one end of its tolerance range having the same value as another resistor at the opposite end of its tolerance range. The values were rounded to 10, 15, 22, 33, 47, 68, and 100, as illustrated in Figure 10-8 where each blue diamond represents the possible range of actual values of a 20% resistor with a theoretical value shown by the white horizontal line at the center of the diamond.

Resistor factors repeat themselves in multiples of 10. Thus, for example, beginning with a resistor of 100Ω, subsequent increasing values will be 150, 220, 330, 470, 680, and 1K, whereas the range of resistors beginning with 1Ω will be 1.5, 2.2, 3.3, 4.7, 6.8, and 10Ω.

Resistance multiplication factors are now expressed as a list of preferred values by the International Electrotechnical Commission (IEC) in their 60063 standard. Intermediate factors have been added to the basic sequence to accommodate better tolerances. A table showing resistor values for tolerances of plus-or-minus 20%, 10%, and 5% appears in Figure 10-9. Resistors with a tolerance of 5% have become increasingly common.

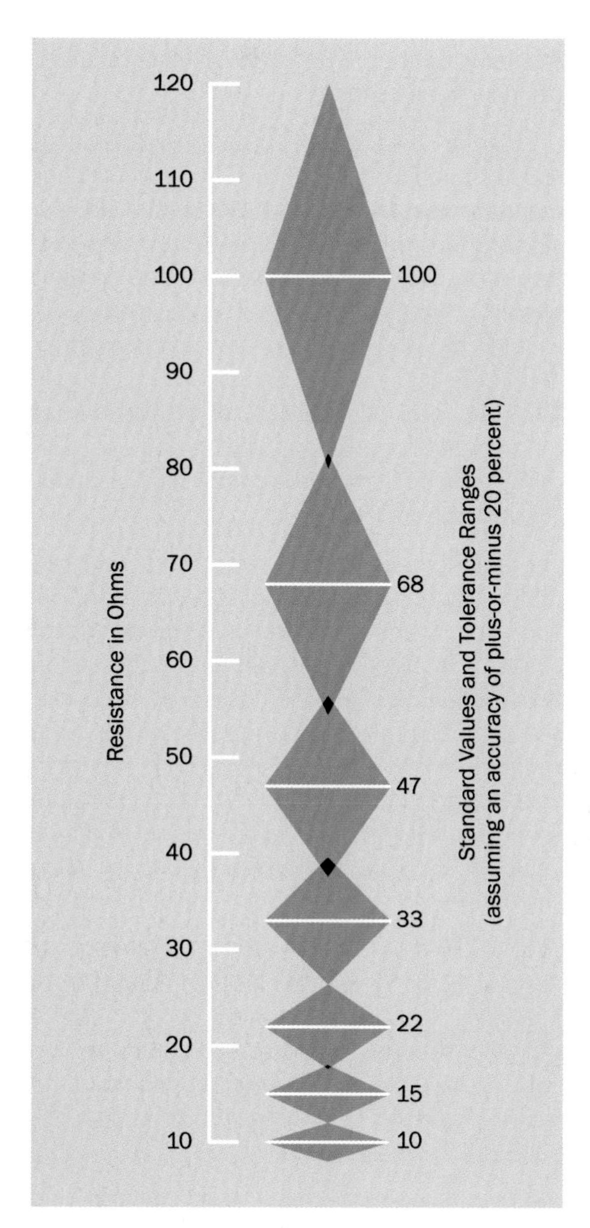

Figure 10-8. *Graphical representation of standard resistor values (white lines) established by the International Electrotechnical Commission, showing the acceptable range of actual values (dark blue areas) assuming precision of plus-or-minus 20%. The overlap, if any, between each range and the next is shown in black.*

The IEC has established 3-digit preferred values for resistors with values accurate to plus-or-minus 0.5%.

Standard Resistor Values

Resistor Tolerances (plus or minus)		
20%	10%	5%
100	100	100
		91
	82	82
		75
68	68	68
		62
	56	56
		51
47	47	47
		43
	39	39
		36
33	33	33
		30
	27	27
		24
22	22	22
		20
	18	18
		16
15	15	15
		13
	12	12
		11
10	10	10

Figure 10-9. *Standard values for resistors of different precisions. For resistors outside the range shown, values can be found by multiplying or dividing (repeatedly, if necessary) by a factor of 10.*

Because many capacitors still have a tolerance no better than 20%, their values also conform with the old original set of resistance values, although the units are expressed in farads or fractions of a farad. See the **capacitors** entry in this encyclopedia for additional information.

Value Coding

Through-hole axial resistors are traditionally printed with a sequence of three colored bands to express the value of the component, each of the first two bands representing a digit from 0 through 9, while the third band indicates the decimal multiplier (the number of zeroes, from 0 to 9, which should be appended to the digits). A fourth band of silver or gold indicates 10% or 5% tolerance respectively. No fourth band would indicate 20% tolerance, although this has become very rare.

Many resistors now have five color bands, to enable the representation of intermediate or fractional values. In this scheme, the first three bands have numeric values (using the same color system as before) while the fourth band is the multiplier. A fifth band, at the opposite end of the resistor, indicates its tolerance.

In Figure 10-10 the numeric or multiplier value of each color is shown as a "spectrum" at the top of the figure. The tolerance, or precision of a resistor, expressed as a plus-or-minus percentage, is shown using silver, gold, and various colors, at the bottom of the figure.

Two sample resistors are shown. The upper one has a value of 1K, indicated by the brown and black bands on the left (representing numeral 1 followed by a numeral 0) and the third red band (indicating two additional zeroes). The gold band at right indicates a precision of 5%. The lower one has a value of 1.05K, indicated by the brown, black, and green bands on the left (representing numeral 1 followed by numeral 0 followed by a numeral 5) and the fourth band brown (indicating one additional zero). The brown band at right indicates a precision of 1%.

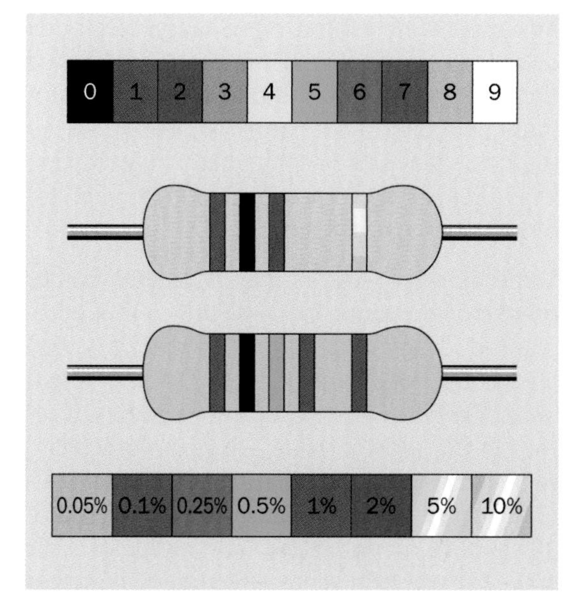

Figure 10-10. *Color coding of through-hole resistors. See text for details.*

In extremely old equipment, resistors may be coded with the body-tip-dot scheme, in which the body color represents the initial digit, the end color represents the second digit, and a dot represents the multiplier. The numeric identities of the colors is the same as in the current color scheme.

In all modern schemes, the three or four bands that show the resistance value are spaced close together, while a larger gap separates them from the band that shows the tolerance. The resistor value should be read while holding the resistor so that the group of closely-spaced numeric bands is on the left.

Confusingly, some resistors may be found where the first three bands define the value, using the old three-band convention; the fourth band indicates tolerance; and a fifth band at the opposite end of the component indicates reliability. However, this color scheme is uncommon.

Other color-coding conventions may be found in special applications, such as military equipment.

It is common for through-hole carbon-film resistors to have a beige body color, while through-hole metal-film resistors often have a blue body color. However, in relatively rare instances, a blue body color may also indicate a *fusible* resistor (designed to burn out harmlessly like a fuse, if it is overloaded) while a white body may indicate a *non-flammable* resistor. Use caution when replacing these special types.

Some modern resistors may have their values printed on them numerically. Surface-mount resistors also have digits printed on them, but they are a code, not a direct representation of resistance. The last digit indicates the number of zeroes in the resistor value, while the preceding two or three numbers define the value itself. Letter R is used to indicate a decimal point. Thus a 3R3 surface-mount resistor has a value of 3.3Ω, while 330 would indicate 33Ω, and 332 indicates 3,300Ω. A 2152 surface-mount resistor would have a value of 21,500Ω.

A surface-mount resistor with a single zero printed on it is a *zero ohm* component that has the same function as a **jumper** wire. It is used for convenience, as it is easily inserted by automated production-line equipment. It functions merely as a bridge between traces on the circuit board.

When resistor values are printed on paper in schematics, poor reproduction may result in omission of decimal points or introduction of specks that look like decimal points. Europeans have addressed this issue by using the letter as a substitute for a decimal point so that a 5.6K resistor will be shown as 5K6, or a 3.3M resistor will be shown as 3M3. This practice is followed infrequently in the United States.

Stability

This term describes the ability of a resistor to maintain an accurate value despite factors such as temperature, humidity, vibration, load cycling, current, and voltage. The *temperature coefficient* of a resistor (often referred to as T_{cr} or T_c, not to be confused with the time constant of a charging capacitor) is expressed in parts per million change in resistance for each degree centigrade deviation from room temperature (usually assumed to be 25 degrees Centigrade). T_c may be a positive or a negative value.

The *voltage coefficient* of resistance—often expressed as V_c—describes the change of a resistor's value that may occur as a function of changes in voltage. This is usually significant only where the resistive element is carbon-based. If V1 is the rated voltage of the resistor, R1 is its rated resistance at that voltage, V2 is 10% of the rated voltage, and R2 is the actual resistance at that voltage, the voltage coefficient, V_c, is given by this formula:

$$V_c = (100 * (R1 - R2)) / (R2 * (V1 - V2))$$

Materials

Resistors are formed from a variety of materials.

Carbon composite. Particles of carbon are mixed with a binder. The density of the carbon determines the end-to-end resistance, which typically ranges from 5Ω to 10M. The disadvantages of this system are low precision (a 10% tolerance is common), relatively high voltage coefficient of resistance, and introduction of noise in sensitive circuits. However, carbon-composite resistors have low inductance and are relatively tolerant of overload conditions.

Carbon film. A cheap and popular type, made by coating a ceramic substrate with a film of carbon compound. They are available in both through-hole and surface-mount formats. The range of resistor values is comparable with carbon-composite types, but the precision is increased, typically to 5%, by cutting a spiral groove in the carbon-compound coating during the manufacturing process. The carbon film suffers the same disadvantages of carbon composite resistors, but to a lesser extent. Carbon film resistors generally should not be substituted for metal film resistors in applications where accuracy is important.

Metal film. A metallic film is deposited on a ceramic substrate, and has generally superior characteristics to carbon-film resistors. During manufacture, a groove may be cut in the metal film to adjust the end-to-end resistance. This may cause the resistor to have higher inductance than carbon-composite types, though it has lower noise. Tolerances of 5%, 2%, and 1% are available. This type of resistor was originally more expensive than carbon-film equivalents, but the difference is now fractional. They are available in both through-hole and surface-mount formats. They are available in lower-wattage variants (1/8 watt is common).

Thick-film resistors are spray-coated, whereas thin-film resistors are sputtered nichrome. Thin-films enjoy a flatter temperature coefficient and are typically used in environments that have a huge operational temperature range, such as satellites.

Bulk metal foil. The type of foil used in metal film resistors is applied to a ceramic wafer and etched to achieve the desired overall resistance. Typically these resistors have axial leads. They can be extremely accurate and stable, but have a limited maximum resistance.

Precision wire-wound. Formerly used in applications requiring great accuracy, but now largely replaced by precision metal foil.

Power wire-wound. Generally used when 1 or 2 watts or more power dissipation is required. Resistive wire is wrapped around a core that is often ceramic. This can cause the resistor to be referred to, inaccurately, as "ceramic." The core may alternatively be fiberglass or some other electrically insulating compound that actively sinks heat. The component is either dipped (typically in vitreous enamel or cement) or is mounted in an aluminum shell that can be clamped to a heat sink. It almost always has the ohm value printed on it in plain numerals (not codes).

Two typical wire-wound resistors are shown in Figure 10-11. The upper resistor is rated at 12W and 180Ω while the lower resistor is rated at 13W and 15K.

Figure 10-11. *Two wire-wound resistors of greatly differing resistance but similar power dissipation capability.*

A larger wire-wound resistor is shown in Figure 10-12, rated for 25W and 10Ω.

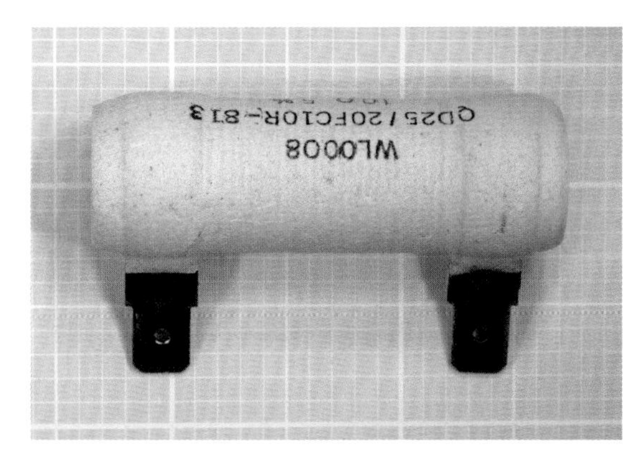

Figure 10-12. *A large wirewound resistor rated to dissipate 25W.*

In Figure 10-13, two resistors encapsulated in cement coatings are shown with the coatings removed to expose the elements. At left is a 1.5Ω 5W resistor, which uses a wire-wound element. At right is a very low-value 0.03Ω 10W resistor.

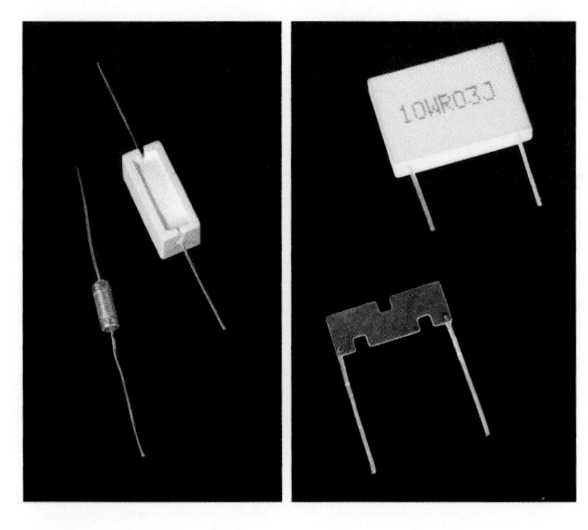

Figure 10-13. *Two low-value resistors with their cement coatings removed to show the resistive elements.*

In Figure 10-14, the resistor at right has an exposed 30Ω element while the resistor at left is rated 10W and 6.5Ω, enclosed in an anodized aluminum shell to promote heat dissipation.

Figure 10-14. *A 30Ω resistor (right) and 6.5Ω resistor (left).*

In power resistors, heat dissipation becomes an important consideration. If other factors (such as voltage) remain the same, a lower-value resistor will tend to pass more current than a higher-value resistor, and heat dissipation is proportional to the square of the current. Therefore power wire-wound resistors are more likely to be needed where low resistance values are required. Their coiled-wire format creates significant inductance, making them unsuitable to pass high frequencies or pulses.

How to Use it

Some of the most common applications for a resistor are listed here.

In Series with LED

To protect an LED from damage caused by excessive current, a *series resistor* is chosen to allow a current that does not exceed the manufacturer's specification. In the case of a single through-hole LED (often referred to as an *indicator*), the forward current is often limited to around 20mA, and the value of the resistor will depend on the voltage being used. (See Figure 10-15.)

When using high-output LEDs (which may contain multiple elements in a single 5mm or 10mm package), or LED arrays that are now being used for domestic lighting, the acceptable current may be much greater, and the LED unit may contain its own current-limiting electronics. A datasheet should be consulted for details.

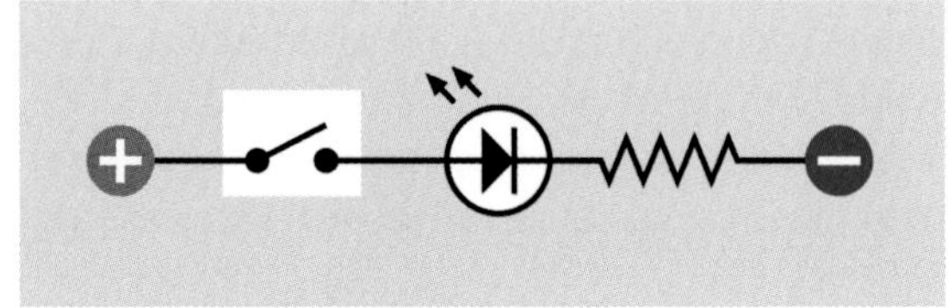

Figure 10-15. *A series resistor is necessary to limit the current that passes through an LED.*

Current Limiting with a Transistor

In Figure 10-16, a transistor is switching or amplifying current flowing from B to C. A resistor is used to protect the base of the transistor from excessive current flowing from point A. Resistors are also commonly used to prevent excessive current from flowing between B and C.

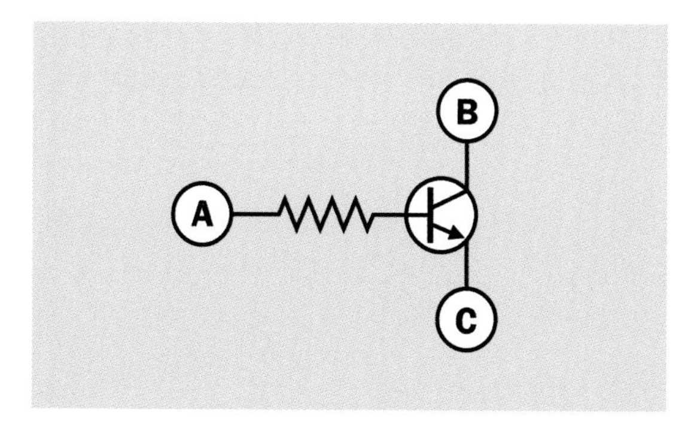

Figure 10-16. *A resistor is typically necessary to protect the base of a transistor from excessive current.*

Pullup and Pulldown Resistors

When a mechanical switch or pushbutton is attached to the input of a logic chip or **microcontroller**, a *pullup* or *pulldown* resistor is used, applying positive voltage or grounding the pin, respectively, to prevent it from "floating" in an indeterminate state when the switch is open. In Figure 10-17, the upper schematic shows a pulldown resistor, whereas the lower schematic shows a pullup resistor. A common value for either of them is 10K. When the pushbutton is pressed, its direct connection to positive voltage or to ground easily overwhelms the effect of the resistor. The choice of pullup or pulldown resistor may depend on the type of chip being used.

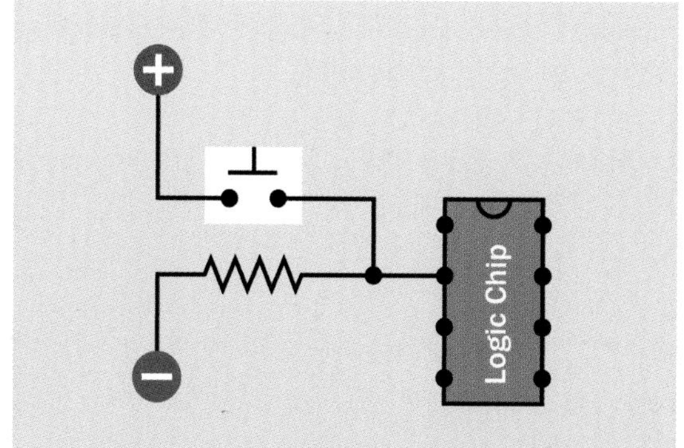

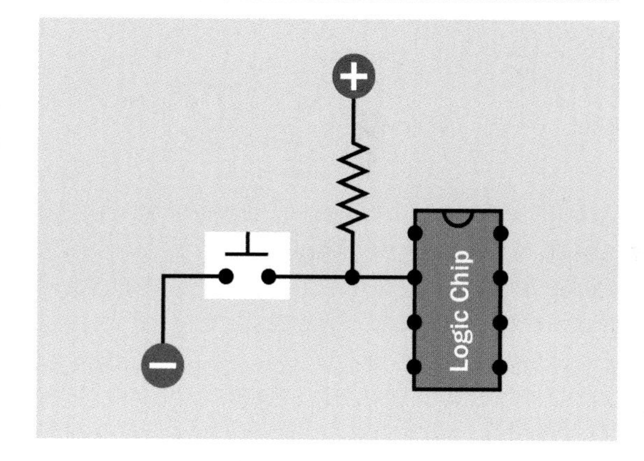

Figure 10-17. *A pulldown resistor (top) or pullup resistor (bottom) prevents an input pin on a logic chip or microcontroller from "floating" in an indeterminate state when the button is not being pressed.*

Audio Tone Control

A resistor-capacitor combination can limit the high-frequency in a simple audio tone-control circuit, as shown in Figure 10-18. Beneath a signal travelling from A to B, a resistor is placed in series with a capacitor that passes high frequencies to ground. This is known as a low-pass filter.

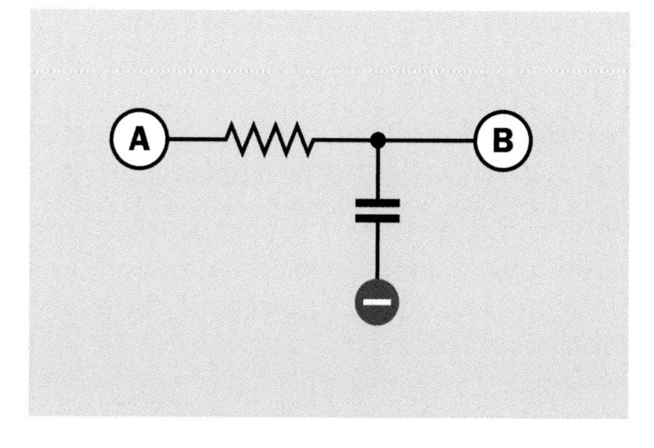

Figure 10-18. *This configuration may be used to remove high frequencies from an audio signal. It is known as a low-pass filter because low frequencies are passed from A to B.*

RC Network

A resistor will adjust the charge/discharge time when placed in series with a capacitor, as in Figure 10-19. When the switch closes, the resistor limits the rate at which the capacitor will charge itself from the power supply. Because a capacitor has an ideally infinite resistance to DC current, the voltage measured at point A will rise until it is close to the supply voltage. This is often referred to as an *RC* (*resistor-capacitor*) network and is discussed in greater detail in the **capacitor** section of this encyclopedia.

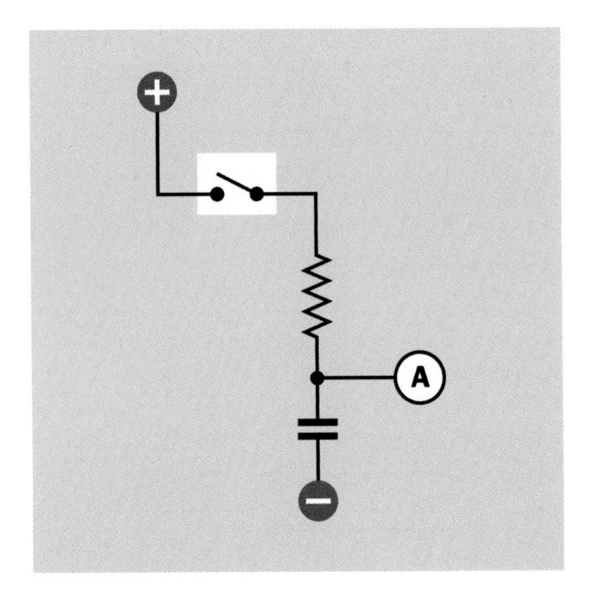

Figure 10-19. *In an RC (resistor-capacitor) network, a resistor limits the rate of increase in potential of the capacitor, measured at A, when the switch is closed.*

Voltage Divider

Two resistors may be used to create a voltage divider (see Figure 10-20). If V_{in} is the supply voltage, the output voltage, V_{out}, measured at point A, is found by the formula:

$$V_{out} = V_{in} * (R2 / (R1 + R2))$$

In reality, the actual value of V_{out} is likely to be affected by how heavily the output is loaded.

If the output node has a high impedance, such as the input to a logic chip or comparator, it will be more susceptible to electrical noise, and lower-value resistors may be needed in the voltage divider to maintain a higher current flow and maintain stability in the attached device.

Resistors in Series

If resistors in series have values R1, R2, R3 . . . the total resistance, R, is found by summing the individual resistances:

```
R = R1 + R2 + R3. . .
```

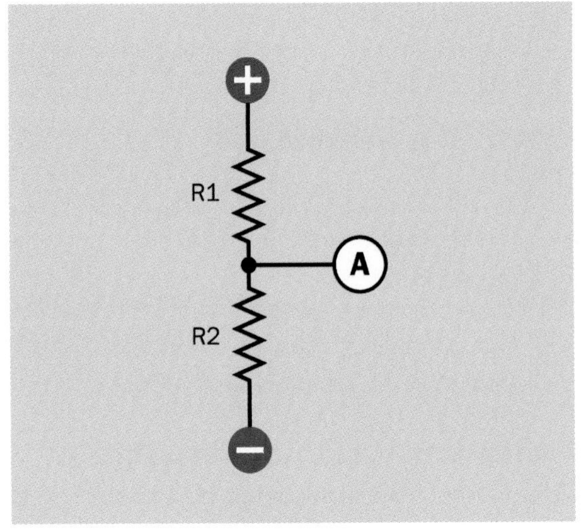

Figure 10-20. *In a DC circuit, a pair of resistors may be placed in series to function as a voltage divider. The voltage measured at A will be lower than the supply voltage, but above ground potential.*

The current through each of the resistors will be the same, whereas the voltage across each of them will vary proportionately with its resistance. If the supply voltage across the series of resistors is VS, and the total of all the resistor values is RT, and the resistance of one resistor is R1, the voltage across that resistor, V1, will be given by the formula:

```
V1 = VS * (R1 / RT)
```

Resistors in Parallel

Where two or more resistors (R1, R2, R3 . . .) are wired in parallel, their total resistance, R, is found from the formula:

```
1/R = ( 1/R1 ) + ( 1/R2 ) + ( 1/R3 ). . .
```

Suppose that R1, R2, R3 . . . all have the same individual resistance, represented by RI, and the number of resistors is N. Their total resistance, RT, when wired in parallel, will be:

```
RT = RI / N
```

If each resistor has an equal resistance and also has an equal individual rating in watts (represented by WI), the total wattage (WT) that they can handle when wired in parallel to share the power will be:

```
WT = WI * N
```

Therefore, if an application requires high-wattage resistors, multiple lower-wattage, higher-value resistors may be substituted if they are wired in parallel—and may even be cheaper than a single high-wattage wire-wound resistor. For example, if a 5W, 50Ω resistor is specified, 10 resistors can be substituted, each rated at 0.5W and 500Ω. Bear in mind that if they are tightly bundled, this will interfere with heat dissipation.

What Can Go Wrong

Heat

Resistors are probably the most robust of all electronic components, with high reliability and a long life. It is difficult to damage a resistor by overheating it with a soldering iron.

The wattage rating of a resistor does not necessarily mean that it should be used to dissipate that amount of power on a constant basis. Small resistors (1/4 watt or less) can overheat just as easily as big ones. Generally speaking, it is safe practice not to exceed 75% of a resistor's power rating on a constant basis.

Overheating is predictably more of a problem for power resistors, where provision must be made for heat dissipation. Issues such as component crowding should be considered when deciding how big a heat sink to use and how much ventilation. Some power resistors may function reliably at temperatures as high as 250 degrees Centigrade, but components near them are likely to be less tolerant and plastic enclosures may soften or melt.

Noise

The electrical *noise* introduced by a resistor in a circuit will vary according to the composition of the resistor, but for any given component, it will be proportional to voltage and current. Low-noise circuits (such as those at the input stage of a high-gain amplifier) should use low-wattage resistors at a low voltage where possible.

Inductance

The coiled wire of a wire-wound resistor will be significantly inductive at low frequencies. This is known as *parasitic inductance*. It will also have a resonant frequency. This type of resistor is unsuitable for applications where frequency exceeds 50KHz.

Inaccuracy

When using resistors with 10% tolerance, imprecise values may cause greater problems in some applications than in others. In a voltage divider, for instance, if one resistor happens to be at the high end of its tolerance range while the other happens to be at the low end, the voltage obtained at the intersection of the resistors will vary from its expected value. Using the schematic shown in Figure 10-20, if R1 is rated for 1K and R2 is rated for 5K, and the power supply is rated at 12VDC, the voltage at point A should be:

```
V = 12 * (( 5 / (5 + 1)) = 10
```

However, if R1 has an actual value of 1.1K and R2 has an actual value of 4.5K, the actual voltage obtained at point A will be:

```
V = 12 * (( 4.5 / (4.5 + 1.1)) = 9.6
```

If the resistors are at opposite ends of their respective tolerance ranges, so that R1 has an actual value of 900Ω while the lower resistor has an actual value of 5.5K, the actual voltage obtained will be:

```
V = 12 * (( 5.5 / (5.5 + 0.9)) = 10.3
```

The situation becomes worse if the two resistors are chosen to be of equal value, to provide half of the supply voltage (6 volts, in this example) at their intersection. If two 5K resistors are used, and the upper one is actually 4.5K while the lower one is 5.5K, the actual voltage will be:

```
V = 12 * (( 5.5 / (4.5 + 5.5)) = 6.6
```

Whether this variation is significant will depend on the particular circuit in which the voltage divider is being used.

Common through-hole resistors may occasionally turn out to have values that are outside their specified tolerance range, as a result of poor manufacturing processes. Checking each resistor with a meter before placing it in a circuit should be a standard procedure.

When measuring the voltage drop introduced by a resistor in an active circuit, the meter has its own internal resistance that will take a proportion of the current. This is known as *meter loading* and will result in an artificially low reading for the potential difference between the ends of a resistor. This problem becomes significant only when dealing with resistors that have a high value (such as 1M), comparable with the internal resistance of the meter (likely to be 10M or more).

Wrong Values

When resistors are sorted into small bins by the user, errors may be made, and different values may be mixed together. This is another reason for checking the values of components before using them. Identification errors may be nontrivial and easily overlooked: the visible difference between a 1 megohm resistor and a 100Ω resistor is just one thin color band.

potentiometer

11

Also known as a *variable resistor*; may be substituted for a *rheostat*.

OTHER RELATED COMPONENTS

- **rotational encoder** (see Chapter 8)
- **resistor** (see Chapter 10)

What It Does

When a voltage is applied across a potentiometer, it can deliver a variable fraction of that voltage. It is often used to adjust sensitivity, balance, input, or output, especially in audio equipment and sensors such as motion detectors.

A potentiometer can also be used to insert a variable resistance in a circuit, in which case it should really be referred to as a *variable resistor*, although most people will still call it a potentiometer.

It can be used to adjust the power supplied to a circuit, in which case it is properly known as a *rheostat*, although this term is becoming obsolete. Massive rheostats were once used for purposes such as dimming theatrical lighting, but solid-state components have taken their place in most high-wattage applications.

A full-size, classic-style potentiometer is shown in Figure 11-1.

Figure 11-1. *A generic or classic-style potentiometer, approximately one inch in diameter.*

Schematic symbols for a potentiometer and other associated components are shown in Figure 11-2, with American versions on the left and European versions on the right in each case. The symbols for a potentiometer are at the top. The correct symbols for a variable resistor or rheostat are shown at center, although a potentiometer symbol may often be used instead. A preset variable resistor is shown at the bottom, often referred to as a *trimmer* or *Trimpot*. In these examples, each has an arbitrary rated resistance of 4,700Ω. Note the European substitution of K for a decimal point.

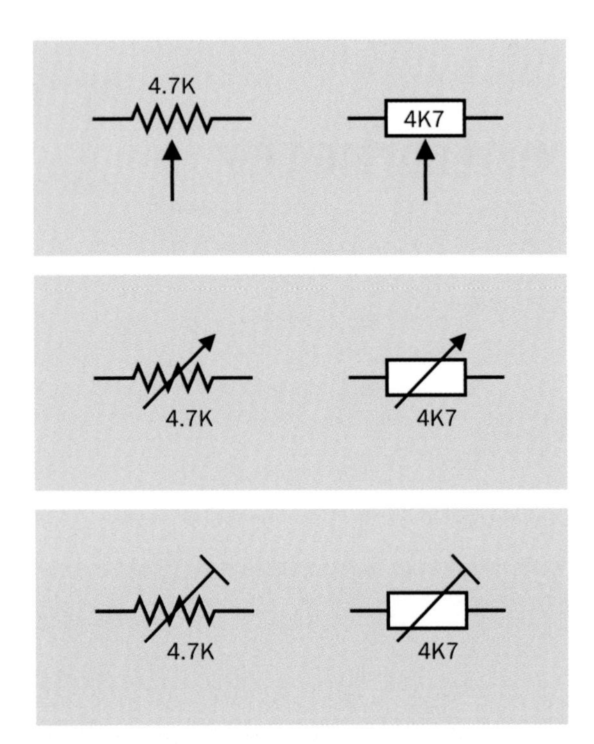

Figure 11-2. *American (left) and European (right) symbols for a potentiometer, a rheostat, and a trimmer potentiometer, reading from top to bottom. The 4.7K value was chosen arbitrarily.*

How It Works

A potentiometer has three terminals. The outer pair connect with the opposite ends of an internal resistive element, such as a strip of conductive plastic, sometimes known as the *track*. The third center terminal connects internally with a contact known as the *wiper* (or rarely, the *pick-off*), which touches the strip and can be moved from one end of it to the other by turning a shaft or screw, or by moving a slider.

If an electrical potential is applied between opposite ends of the resistive element, the voltage "picked off" by the wiper will vary as it moves. In this mode, the potentiometer works as a resistive voltage divider. For example, in a potentiometer with a linear taper (see "Variants," coming up), if you attach the negative side of a 12V battery to the right-hand end terminal and the positive side to the left-hand end terminal, you will find an 8V potential at the center terminal when the potentiometer has rotated clockwise through one-third of its range. In Figure 11-3, the base of the shaft (shown in black) is attached to an arm (shown in green) that moves a wiper (orange) along a resistive element (brown). The voltages shown assume that the resistive element has a linear taper and will vary slightly depending on wire resistance and other factors.

Because a potentiometer imposes a voltage reduction, it also reduces current flowing through it, and therefore creates waste heat which must be dissipated. In an application such as an audio circuit, small currents and low voltages generate negligible heat. If a potentiometer is used for heavier applications, it must be appropriately rated to handle the wattage and must be vented to allow heat to disperse.

To use a potentiometer as a *variable resistor* or *rheostat*, one of its end terminals may be tied to the center terminal. If the unused end terminal is left unconnected, this raises the risk of picking up stray voltages or "noise" in sensitive circuits. In Figure 11-4, a potentiometer is shown adjusting a series resistance for an *LED* for demonstration purposes. More typically, a trimmer would be used in this kind of application, since a user is unlikely to need to reset it.

Variants

Linear and Log Taper

If the resistive element in a potentiometer is of constant width and thickness, the electrical potential at the wiper will change in ratio with the rotation of the wiper and shaft (or with movement of a slider). This type of potentiometer is said to have a linear taper even though its element does not actually taper.

For audio applications, because human hearing responds nonlinearly to sound pressure, a potentiometer that has a linear taper may seem to have a very slow action at one end of its scale and an abrupt effect at the other. This problem used to be solved with a non-uniform or tapered re-

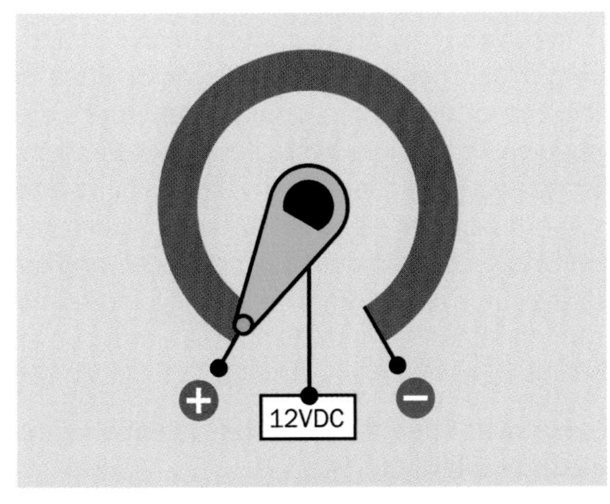

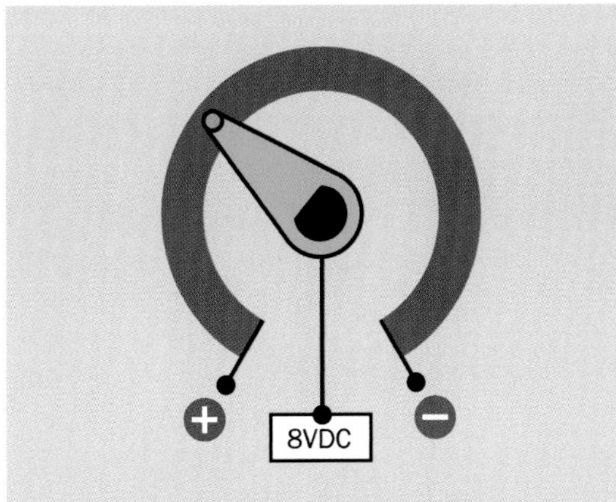

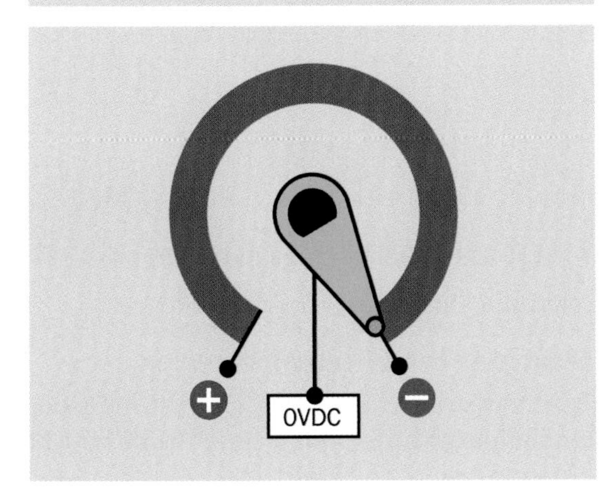

Figure 11-3. *Inside a potentiometer. See text for details.*

sistive element. More recently, a combination of resistive elements has been used as a cheaper option. Such a potentiometer is said to have an audio taper or a log taper (since the resistance

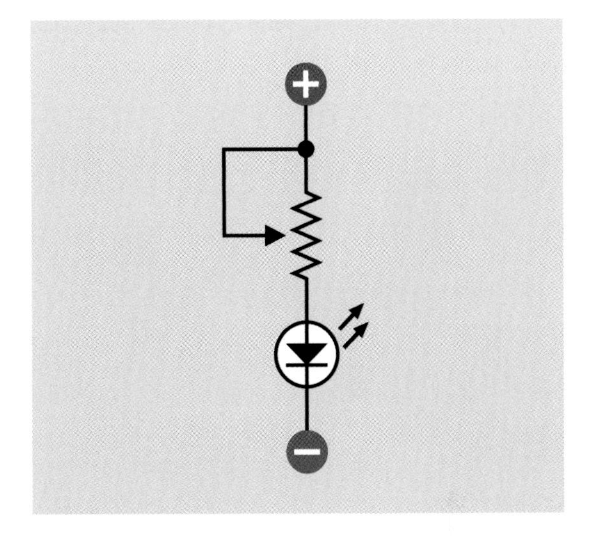

Figure 11-4. *A potentiometer can be used to adjust a series resistance, as shown in this schematic. Tying the wiper to one of the end terminals reduces the risk of picking up electrical noise.*

may vary as a logarithm of the angle of rotation). A reverse audio taper or antilog taper varies in the opposite direction, but this type has become very uncommon.

Classic-style Potentiometer

This consists of a sealed circular can, usually between 0.5″ and 1″ in diameter, containing a resistive strip that is shaped as a segment of a circle. A typical example is shown in Figure 11-1, although miniaturized versions have become more common. A shaft mounted on the can turns the internal wiper that presses against the strip. For panel-mount applications, a threaded bushing at the base of the shaft is inserted through a hole in the front panel of the electronics enclosure, and a nut is tightened on the bushing to hold the potentiometer in place. Often there is also a small offset index pin that, when paired with a corresponding front panel hole, will keep the pot from spinning freely.

Many modern potentiometers are miniaturized, and may be packaged in a box-shaped plastic enclosure rather than a circular can. Their power ratings are likely to be lower, but their principle of operation is unchanged. Two variants are shown in Figure 11-5.

Figure 11-5. *Two modern miniaturized potentiometers. At left: 5K. At right: 10K. Both are rated to dissipate up to 50mW.*

The three terminals on the outside of a potentiometer may be solder lugs, screw terminals, or pins for direct mounting on a circuit board. The pins may be straight or angled at 90 degrees.

The resistive element may use carbon film, plastic, cermet (a ceramic-metal mixture), or resistive wire wound around an insulator. Carbon-film potentiometers are generally the cheapest, whereas wire-wound potentiometers are generally the most expensive.

Wire-wound potentiometers may handle more power than the other variants, but as the wiper makes a transition from one turn of the internal wire element to the next, the output will tend to change in discrete steps instead of varying more smoothly.

In a potentiometer with detents, typically a spring-loaded lever in contact with notched internal wheel causes the shaft to turn in discrete steps that create a stepped output even if the resistive element is continuous.

The shaft may be made of metal or plastic, with its length and width varying from one component to another. A control knob can be fitted to the end of the shaft. Some control knobs are push-on, others have a set screw to secure them. Shafts may be splined and split, or round and smooth, or round with a flat surface that matches the shape of a socket in a control knob and reduces the risk of a knob becoming loose and turning freely. Some shafts have a slotted tip to enable screwdriver adjustment.

Some shaft options for full-size potentiometers are shown in Figure 11-6.

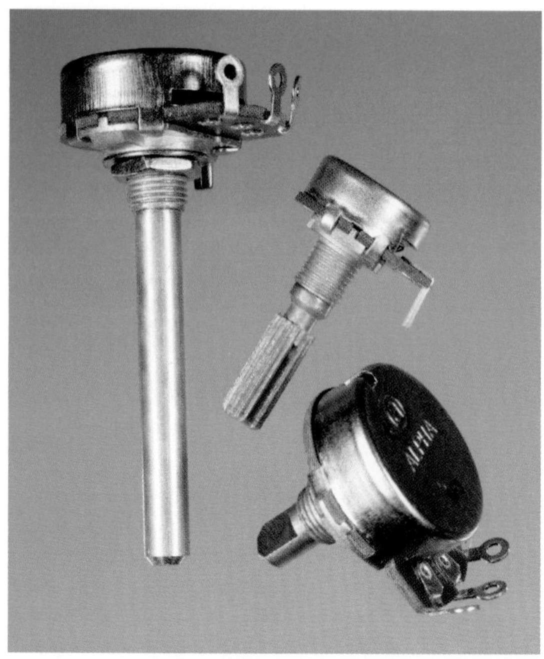

Figure 11-6. *Three shaft options for potentiometers.*

Multiple-Turn Potentiometer

To achieve greater precision, a track inside a potentiometer may be manufactured in the form of a helix, allowing the wiper to make multiple turns on its journey from one end of the track to the other. Such multiple-turn potentiometers typically allow 3, 5, or 10 turns to move the wiper from end to end. Other multiple-turn potentiometers may use a screw thread that advances a wiper along a linear or circular track. The latter is comparable with a trimmer where multiple turns of a screwdriver are used to rotate a worm gear that rotates a wiper between opposite ends of a circular track.

Ganged Potentiometer

Two (or rarely, more) potentiometers can be stacked or combined so that their resistive elements and wipers share the same shaft but can use different voltages or have different taper. Each resistance-wiper assembly is known as a *cup*, and the potentiometers are said to be *ganged*.

Flat ganged potentiometers combine two resistive elements in one enclosure. Some *dual ganged* potentiometers are concentric, meaning that the pots are controlled separately by two shafts, one inside the other. Suitable concentric knobs must be used. You are unlikely to find these potentiometers sold as components in limited quantities.

Switched Potentiometer

In this variant, when the shaft is turned clockwise from an initial position that is fully counterclockwise, it flips an internal switch connected to external terminals. This can be used to power-up associated components (for example, an audio amplifier). Alternatively, a switch inside a potentiometer may be configured so that it is activated by pulling or pushing the shaft.

Slider Potentiometer

Also known as a *slide potentiometer*. This uses a straight resistive strip and a wiper that is moved to and fro linearly by a tab or lug fitted with a plastic knob or finger-grip. Sliders are still found on some audio equipment. The principle of operation, and the number of terminals, are identical to the classic-style potentiometer. Sliders typically have solder tabs or PC pins. In Figure 11-7, the large one is about 3.5" long, designed for mounting behind a panel that has a slot to allow the sliding lug to poke through. Threaded holes at either end will accept screws to fix the slider behind the panel. A removable plastic finger-grip (sold separately, in a variety of styles) has been pushed into place. Solder tabs underneath the slider are hidden in this photo. The smaller slider is designed for through-hole mounting on a circuit board.

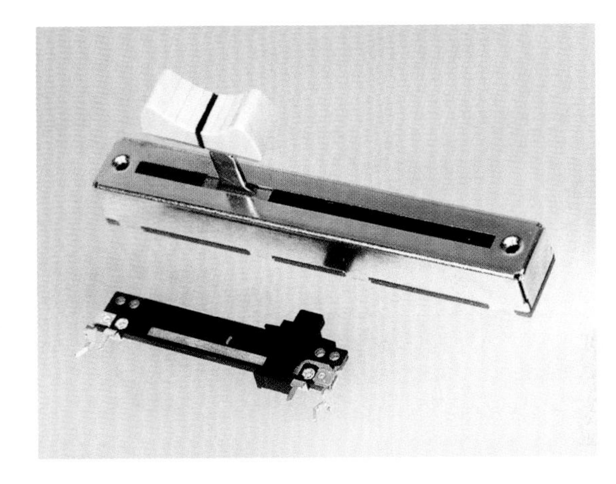

Figure 11-7. *Slider potentiometers.*

Trimmer Potentiometer

Often referred to as Trimpots, this is actually a proprietary brand name of Bourns. They are usually mounted directly on circuit boards to allow fine adjustment or trimming during manufacturing and testing to compensate for variations in other components. Trimmers may be single-turn or multi-turn, the latter containing a worm gear that engages with another gear to which the wiper is attached. Trimmers always have linear taper. They may be designed for screwdriver adjustment or may have a small knurled shaft, a thumb wheel, or a knob. They are not usually accessible by the end user of the equipment, and their setting may be sealed or fixed when the equipment is assembled. In Figure 11-8, the beige Spectrol trimmer is a single-turn design, whereas the blue trimmer is multi-turn. A worm gear inside the package, beneath the screw head, engages with an interior gear wheel that rotates the wiper.

In Figure 11-9, a 2K trimmer potentiometer has a knurled dial attached to allow easy finger adjustment, although the dial also contains a slot for a flat-blade screwdriver.

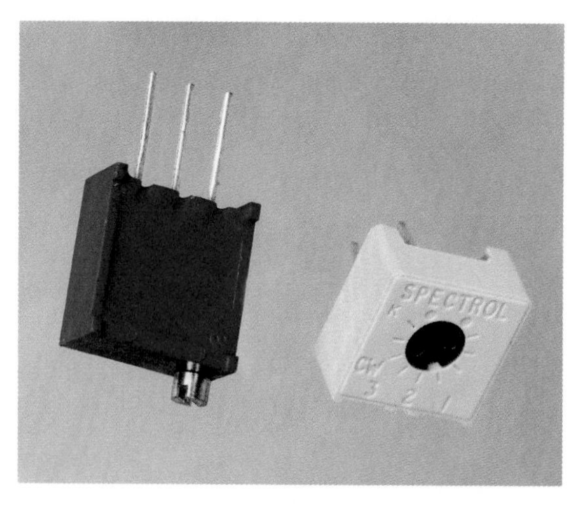

Figure 11-8. *Like most trimmers, these are designed for through-hole mounting on a circuit board.*

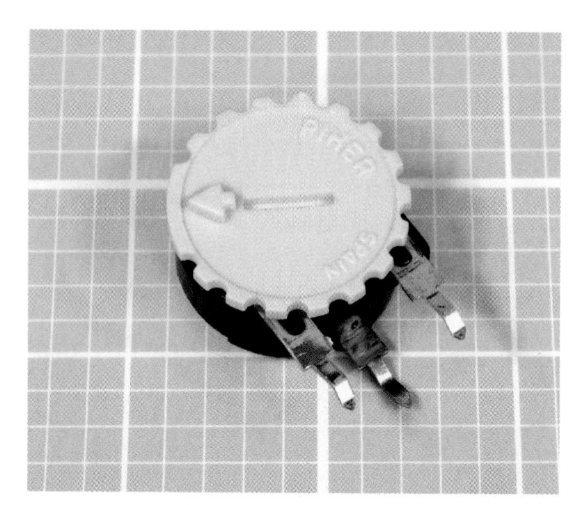

Figure 11-9. *A trimmer potentiometer with a knurled dial to facilitate finger adjustment.*

How to Use it

The classic-style potentiometer was once used universally to control volume, bass, and treble on audio equipment but has been replaced increasingly by digital input devices such as **tactile switches** (see "Tactile Switch" on page 34) or **rotational encoders** (see Chapter 8), which are more reliable and may be cheaper, especially when assembly costs are considered.

Potentiometers are widely used in lamp dimmers and on cooking stoves (see Figure 11-10). In these applications, a solid-state switching device such as a **triac** (described in Volume 2) does the actual work of moderating the power to the lamp or the stove by interrupting it very rapidly. The potentiometer adjusts the duty cycle of the power interruptions. This system wastes far less power than if the potentiometer controlled the lighting or heating element directly as a rheostat. Since less power is involved, the potentiometer can be small and cheap, and will not generate significant heat.

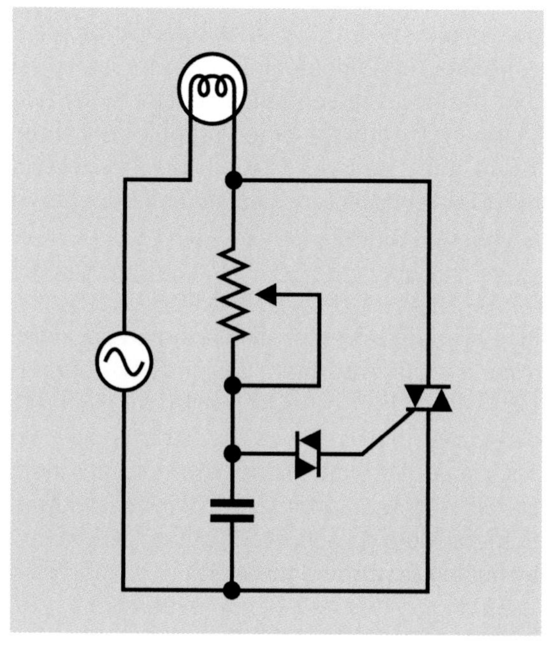

Figure 11-10. *Typical usage of a potentiometer in conjunction with a diac, triac, and capacitor to control the brightness of an incandescent bulb, using an AC power supply. Diacs and triacs are discussed in Volume 2.*

Because true logarithmic potentiometers have become decreasingly common, a linear potentiometer in conjunction with a fixed resistor can be used as a substitute, to control audio input. See Figure 11-11.

A potentiometer may be used to match a sensor or analog input device to an analog-digital con-

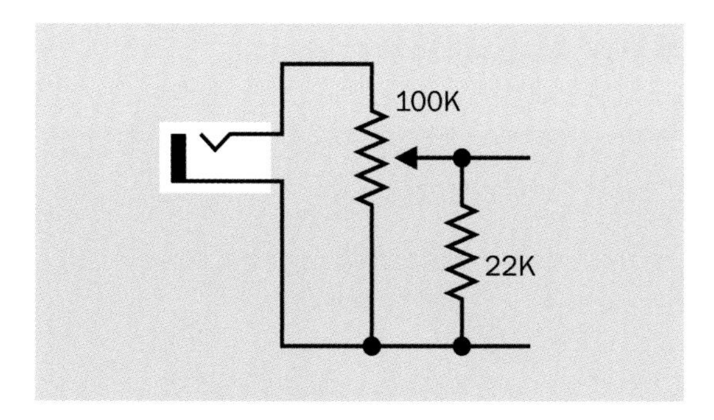

Figure 11-11. *In this circuit, a 100K linear potentiometer is used in conjunction with a 22K resistor to create an approximately logarithmic volume control for an audio system with input coming from a mono jack socket at left.*

verter, or it can calibrate a device such as a temperature or motion sensor.

What Can Go Wrong

Wear and Tear

Since classic-style potentiometers are electromechanical devices, their performance will deteriorate as one part rubs against another. The long open slot of a slider potentiometer makes it especially vulnerable to contamination with dirt, water, or grease. Contact-cleaner solvent, lubricant-carrying sprays, or pressurized "duster" gas may be squirted into a potentiometer to try to extend its life. Carbon-film potentiometers are the least durable and in audio applications will eventually create a "scratchy" sound when they are turned, as the resistive element deteriorates.

If the wiper deteriorates to the point where it no longer makes electrical contact with the track, and if the potentiometer is being used as a variable resistor, two failure modes are possible, shown in Figure 11-12. Clearly the right-hand schematic is a better outcome. This is an argument for always tying the wiper to the "unused" end of the track.

If you are designing a circuit board that will go through a production process, temperature variations during wave soldering, and subsequent washing to remove flux residues, create hostile conditions for potentiometers, especially sliders where the internal parts are easily contaminated. It will be safer to hand-mount potentiometers after the automated process.

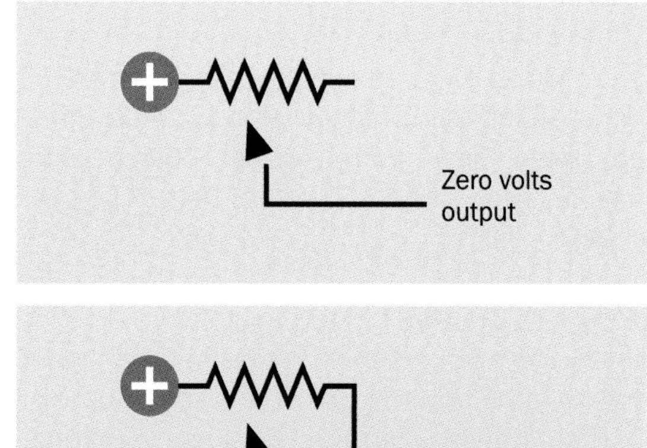

Figure 11-12. *If the wiper of a potentiometer breaks (indicated by the loose arrow head) as a result of wear and tear, and the potentiometer is being used as a variable resistor, the voltage from it will drop to zero (top schematic) unless the wiper has been tied to one end of the track (bottom schematic).*

Knobs that Don't Fit

Control knobs are almost always sold separately from potentiometers. Make sure the shaft of the potentiometer (which may be round, round-with-flat, or knurled) matches the knob of your choice. Note that some shaft diameters are expressed in inches, while others are metric.

Nuts that Get Lost

For panel-mounted potentiometers, a nut that fits the thread on the bushing is almost always included with the potentiometer; an additional nut and lock washer may also be supplied. Because there is no standardization of threads on potentiometers, if you lose a nut, you may have some difficulty finding an exact replacement.

A Shaft that Isn't Long Enough

When choosing a shaft length, if in doubt, buy a potentiometer with a long shaft that you can cut to the desired length.

Sliders with No Finger Grip

Slider potentiometers are often sold without a knob or plastic finger-grip, which must be ordered separately and may be available in different styles. The finger-grip usually push-fits onto the metal or plastic tab or lug that moves the slider to and fro.

Too Big to Fit

Check the manufacturer's datasheet if you need to know the physical size of the potentiometer. Photographs may be misleading, as a traditional-style potentiometer that is 0.5" in diameter looks much the same as one that is 1" in diameter. High-wattage potentiometers will be more costly and physically large (2 to 3 inches in diameter). See Figure 11-13.

Overheating

Be sure to leave sufficient air space around a high-wattage potentiometer. Carefully calculate the maximum voltage drop and current that you may be using, and choose a component that is appropriately rated. Note that if you use the potentiometer as a rheostat, it will have to handle more current when its wiper moves to reduce its resistance. For example, if 12VDC are applied through a 10-ohm rheostat to a component that has a resistance of 20 ohms, current in the circuit will vary from 0.4 amps to 0.6 amps depending on the position of the rheostat. At its maximum setting, the rheostat will impose a 4V voltage drop and will therefore dissipate 1.6 watts from the full length of its resistive element. If the rheostat is reset to impose only a 4-ohm resistance, the voltage drop that it imposes will be 2V, the current in the circuit will be 0.5 amps, and the rheostat will therefore dissipate 1 watt from 4/10ths of the length of its resistive element. A wire-wound potentiometer will be better able to handle high dissipation from a short segment of its element than other types of rheostat. Add a fixed resistor in series with a rheostat if necessary to impose a limit on the current.

Figure 11-13. *The large potentiometer is approximately 3" in diameter, rated at 5 ohms, and able to handle more than 4 amps. The small potentiometer is 5/8" diameter, rated at 2K and 1/4 watt, with pins designed for through-hole insertion in a circuit board, and a grooved shaft that accepts a push-on knob. Despite the disparity in size, the principle of operation and the basic features are identical.*

When using a trimmer potentiometer, limit the current through the wiper to 100mA as an absolute maximum value.

The Wrong Taper

When buying a potentiometer, remember to check the specification to find out whether it has linear or audio/log taper. If necessary, attach a meter, with the potentiometer set to its center position, to verify which kind of taper you have. While holding the meter probes in place, rotate the potentiometer shaft to determine which way an audio/log taper is oriented.

capacitor 12

Quite often referred to as a *cap*. Formerly known (primarily in the United Kingdom) as a *condenser*, but that term has become obsolete.

OTHER RELATED COMPONENTS

- **variable capacitor** (See Chapter 13)
- **battery** (See Chapter 2)

What It Does

A capacitor connected across a DC power source will accumulate a charge, which then persists after the source is disconnected. In this way, the capacitor stores (and can then discharge) energy like a small rechargeable battery. The charge/discharge rate is extremely fast but can be limited by a series resistor, which enables the capacitor to be used as a timing component in many electronic circuits.

A capacitor can also be used to block DC current while it passes pulses, or electrical "noise," or alternating current, or audio signals, or other wave forms. This capability enables it to smooth the output voltage provided by **power supplies**; to remove spikes from signals that would otherwise tend to cause spurious triggering of components in digital circuits; to adjust the frequency response of an audio circuit; or to couple separate components or circuit elements that must be protected from transmission of DC current.

Schematic symbols for capacitors are shown in Figure 12-1. At top-left is a nonpolarized capacitor, while the other two indicate that a polarized capacitor must be used, and must be oriented as shown. The variant at the bottom is most commonly used in Europe. Confusingly, the nonpolarized symbol may also be used to identify a polarized capacitor, if a + sign is added. The polarized symbols are sometimes printed without + signs, but the symbols still indicate that polarity must be observed.

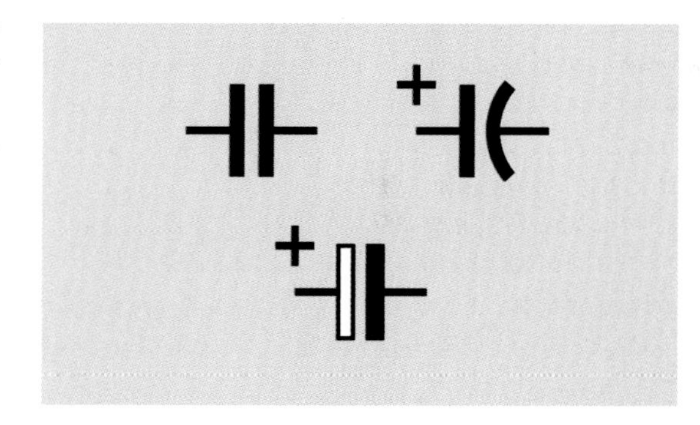

Figure 12-1. *Schematic symbols for polarized and nonpolarized capacitors. See text for details.*

How It Works

In its simplest form, a capacitor consists of two plates, each with a lead attached to it for connection with a DC power source. The plates are separated by a thin, insulating layer known as the *dielectric*, which is usually a solid or a paste but may be liquid, gel, gaseous, or vacuum.

The plates in most capacitors are made from thin metal film or metallized plastic film. To minimize the size of the component, the film may be rolled up to form a compact cylindrical package, or multiple flat sections may be interleaved.

Electrons from the power source will migrate onto the plate attached to the negative side of the source, and will tend to repel electrons from the other plate. This may be thought of as creating *electron holes* in the other plate or as attracting *positive charges*, as shown in Figure 12-2. When the capacitor is disconnected from the power source, the opposite charges on its plates will persist in equilibrium as a result of their mutual attraction, although the voltage will gradually dissipate as a result of *leakage*, either through the dielectric or via other pathways.

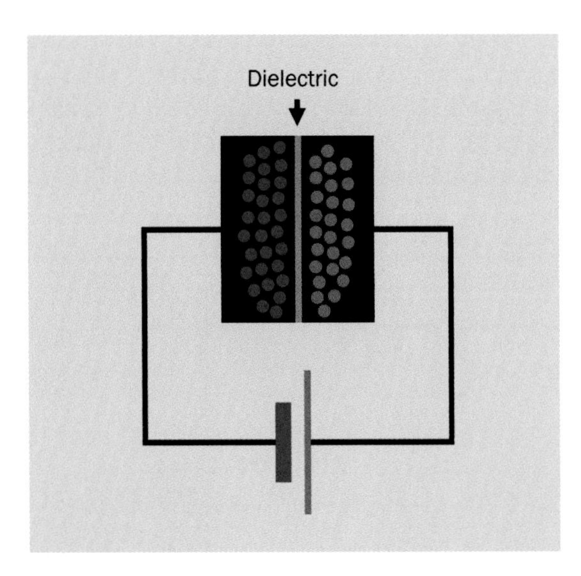

Figure 12-2. *Because the plates of a capacitor are electrically conductive, they will become populated with positive and negative charges when connected with a DC power source. As opposite charges attract each other, they will tend to congregate on either side of the dielectric, which is an insulating layer. The battery symbol is shown here colored for clarity.*

When a **resistor** is placed across the leads of a charged capacitor, the capacitor will discharge itself through the resistor at a rate limited by the resistance. Conversely, if a capacitor is charged through a resistor, the resistor will limit the charging rate.

A resistor in series with a capacitor is known as an *RC network* (Resistor-Capacitor network). In Figure 12-3, an RC circuit is shown with a SPDT switch that charges or discharges the capacitor via a series resistor. The voltage at point A increases nonlinearly (relative to the negative side of the power supply) while the capacitor is charging, and decreases nonlinearly while the capacitor is discharging, as suggested by the graphs. At any moment, the time that the capacitor takes to acquire 63% of the difference between its current charge and the voltage being supplied to it is known as the *time constant* for the circuit. See "The Time Constant" on page 99 for additional information.

When a capacitor is connected across an AC voltage source, each surge of electrons to one plate induces an equal and opposite positive surge to the other plate, and when polarity of the power supply reverses, the charges on the plates switch places. These surges may make it seem that the capacitor is conducting AC current, even though the dielectric that separates the plates is an insulator. See Figure 12-4. Often a capacitor is said to "pass" AC, even though this is not really happening. For convenience, and because the concept is widely established, this encyclopedia refers to capacitors as "passing" AC.

Depending on the size of the capacitor, it will block some AC frequencies while passing others. Generally speaking, a smaller capacitor will pass high frequencies relatively efficiently, as each little surge of current fills each plate. However, the situation is complicated by the *inductive reactance* (which creates the *effective series resistance*) of a capacitor, as discussed below. See "Alternating Current and Capacitive Reactance" on page 100.

Variants

Format

The three most common packages for capacitors are cylindrical, disc, and rectangular tablet.

A *cylindrical capacitor* may have *axial leads* (a wire attached to each end) or *radial leads* (both wires emerging from one end). Radial capacitors are

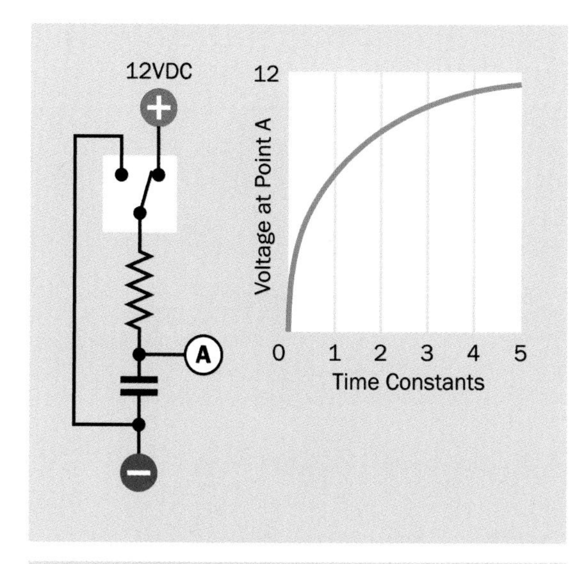

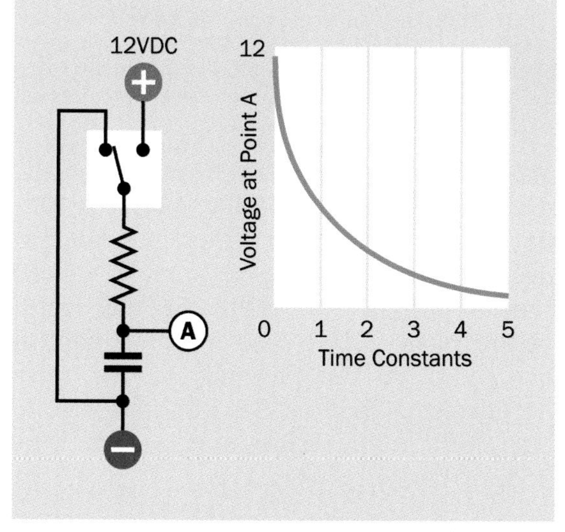

Figure 12-3. *An RC (Resistor-Capacitor) network with a switch to control charge and discharge of a capacitor. At top, the curve gives an approximate idea of the charging behavior of the capacitor. At bottom, the curve illustrates its discharging behavior.*

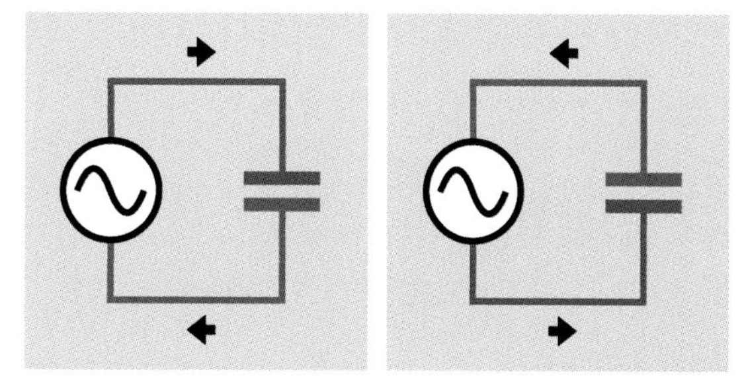

Figure 12-4. *In the left diagram, a source of alternating current charges the upper plate of a capacitor positively and the lower plate negatively. This process entails a flow of conventional current shown by the arrows. A moment later, when the AC current flow reverses, the flow also reverses, creating the impression that the capacitor "passes" AC current.*

Figure 12-5. *Cylindrical capacitors with radial leads. All are electrolytic.*

more widely used as they allow easy insertion into a circuit board. The capacitor is usually packaged in a small aluminum can, closed at one end, capped with an insulating disc at the other end, and wrapped in a thin layer of insulating plastic. Some samples are shown in Figure 12-5 and Figure 12-6.

A *disc capacitor* (sometimes referred to as a *button capacitor*) is usually encased in an insulating ceramic compound, and has radial leads. Modern small-value ceramic capacitors are more likely to be dipped in epoxy, or to be square tablets. Some samples are shown in Figure 12-7.

A *surface-mount capacitor* is square or rectangular, usually a few millimeters in each dimension,

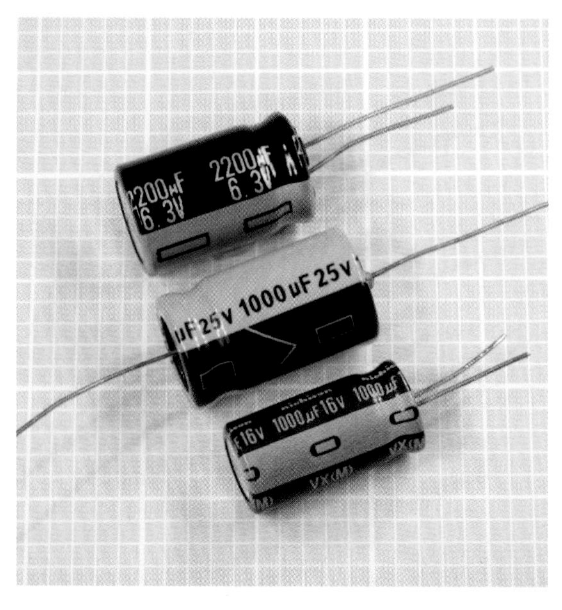

Figure 12-6. *Cylindrical capacitors with radial leads (top and bottom) and axial leads (center). All are electrolytic.*

Figure 12-7. *Generic ceramic capacitors. Left: rated for 0.1µF at 50V. Center: 1µF at 50V. Right: 1µF at 50V.*

with two conductive pads or contacts at opposite ends. It appears almost identical to a surface-mount resistor. Larger-value capacitors are inevitably bigger but can still be designed for surface-mount applications. See Figure 12-8.

Figure 12-8. *Most surface-mount capacitors are as tiny as other surface-mount components, but this 4,700µF electrolytic (at 10V) has a base approximately 0.6" square. A solder tab is visible at the center of the nearest edge.*

Many capacitors are nonpolarized, meaning that they are insensitive to polarity. However, electrolytic and tantalum capacitors must be connected "the right way around" to any DC voltage source. If one lead is longer than the other, it must be the "more positive" lead. A mark or band at one end of the capacitor indicates the "more negative" end. Tantalum capacitors are likely to indicate the positive lead by using a + sign on the body of the component.

An arrow printed on the side of a capacitor usually points to the "more negative" terminal. In an aluminum can with axial leads, the lead at one end will have an insulating disc around it while the other lead will be integral with the rounded end of the can. The wire at the insulated end must be "more positive" than the wire at the other end.

A *capacitor array* contains two or more capacitors that are isolated from each other internally and accessed by external contacts. They are sold in surface-mount format and also in *through-hole* chips of DIP (dual-inline package) or SIP (single-inline package) format. The internal components may be connected in one of three configurations:

isolated, common-bus, or dual-ended common bus. Technically the isolated configuration should be referred to as a capacitor array, but in practice, all three configurations are usually referred to as *capacitor networks*. See Figure 12-9 and Figure 12-10.

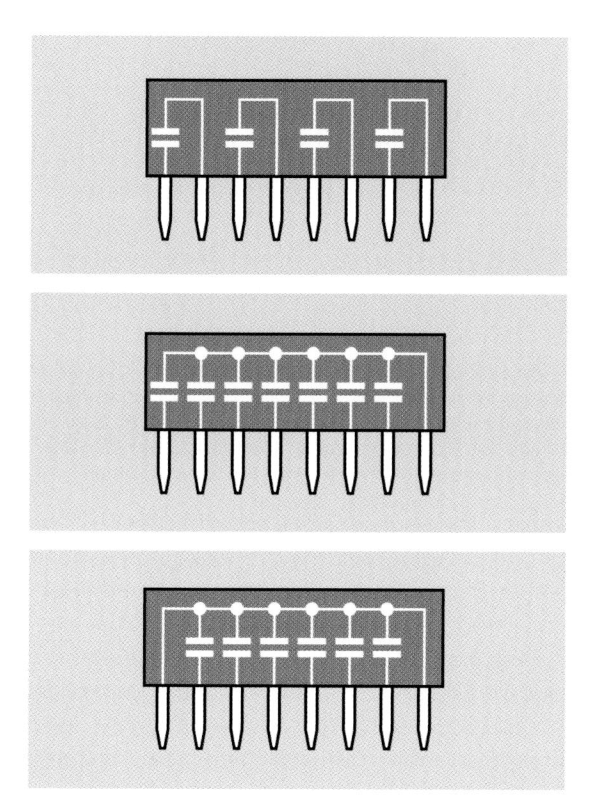

Figure 12-9. *A capacitor network most often consists of a single-inline package (SIP) chip containing multiple capacitors in one of three configurations shown here. Top: Isolated. Center: Common bus. Bottom: Dual-ended common bus. Individual capacitor values ranging from 0.001μF to 0.1μF are common.*

Capacitor networks can reduce the component count in circuits where digital logic chips require bypass capacitors. They are comparable in concept to *resistor arrays*.

Chips containing *RC circuits* (multiple resistor-capacitor pairs) are available, although uncommon.

Figure 12-10. *A capacitor array in through-hole, SIP format.*

Principal Types

Electrolytic capacitors are relatively cheap, compact, and available in large values. These attributes have made them a popular choice in consumer electronics, especially for power supplies. The capacitive capability of an electrolytic is refreshed by periodic application of voltage. A moist paste inside the capacitor is intended to improve the dielectric performance when voltage is applied, but can dry out during a period of years. If an electrolytic is stored for 10 years or so, it may allow a short circuit between its leads when power is applied to it. The capacitors in Figure 12-5 and Figure 12-6 are all electrolytic. The capacitor in Figure 12-11 is at the high end of the scale.

A *bipolar electrolytic* is a single package containing two electrolytic capacitors in series, end-to-end, with opposed polarities, so that the combination can be used where the voltage of a signal fluctuates above and below 0VDC. See Figure 12-12 and Figure 12-13. This type of component is likely to have "BP" (bipolar) or "NP" (nonpolarized) printed on its shell. It may be used in audio circuits where polarized capacitors are normally unsuitable, and is likely to be cheaper than non-electrolytic alternatives. However, it

Figure 12-11. *This 13,000µF electrolytic capacitor is larger than would be required in most everyday applications.*

suffers from the same weaknesses as all electrolytics.

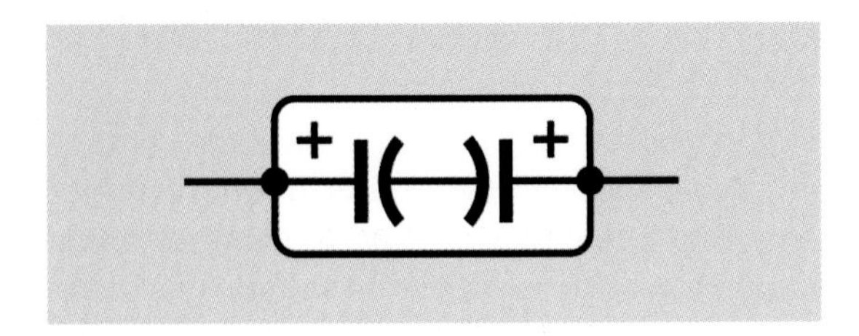

Figure 12-12. *Schematic view of the internal configuration of a bipolar electrolytic capacitor, also known as nonpolarized electrolytic capacitor. It consists of two electrolytics in series, with opposing polarities.*

Tantalum capacitors are compact but relatively expensive, and can be vulnerable to voltage

Figure 12-13. *Bipolar electrolytic capacitors. The larger size of the one at top-left is a consequence of its higher voltage rating. "BP" on the other two capacitors is an acronym for "bipolar," meaning that they have no polarity, even though one lead may be shorter than the other.*

spikes. They are sensitive to application of the wrong polarity. Typically they are epoxy-dipped rather than mounted inside a small aluminum can like electrolytics, and consequently the electrolyte may be less likely to evaporate and dry out. In Figure 12-14, two tantalum capacitors (rated 330µF at 6.3V, left, and 100µF at 20V, right) are shown above a polyester film capacitor (rated 10µF at 100V). Surface-mount tantalum capacitors are decreasing in popularity as large-value ceramic capacitors are becoming available, with smaller dimensions and lower equivalent series resistance.

Plastic-film capacitors are discussed in the following section.

Single-layer *ceramic* capacitors are often used for bypass, and are suitable for high-frequency or audio applications. Their value is not very stable with temperature, although the "NPO" variants are more stable. Multilayer ceramic capacitors are more compact than single-layer ceramic, and consequently are becoming increasingly popular. Three multilayer ceramic capacitors are

Figure 12-14. *Two tantalum capacitors are shown above a polyester film capacitor. The polarity of the tantalum capacitors is indicated by the plus signs adjacent to the longer lead, in each case. The polyester capacitor is nonpolarized.*

shown in Figure 12-15. At bottom-right, even the largest (rated at 47µF at 16V) is only 0.2" square.

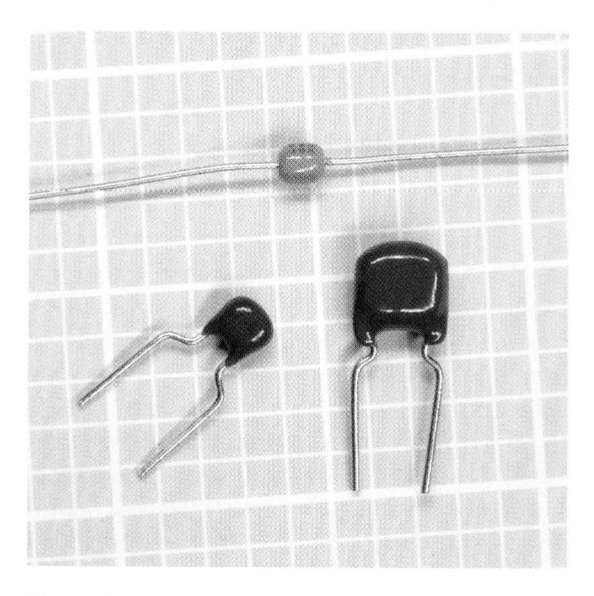

Figure 12-15. *Multilayer ceramic capacitors are extremely compact, and are nonpolarized. Top: 1,000pF (i.e. 1nF) at 100V. Bottom left: 1µF at 25V. Bottom right: 47µF at 16V.*

Dielectrics

The dielectric used in a capacitor most often consists of an electrolytic layer, a ceramic compound, a plastic film (polycarbonate, polypropylene, or polystyrene), or paper.

An *electrolytic layer* in an electrolytic capacitor traditionally consists of paper soaked in an electrolyte. It is interleaved with a thin film of aluminum on which is deposited a layer of aluminum oxide. The layers are rolled up to create a cylindrical component. The functioning dielectric is created when voltage is applied.

Polyester
: This is the most common type of plastic film, with the highest dielectric constant, enabling highest capacitance per unit volume. Widely used in DC applications, but the rolled layers create parasitic inductance. Often used in decoupling, coupling, and bypass, but not so suitable for situations requiring stability and low leakage. May not be suitable for high current.

Polycarbonate
: Thermally very stable, often specified for filters and timing circuits that require a fixed frequency. An excellent type of capacitor, compatible for mil-spec applications, but expensive.

Mylar, Polyester, and other plastic-film types are often used in audio circuits, where their voltage limitation (typically less than 100VDC) is not a problem, and their nonpolarized attribute is an advantage.

Polypropylene
: Vulnerable to heat (a maximum of 85 degrees Centigrade is common), and less thermally stable than polycarbonate. A very low power dissipation factor allows it to handle higher power at higher frequencies. Available with tolerances down to 1%. These capacitors are a popular choice in crossover networks for loudspeaker combinations, and are used in switching power supplies.

They tend to be physically larger than other capacitors using film dielectric.

Values

Farads

The electrical storage capacity of a capacitor is measured in *farads,* universally represented by the letter F. A capacitor that can be charged with a potential difference between its plates of 1 volt, in a time of 1 second, during which it draws 1 amp, has a capacitance of 1 farad.

Because the farad is a large unit, capacitors in electronic circuits almost always have fractional values: *microfarads* (µF), *nanofarads* (nF), and *picofarads* (pF). The Greek letter µ (mu) should be used in the µF abbreviation, but a lowercase letter u is often substituted. Thus, for example, 10uF means the same as 10µF.

1F = 1,000,000µF, and 1µF = 1,000,000pF. Therefore, 1 farad is equivalent to 1 trillion picofarads —a very wide range of possible values. See Figure 12-16 and Figure 12-17 for charts showing equivalent values in different units.

pF	nF	µF
1	0.001	0.000001
10	0.01	0.00001
100	0.1	0.0001
1,000	1	0.001
10,000	10	0.01
100,000	100	0.1
1,000,000	1,000	1

Figure 12-16. *Equivalent values for picofards, nanofarads, and microfarads. The nF unit is used primarily in Europe.*

The nF unit is more common in Europe than in the United States. A 1nF capacitance is often expressed in the US as 0.001µF or 1,000pF. Similarly, a 10nF capacitance is almost always expressed as a 0.01µF, and a 0.1nF capacitance is more likely to be expressed as 100pF.

µF	F
1	0.000001
10	0.00001
100	0.0001
1,000	0.001
10,000	0.01
100,000	0.1
1,000,000	1

Figure 12-17. *Equivalent values for microfarads and farads. Because the farad is such a large unit, electronic circuits almost always use fractional values.*

European schematics may use value-symbols as a substitute for decimal points. For example, a 4.7pF capacitor may be shown as 4p7, a 6.8nF capacitor may be shown as 6n8, and a 3.3µF capacitor may be shown as 3µ3.

Commonly Used Values

The traditional range of capacitor values was established on the same basis as the traditional range of resistor values, by assuming an accuracy of plus-or-minus 20% and choosing factors that would minimize the possible overlap between adjacent tolerance ranges. The factors 1.0, 1.5, 2.2, 3.3, 4.7, 6.8, and 10 satisfy this requirement. See Chapter 10 for a more detailed explanation, including a graphical representation of values and overlaps in Figure 10-8. While many resistors are now manufactured with high precision, 20% tolerance is still common for electrolytic capacitors. Other types of capacitors are available with an accuracy of 10% or 5%, but are more expensive.

While large-value capacitors are likely to have their actual value printed on them, smaller capacitors are identified by a variety of different codes. These codes are not standardized among manufacturers, and exist in various colors and abbreviations. A multimeter that can measure capacitance is a quicker, easier, and more reliable method of determining the value of a component than trying to interpret the codes.

In addition to capacitance, a large capacitor is likely to have its working voltage printed on it. Exceeding this value increases the risk of damaging the dielectric. In the case of electrolytic capacitors, a voltage that is much lower than the rated value should also be avoided, because these capacitors require an electrical potential to maintain their performance.

In common electronics applications, values larger than 4,700µF or smaller than 10pF are unusual.

Electrolytics are available at a moderate price in a wider range of values than other commonly used capacitors. They range from 1µF to 4,700µF and sometimes beyond. Working voltages typically range from 6.3VDC to 100VDC, but can be as high as 450VDC.

Tantalum capacitors are usually unavailable in sizes above 150µF or for voltages above 35VDC.

Single-layer ceramic capacitors have small values ranging from 0.01µF to 0.22µF, with working voltages usually not exceeding 50VDC, although very small-value capacitors may be rated much higher for special applications. Poor tolerances of +80% to -20% are common.

Some variants of multi-layer ceramic capacitors are capable of storing up to 47µF, although 10µF is a more common upper limit. They are seldom rated above 100VDC. Some are accurate to plus-or-minus 5%.

Dielectric Constant

If A is the area of each plate in a capacitor (measured in square centimeters), and T is the thickness of the dielectric (measured in centimeters), and K is the *dielectric constant* of the capacitor, the capacitance, C (measured in farads) will be obtained from the formula:

```
C = (0.0885 * K * A) / T
```

The dielectric constant of air is 1. Other dielectrics have different standard values. Polyethylene, for instance, has a constant of approximately 2.3. Thus a capacitor of 1 square centimeter plate area and polyethylene dielectric 0.01 centimeters thick would have a capacitance of about 20pF. A tantalum capacitor of equal plate area and dielectric thickness would have capacitance closer to 100pF, since the dielectric constant of tantalum oxide is much higher than that of polyethylene.

The Time Constant

When a capacitor is charged in series through a resistor (it is used in an RC network), and it begins with no charge on its plates, the *time constant* is the time, in seconds, required to charge the capacitor to 63% of the supply voltage. After an additional, identical interval of time, the capacitor will acquire 63% of the remaining difference between itself and the power supply. In theory the capacitor gets closer and closer to a full charge, but never quite reaches 100%. However, five time constants are sufficient for the capacitor to reach 99%, which is regarded as close enough to being fully charged for all practical purposes.

Refer to Figure 12-3 for a schematic of an RC network.

The time constant is a simple function of the resistance and the capacitance. If R is the value of the resistor (in ohms), and C is the value of the capacitor (in farads), the time constant, TC, will be obtained by the formula:

```
TC = R * C
```

If we multiply the R value by 1,000 while dividing the C value by 1,000, the time constant remains the same, and we can use the more convenient values of kilohms for the resistance and µF for the capacitance. In other words, the formula tells us that a 1K resistor in series with a 1,000µF capacitor has a time constant of 1 second.

The formula suggests that if the value of R diminishes to zero, the capacitor will charge instantly. In reality, the charging time will be rapid but finite, limited by factors such as the electrical resistance of the materials used.

Multiple Capacitors

When two or more capacitors are wired in parallel, their total capacitance is the sum of their separate capacitances. When two or more capacitors are wired in series, the relationship between their total capacitance © and their individual capacitances (C1, C2, C3 . . .) is given by this formula:

```
1 / C = (1/C1) + (1/C2) + (1/C3). . .
```

The formula to calculate the total capacitance of capacitors connected in series resembles the one used to calculate the total resistance of resistors connected in parallel. See Chapter 10.

Alternating Current and Capacitive Reactance

The apparent resistance of a capacitor to AC is properly known as *capacitive reactance*. In the following formula, capacitive reactance (X_C, in ohms) is derived as a function of capacitance (C, in farads) and AC frequency (f, measured in hertz):

$$X_C = 1 / (2 * \pi * f * C)$$

The formula shows that when frequency becomes zero, capacitive reactance becomes infinite; in other words, a capacitor has theoretically infinite resistance when DC current tries to flow through it. In reality, a dielectric has a finite resistance, and thus always allows some leakage.

The formula also shows that capacitive reactance diminishes when the size of the capacitor increases and/or the frequency being applied to it increases. From this it appears that an AC signal will be attenuated less at higher frequencies, especially if we use a small capacitor. However, a real-world capacitor also exhibits some degree of *inductive reactance*. This value will depend on its configuration (cylindrical vs. multiple flat plates), its physical length, the materials from which it is fabricated, the lengths of its leads, and other factors. Inductive reactance tends to *increase* with frequency, and since capacitive reactance tends to *decrease* with frequency, at some point the curves for the two functions intersect. This point represents the capacitor's *self-resonant frequency*, which is often referred to simply as its *resonant frequency*. See Figure 12-18.

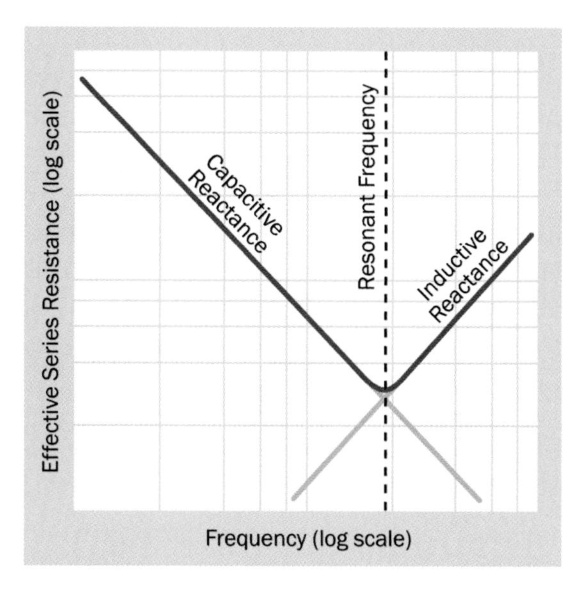

Figure 12-18. *As an AC current applied to a capacitor increases in frequency, the capacitive reactance of the component decreases, while its inductive reactance increases. The resonant frequency of the capacitor is found where the two functions intersect.*

Equivalent Series Resistance

A theoretically ideal capacitor would be purely reactive, without any resistance. In reality, capacitors are not ideal, and they have *equivalent series resistance,* or *ESR*. This is defined as the resistor that you would have to place in series with an ideal version of the capacitor, so that the combination would behave like the real version of the capacitor on its own.

If X_c is the reactance of the capacitor, then its Q factor (which means its *quality factor*) is given by the simple formula:

$$Q = X_C / ESR$$

Thus, the quality factor is higher if the ESR is relatively low. However, the reactance of the capacitor will vary significantly with frequency, and this simple formula is only an approximate guide.

The Q-factor for capacitors should not be confused with the Q-factor for inductors, which is calculated quite differently.

How to Use it

The figures illustrate some simplified schematics for common applications.

Bypass Capacitor

In Figure 12-19, a low-value capacitor (often 0.1μF) is placed near the power input pin of a sensitive digital chip to divert high-frequency spikes or noise to negative ground. This *bypass capacitor* may also be described as a *decoupling capacitor*.

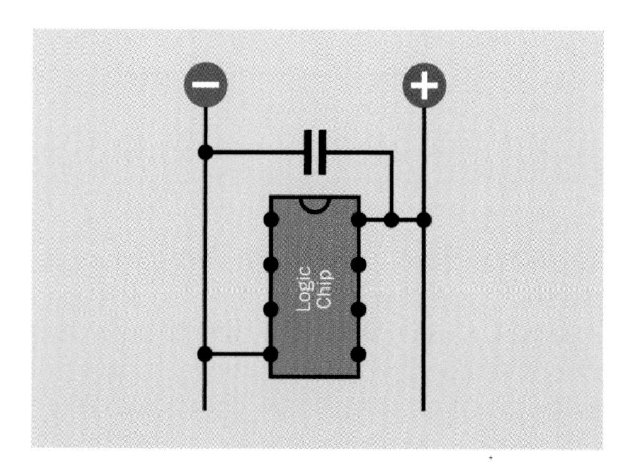

Figure 12-19. *A bypass capacitor (typically 0.1μF) configured to protect an integrated circuit logic chip from voltage spikes and noise in the power supply.*

Coupling Capacitor

In Figure 12-20, a 1μF *coupling capacitor* transmits a pulse from one section of a circuit to another, while blocking the DC voltage. Some reshaping of the waveform may occur.

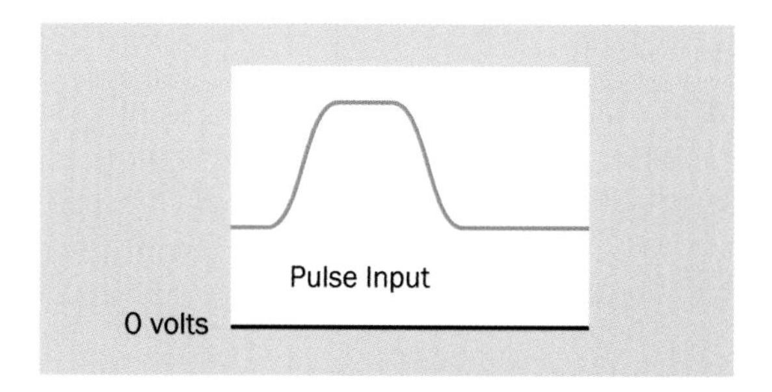

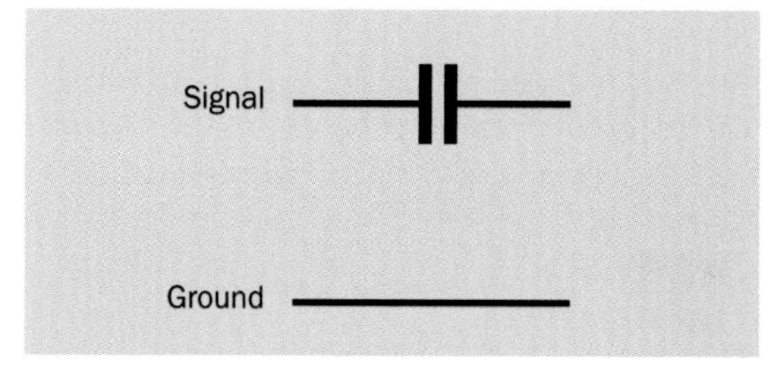

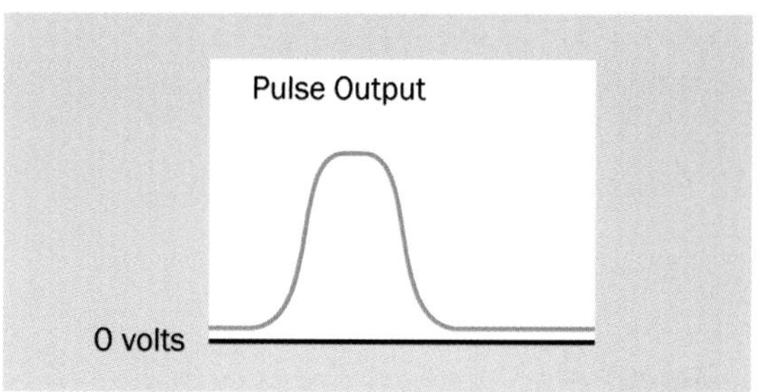

Figure 12-20. *A coupling capacitor (typically around 1μF) preserves DC isolation of one section of a circuit from another, while allowing a pulse to be transmitted.*

High-Pass Filter

In Figure 12-21, a 0.1μF capacitor blocks the low-frequency component of a complex waveform and transmits only the higher frequency that was superimposed on the low frequency.

Low-Pass Filter

In Figure 12-22, a 0.1μF decoupling capacitor diverts the higher frequency component of a complex waveform to negative ground, allowing only the lower frequency to be preserved. A lower-value capacitor (such as 0.001μF) will bleed away high-frequency noise from an AM radio source without affecting audio frequencies.

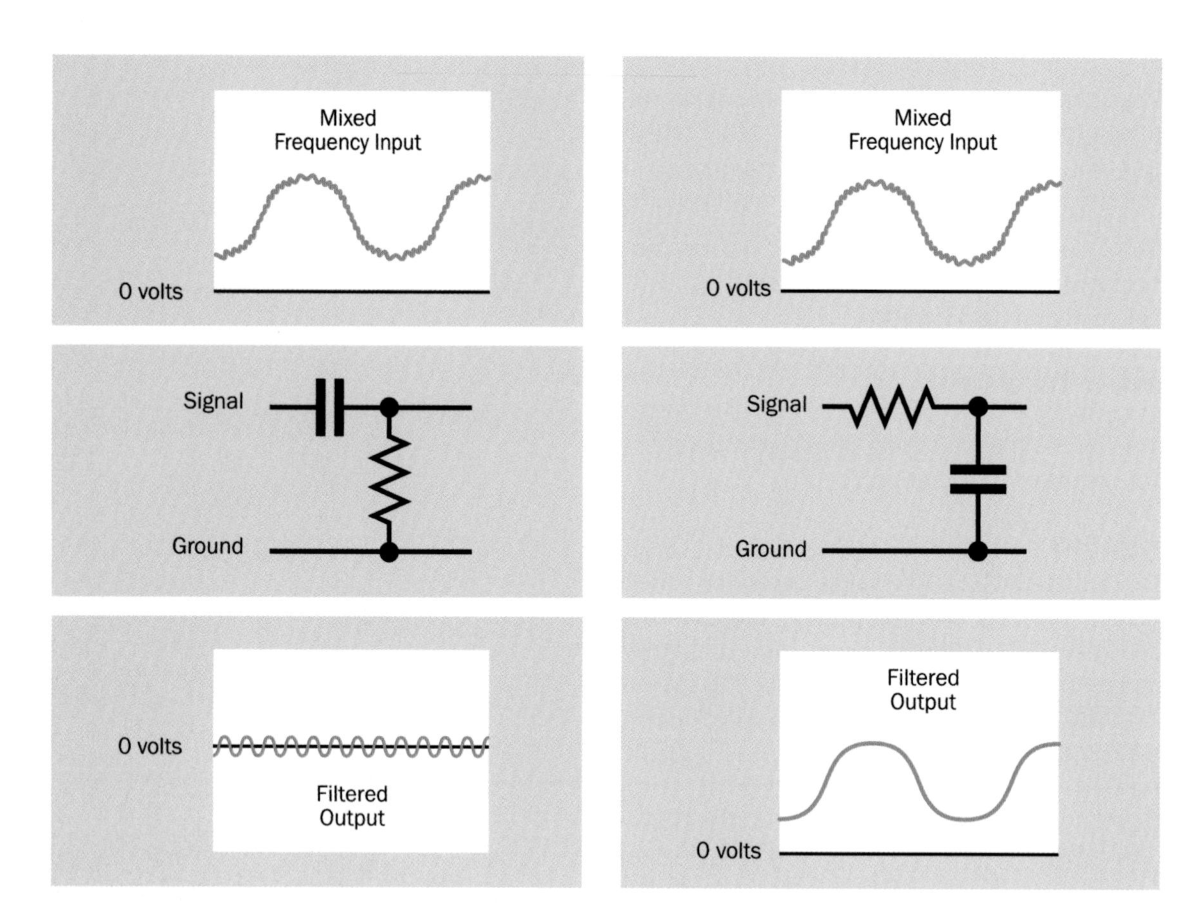

Figure 12-21. *A small capacitor (typically 0.1μF) can be used to create a high-pass filter, passing high frequencies while blocking low frequencies.*

Figure 12-22. *A small capacitor (typically 0.1μF) in this configuration routes high frequencies to negative ground, filtering them out of an analog signal.*

Smoothing Capacitor

In Figure 12-23, a 100μF capacitor charges and discharges to smooth an AC signal after a diode has removed the negative portion.

Snubber

In Figure 12-24, an RC network (inside a white dashed line) is known as a *snubber* when used to protect a switch from the problem of *arcing* (pronounced "arking")—that is, a sustained spark that can quickly erode the switch contacts. Arcing may occur in switches, pushbuttons, or relays that control an inductive load, such as a large motor. This problem can become significant at high DC currents (10A or more) or relatively high AC or DC voltages (100V or more).

When the switch is opened, the magnetic field that has been sustained by the inductive load collapses, causing a surge of current, or *forward EMF*. The capacitor in the snubber absorbs this surge, thus protecting the switch contacts. When the switch is closed again, the capacitor discharges itself, but the resistor limits the outrush of current—again, protecting the switch.

A snubber placed around the switch in a DC circuit could typically use a 0.1μF capacitor (polypropylene or polyester) rated for 125VAC/200VDC, and a 100-ohm carbon resistor rated 0.5 watt or higher. Prepackaged snubbers containing appropriate capacitor-resistor pairs are available from some parts suppliers, primarily for industrial use.

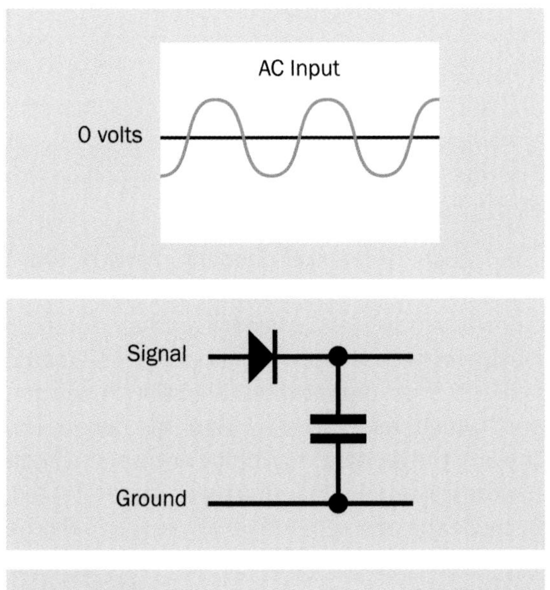

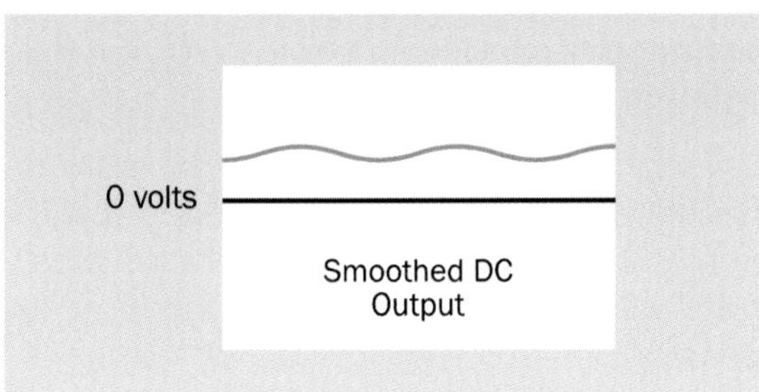

Figure 12-23. *A capacitor of 100µF or more smooths the upper half of an AC signal that has passed through a diode. The capacitor charges during each positive pulse and discharges to "fill the gaps" between them.*

In an AC circuit, a snubber can be placed around the inductive load itself. Although a **diode** is often used this way in a DC circuit, it cannot be used with AC.

Although solid-state switching devices such as a **solid state relay** contain no mechanical contacts, they may still be damaged by substantial pulses of back-EMF, and can be protected by a snubber where they are controlling inductive loads that take 10A or more at 100V or more.

Capacitor as a Battery Substitute

A capacitor may be substituted for a battery for some applications, although it has a lower energy density and will be more expensive to manufacture. A capacitor charges and discharges

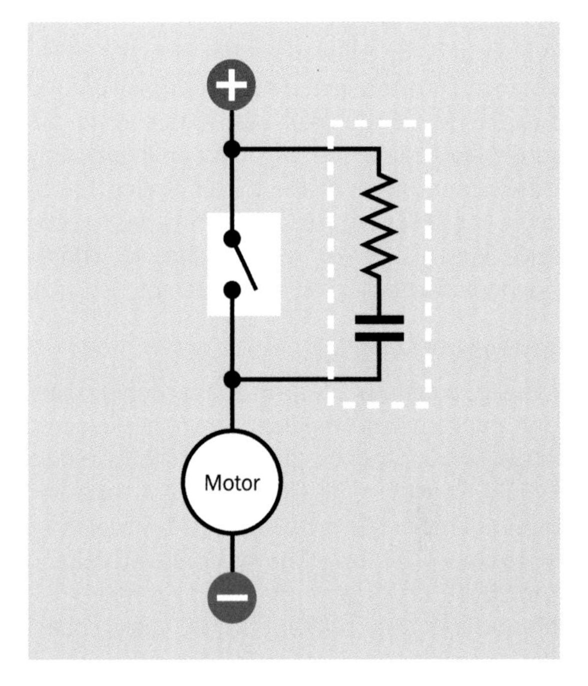

Figure 12-24. *An RC network (outlined with a white dashed line) protects a switch that controls a high inductive load. Used in this way, the RC network is known as a snubber.*

much more rapidly than a battery because no chemical reactions are involved, but a battery sustains its voltage much more successfully during the discharge cycle.

Capacitors that can store a very large amount of energy are often referred to as *supercapacitors*.

What Can Go Wrong

Common problems associated with capacitors are age-related deterioration (especially in electrolytics), inductive reactance (especially in cylindrical formats), nonlinear response, resistivity, excessive current leakage, and dielectric memory. Some of these problems are discussed below. A manufacturer's datasheet should be consulted carefully in conjunction with the notes regarding compositions in the preceding Variants section before making a commitment to a particular type of capacitor.

Wrong Polarity

A polarized capacitor may offer virtually no resistance if it is connected the wrong way around to a DC power source. A very high current can result, damaging the capacitor and probably other components in the circuit. Failing to observe the polarity of a tantalum capacitor can have destructive or even explosive consequences, depending on the amperage.

Voltage Overload

If the DC working voltage of a capacitor is exceeded, there is a risk of breaking down the dielectric and allowing a spark, or arc, that will form a short circuit. Note that the DC rating of a capacitor does not mean that it can be used safely with an equivalent AC voltage. The maximum AC voltage should be no greater than approximately 0.7 times the DC rated voltage. If a DC-rated capacitor is used directly across an AC power line, it will create an effective short circuit.

If capacitors are connected in series or in parallel, ideally the voltage rating for each capacitor should be the same, and certainly no less than the supply voltage.

Tantalum capacitors are easily damaged by current spikes that exceed their maximum working voltage, and are unsuitable for high-frequency coupling because of their inductance.

Leakage

Charge leakage is a problem especially associated with electrolytic capacitors, which are not suitable for storing a charge over a significant interval. Polypropalene or polystyrene film capacitors are a better choice.

Dielectric Memory

Also known as *dielectric absorption*, this is a phenomenon in which a capacitor's electrolyte displays some percentage of its former voltage after the capacitor has been discharged and then disconnected from the circuit. Single-layer ceramic capacitors especially tend to suffer from this problem.

Specific Electrolytic Issues

Electrolytic capacitors have high inductive reactance, are not manufactured to close tolerances, and deteriorate significantly with age. While other components may be stockpiled and used over a period of years, this is not a sensible policy with electrolytics.

The "capacitor plague" affecting many of these capacitors manufactured from 1999 onward provided a salutary lesson regarding their potential weaknesses. Faulty composition of the dielectric allowed it to deteriorate, liberating hydrogen gas, which eventually caused the aluminum shells of the capacitors to bulge and burst. Circuit boards from major manufacturers were affected. Because the problem took two years to become apparent, literally millions of boards with faulty capacitors had been sold before the fault was diagnosed and eventually corrected.

Unfortunately electrolytics cannot be easily replaced with other types of capacitors in applications such as power supplies, because substitutes will be considerably larger and more expensive.

Heat

The equivalent series resistance (ESR) of a large capacitor inevitably means that it must dissipate some power as heat during use. Ripple current can also create heat. Capacitor performance will change as the temperature increases. A common maximum component temperature for electrolytic capacitors is 85 degrees Centigrade.

Vibration

In a high-vibration environment, electrolytics should be protected by clamping them mechanically in place, using a *capacitor clamp*, also known as a *c-clamp*.

Misleading Nomenclature

Rarely, in the United States, the term "mF" may be used as a probable alternative to μF. This can be a source of confusion and risk because mF is properly (but very rarely) used to mean "millifarads." The term should always be avoided.

variable capacitor

13

Formerly known (primarily in the United Kingdom) as a *variable condenser*. The term is now obsolete.

OTHER RELATED COMPONENTS

- **capacitor** (See Chapter 12)

What It Does

A variable capacitor allows adjustment of capacitance in much the same way that a **potentiometer** allows adjustment of resistance.

Large variable capacitors were developed primarily to tune radio receivers, in which they were known as *tuning capacitors*. Cheaper, simpler, and more reliable substitutes gradually displaced them, beginning in the 1970s. Today, they are still used in semiconductor fabrication, in RF plastic welding equipment, in surgical and dental tools, and in ham radio equipment.

Small *trimmer capacitors* are widely available and are mostly used to adjust high-frequency circuits. Many of them look almost indistinguishable from *trimmer potentiometers*.

The schematic symbols commonly used to represent a variable capacitor and a trimmer capacitor are shown in Figure 13-1.

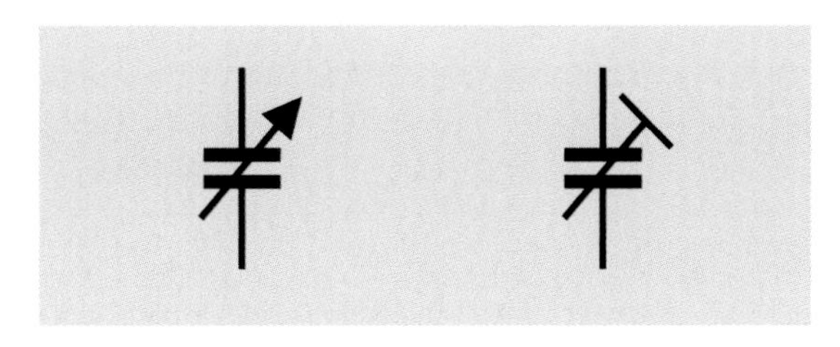

Figure 13-1. *Typical schematic symbols for variable capacitor (left) and trimmer capacitor (right).*

A *varactor* is a form of diode with variable capacitance, controlled by reverse voltage. See "Varactor Diode" on page 219 for this component.

How It Works

The traditional form of variable capacitor consists of two rigid semicircular plates separated by an air gap of 1mm to 2mm. To create more capacitance, additional interleaved plates are added to form a stack. One set of plates is known as the *rotor*, and is mounted on a shaft that can be turned, usually by an externally accessible knob. The other set of plates, known as the *stator*, is mounted on the frame of the unit with ceramic insulators. When the sets of plates completely overlap, the capacitance between them is maximized. As the rotor is turned, the sets of plates gradually disengage, and the capacitance diminishes to near zero. See Figure 13-2.

The air gaps between the sets of plates are the *dielectric*. Air has a dielectric constant of approximately 1, which does not vary significantly with temperature.

The most common shape of plate is a semicircle, which provides a linear relationship between capacitance and the angle of rotation. Other shapes have been used to create a nonlinear response.

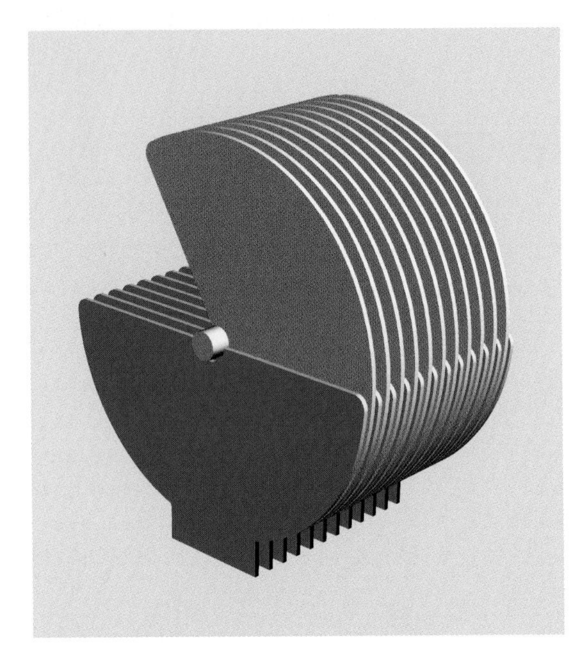

Figure 13-2. *In this simplified view of a variable capacitor, the brown plates constitute the rotor, attached to a central shaft, while the blue plates are the stator. The colors have no electrical significance and are added merely for clarity. The area of overlap between rotor and stator determines the capacitance.*

Reduction gears may be used to enable fine tuning of a variable capacitor, which means multiple turns of a knob can produce very small adjustments of the capacitor. At the peak of variable capacitor design, units were manufactured with high mechanical precision and included *anti-backlash gears*. These consisted of a pair of equal-sized gears mounted flat against each other with a spring between them that attempted to turn the gears in opposite directions from each other. The pair of gears meshed with a single pinion, eliminating the looseness, or backlash, that normally exists when gear teeth interlock. A vintage capacitor with a spring creating anti-backlash gearing (circled) is shown in Figure 13-3. This is a two-gang capacitor—it is divided into two sections, one rated 0 to 35pF, the other rated 0 to 160pF.

Figure 13-3. *A "traditional style" variable capacitor of the type designed to tune radio frequencies. The spring, circled, enables anti-backlash gearing.*

Variants

The traditional variable capacitor, with exposed, air-spaced, rigid, rotating vanes, is becoming hard to find. Small, modern variable capacitors are entirely enclosed, and their plates, or vanes, are not visible. Some capacitors use a pair of concentric cylinders instead of plates or vanes, with an external thumb screw that moves one cylinder up or down to adjust its overlap with the other. The overlap determines the capacitance.

Trimmer capacitors are available with a variety of dielectrics such as mica, thin slices of ceramic, or plastic.

Values

A large traditional capacitor can be adjusted down to a near-zero value; its maximum will be no greater than 500pF, limited by mechanical factors. (See Chapter 12 for an explanation of capacitance units.)

A maximum value for a trimmer capacitor is seldom greater than 150pF. Trimmers may have their values printed on them or may be color-coded, but there is no universal set of codes. Brown, for example, may indicate either a maximum value around 2pF or 40pF, depending on the manufacturer. Check datasheets for details.

The upper limit of a trimmer's rated capacitance is usually no less than the rated value, but can often be 50% higher.

Formats

All trimmer capacitors are designed for mounting on circuit boards. Many are surface-mount, with a minority being through-hole. Surface-mount units may be 4mm × 4mm or smaller. Through-hole are typically 5mm × 5mm or larger. Superficially, trimmer capacitors resemble single-turn trimmer potentiometers with a screw head in the center of a square package. A through-hole example is shown in Figure 13-4.

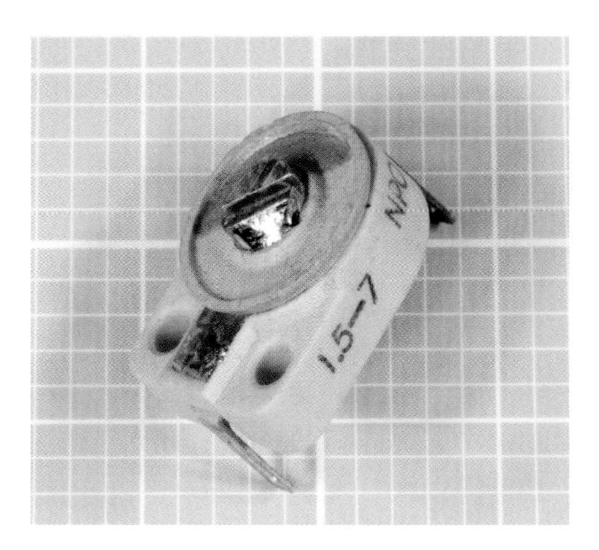

Figure 13-4. *A trimmer capacitor rated 1.5pF to 7.0pF.*

How to Use it

A variable capacitor is often used to tune an *LC circuit,* so called because a coil (with reactance customarily represented by letter L) is wired in parallel with a variable capacitor (represented by letter C). The schematic in Figure 13-5 shows an imaginary circuit to illustrate the principle. When the switch is flipped upward, it causes a large fixed-value capacitor to be charged from a DC power source. When the switch is flipped down, the capacitor tries to pass current through the coil—but the coil's *reactance* blocks the current and converts the energy into a magnetic field. After the capacitor discharges, the magnetic field collapses and converts its energy back into electricity. This flows back to the capacitor, but with inverted polarity. The cycle now repeats with current flowing in the opposite direction. A low-current **LED** across the circuit would flash as the voltage oscillates, until the energy is exhausted.

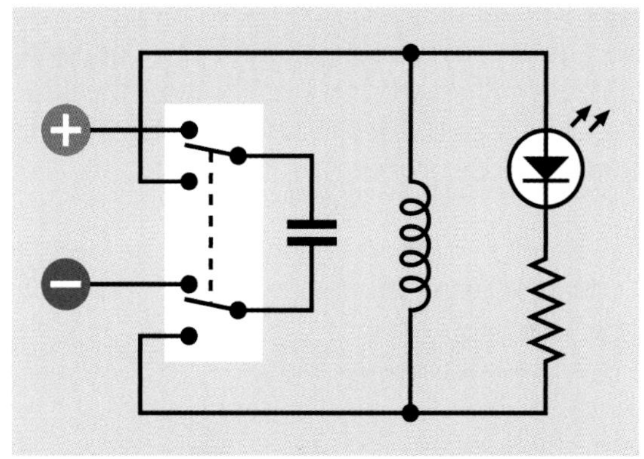

Figure 13-5. *In this imaginary circuit, the capacitor is charged through the double-pole switch in its upper position. When the switch is turned, the capacitor forms an LC (inductance-capacitance) circuit with the coil, and resonates at a frequency determined by their values. In reality, extremely high values would be needed to obtain a visible result from the LED.*

Because the oscillation resembles water sloshing from side to side in a tank, an LC circuit is sometimes referred to as a *tank circuit.*

In reality, unrealistically large values would be required to make the circuit function as described. This can be deduced from the following formula, where f is the frequency in Hz, L is inductance in Henrys, and C is capacitance in Farads:

```
f = 1 / (2π * √(L * C) )
```

For a frequency of 1Hz, a massive coil opposite a very large capacitor of at least 0.1F would be needed.

However, an LC circuit is well-suited to very high frequencies (up to 1,000MHz) by using a very small coil and variable capacitor. The schematic in Figure 13-6 shows a high-impedance earphone and a diode (right) substituted for the LED and the resistor in the imaginary circuit, while a variable capacitor takes the place of the fixed capacitor. With the addition of an antenna at the top and a ground wire at the bottom, this LC circuit is now capable of receiving a radio signal, using the signal itself as the source of power. The resonant frequency of the circuit is tuned by the variable capacitor. The impedance peaks at the resonant frequency, causing other frequences to be rejected by passing them to ground. With suitable refinement and amplification, the basic principle of an LC circuit is used in AM radios and transmitters.

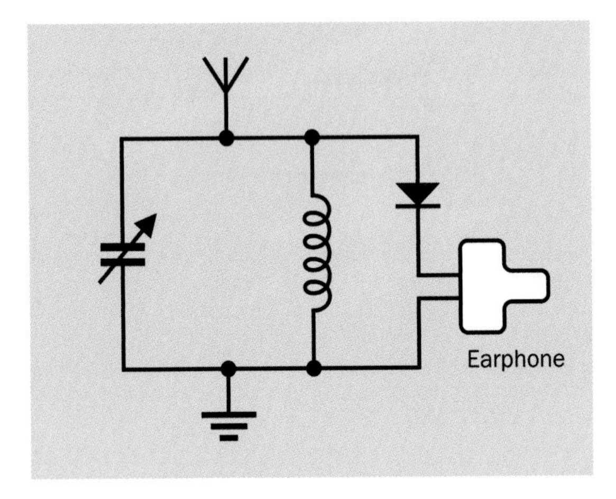

Figure 13-6. *The principle of an LC circuit is used here in a basic circuit that can tune in to a radio station and create barely audible sound through the earphone at right, using only the broadcast signal for power. The variable capacitor adjusts the frequency of the circuit to resonate with the carrier wave of the radio signal.*

Because variable capacitors are so limited in size, they are unsuitable for most timing circuits.

Trimmer capacitors are typically found in high-power transmitters, cable-TV transponders, cellular base stations, and similar industrial applications.

They can be used to fine-tune the resonant frequency of an oscillator circuit, as shown in Figure 13-7.

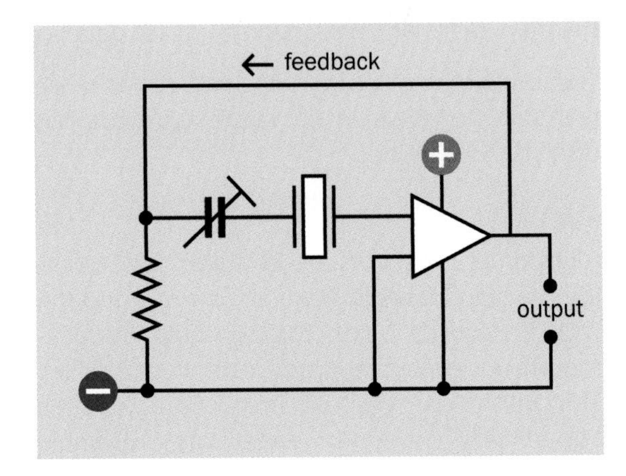

Figure 13-7. *A trimmer capacitor in series with a crystal fine-tunes the frequency of this basic circuit using an op-amp.*

In addition to tuning a circuit frequency, a trimmer capacitor can be used to compensate for changes in capacitance or inductance in a circuit that are caused by the relocation of wires or rerouting of traces during the development process. Readjusting a trimmer is easier than swapping fixed-value capacitors. A trimmer may also be used to compensate for capacitance in a circuit that gradually drifts with age.

What Can Go Wrong

Failure to Ground Trimmer Capacitor While Adjusting it

Although trimmer capacitors are not polarized, the manufacturer may mark one terminal with a plus sign and/or the other with a minus sign. If the capacitor is adjusted while its negative terminal is floating or ungrounded, a metal screwdriver blade will create erroneous readings. Always ground the appropriate side of a trimmer capacitor before fine-tuning it, and preferably use a plastic-bladed screwdriver.

Application of Overcoat Material or "Lock Paint"

Overcoat material is a rubbery adhesive that may be spread over assembled components to immunize them against moisture or vibration. *Lock paint* is a dab of paint that prevents a screw adjustment from turning after it has been set. Most manufacturers advise against applying these materials to a trimmer capacitor, because if penetration occurs, the capacitor can fail.

Lack of Shielding

Variable capacitors should be shielded during use, to protect them from external capacitive effects. Merely holding one's hand close to a variable capacitor will change its value.

inductor

14

The term **inductor** is used here to describe a coil that has the purpose of creating *self-inductance* in an electronic circuit, often while passing alternating current in combination with resistors and/or capacitors. A *choke* is a form of inductor. By comparison, the **electromagnet** entry in this encyclopedia describes a coil containing a center component of ferromagnetic material that does not move relative to the coil, and has the purpose of attracting or repelling other parts that respond to a magnetic field. A coil containing a center component of ferromagnetic material that moves as a result of current passing through the coil is considered to be a **solenoid** in this encyclopedia, even though that term is sometimes more broadly applied.

OTHER RELATED COMPONENTS

- **solenoid** (See Chapter 21)
- **electromagnet** (See Chapter 20)

What It Does

An inductor is a coil that induces a magnetic field in itself or in a *core* as a result of current passing through the coil. It may be used in circuits to block or reshape AC current or a range of AC frequencies, and in this role can "tune" a simple radio receiver or various types of oscillators. It can also protect sensitive equipment from destructive voltage spikes.

The schematic symbol for an inductor includes a coil that can be drawn in two basic styles, shown at the top and at the bottom of Figure 14-1. The style at the bottom has become more common, but the style at the top is not obsolete. In each vertical section of the diagram, the functionality of the symbols is identical.

One or two parallel lines alongside the coil indicate that it is wound around a solid core of material that can be magnetized, while one or two dotted lines indicate that it is wound around a core containing metal particles, such as iron filings. Where no core is shown, this indicates an air core.

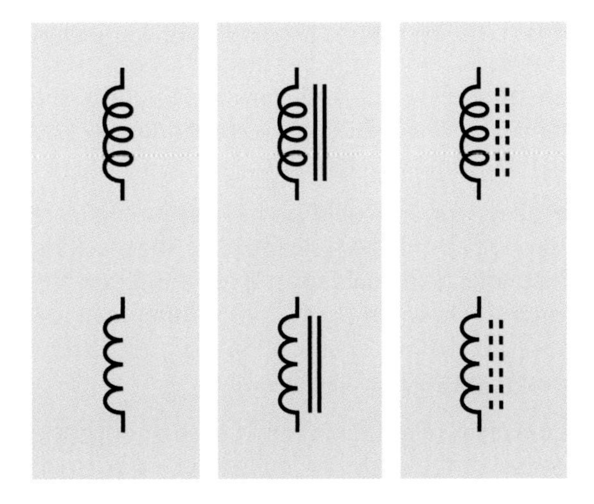

Figure 14-1. *The coil symbol for an inductor may be drawn in two styles which are functionally identical. Line(s) beside the coil indicate a solid core. Dotted line(s) indicate a core containing metal particles.*

A selection of inductors designed for through-hole mounting is shown in Figure 14-2.

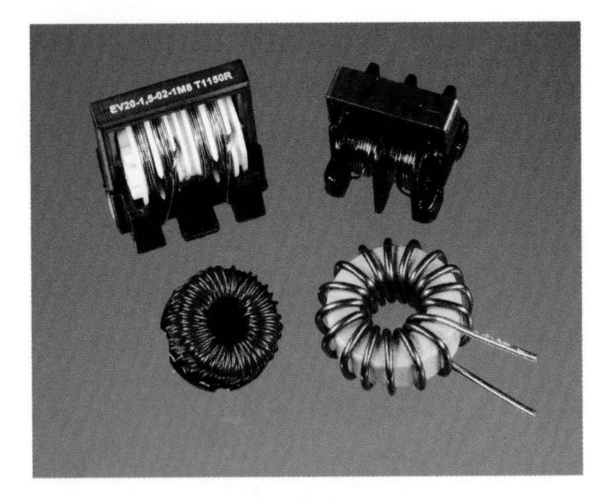

Figure 14-2. *Four inductors designed for through-hole insertion into printed circuit boards.*

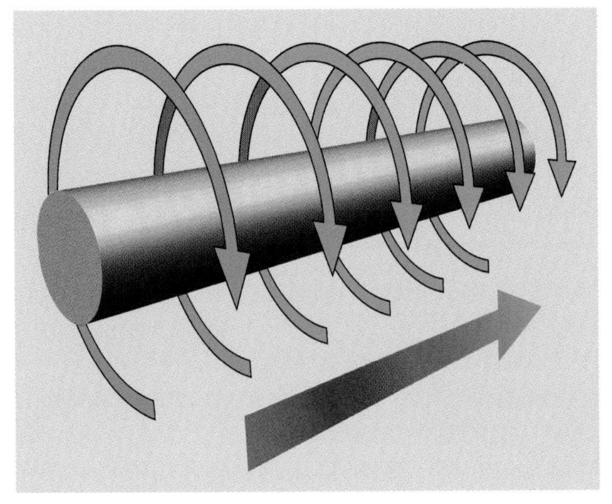

Figure 14-3. *Conventional current passing through a wire from left to right (as indicated by the red/blue arrow) induces a magnetic field around the wire (shown by the green arrows).*

How It Works

Direct current passing through an electrical conductor, such as a wire, creates a magnetic field around the conductor. In Figure 14-3, *conventional current* (flowing from positive to negative) is passing through a straight wire from left to right, as indicated by the red/blue arrow. The resulting magnetic field is indicated by the green arrows. If the wire is now bent into a curve, as shown in Figure 14-4, the magnetic field exerts an aggregate force downward through the curve. This magnetic force is conventionally said to flow from *south* to *north*.

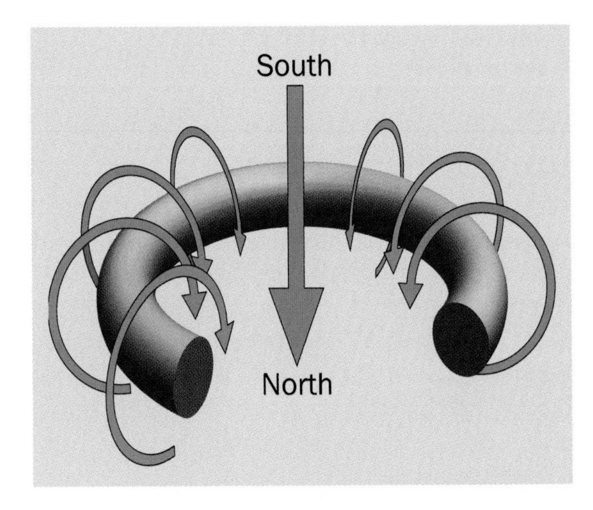

Figure 14-4. *If the wire is bent into a curve, the magnetic fields can create a net force shown by the large green arrow.*

If direct current could be induced to circulate through an unbroken circle of wire, the resulting magnetic field could exert a force through the circle as shown in Figure 14-5, assuming clockwise circulation of conventional current as suggested by the red/blue arrows.

Conversely, if a magnet was pushed through the center of the circle, it would induce a pulse of electric current in the circle. Thus, electricity passing through a wire can induce a magnetic field around the wire, and conversely, a magnet moving near a wire can induce an electric current in the wire. This principle is used in an electrical *generator*, and also in a **transformer**, where alternating current in the primary coil induces a fluctuating magnetic field in the core, and the field in the core is turned back into alternating current in the secondary coil.

Note that a static or unchanging magnetic field will not induce a flow of electricity.

Figure 14-5. *Hypothetically, if conventional current flows around a circular conductor (as suggested by the red/blue arrows), it will create a magnetic field that can create a force as shown by the green arrow.*

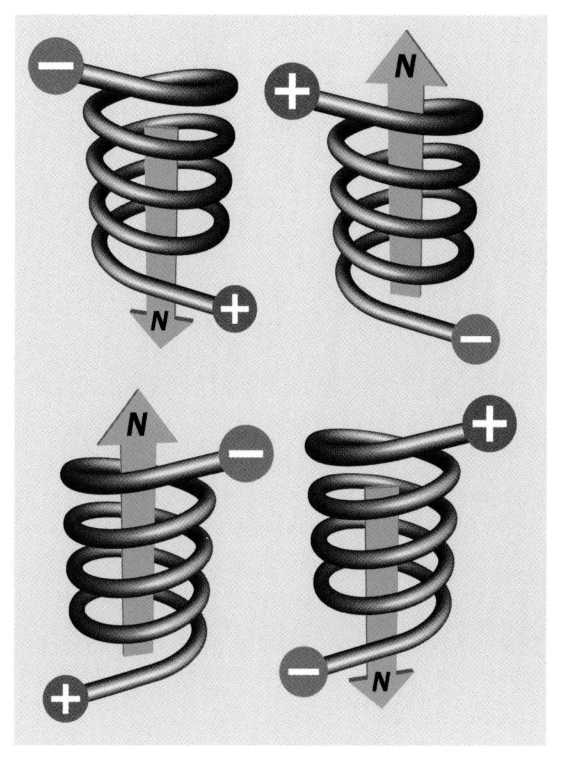

Figure 14-6. *When DC current flows through a coil, it creates magnetic fields that will exert a force whose direction depends on the direction of the current and on whether the coil is wound clockwise or counterclockwise. The force is shown by the green arrow in each case.*

DC Through a Coil

If the wire is formed into a helix (a series of approximate circles) as shown in Figure 14-6, and if DC current is passed through the wire, the aggregate of the magnetic fields can create a force in the direction of the green arrow in each example, depending whether the wire is wound clockwise or counter-clockwise, and depending on the direction of the current. The helix is usually referred to as a *coil* or a *winding*.

In actuality, a magnetic field is not open-ended, and its lines of force are completed by circling around outside the inductor, to complete a *magnetic circuit*. This completion of the field can be demonstrated by the traditional high-school experiment of positioning a compass or scattering iron filings on a sheet of paper above a magnet. A simplified depiction of lines of force completing a magnetic circuit is shown in Figure 14-7, where a coil is inducing the magnetic field. Note that throughout this encyclopedia, the color green is used to indicate the presence of magnetic force.

The completion of a magnetic field is not relevant to the primary function of the inductor. In fact the external part of the magnetic field is mostly a source of trouble in electronics applications, since it can interact with other components, and may necessitate the use of magnetic shielding. In addition, the field is weakened by completing itself through air, as air presents much greater *reluctance* (the magnetic equivalent of resistance) than the core of an electromagnet.

The polarity of a magnetic field created by a coil can be demonstrated by moving a small permanent magnet toward the coil, as shown in Figure 14-8. If the magnet has opposite polarity to the coil, it will tend to be repelled, as like poles

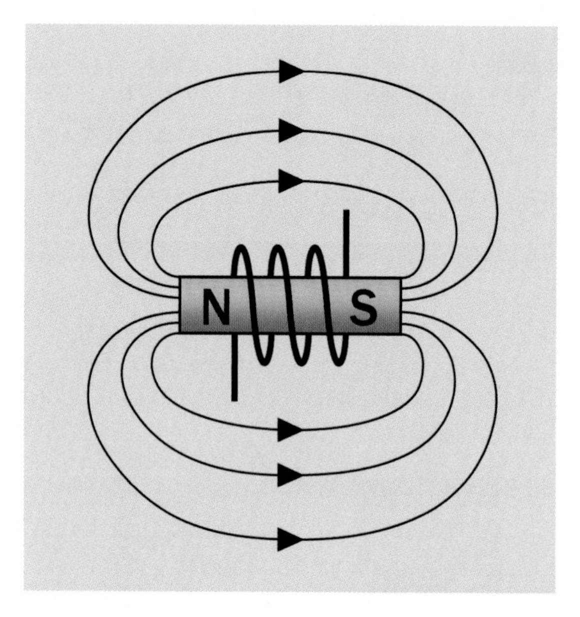

Figure 14-7. *A magnetic field in reality is not open-ended, and each line of force traveling through a rod-shaped magnet or electromagnet is completed outside of the magnet. The completion of magnetic fields has been omitted from other diagrams here for clarity.*

repel. If it has the same polarity, it will tend to be attracted, because opposite poles attract. This principle may be used in **solenoids**.

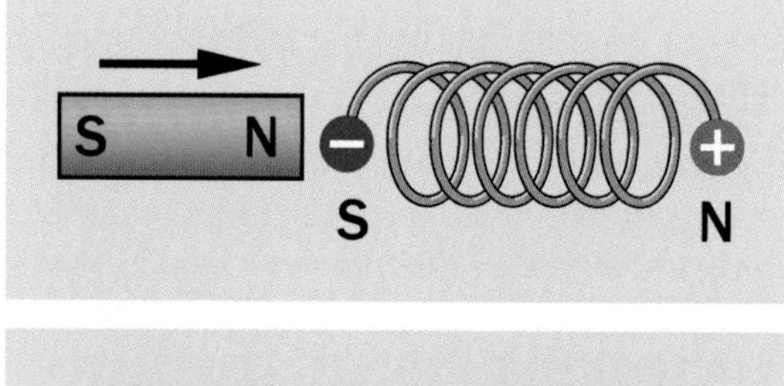

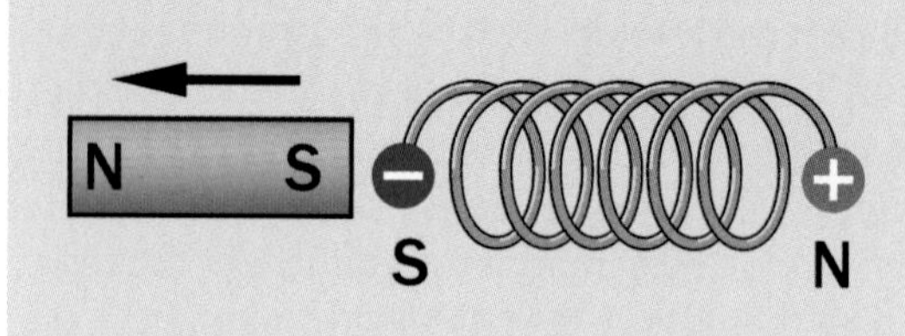

Figure 14-8. *A permanent magnet (left) will either be drawn toward a DC-energized coil or repelled from it, depending on the polarity of the two magnetic fields.*

Magnetic Core

The inductive power of a coil will be enhanced, and the saturation point will be reduced, by using a *magnetic core*. The term "magnetic" in this context does not mean that the core is a permanent magnet; it means that the core can be magnetized briefly by a transient pulse of electricity through the surrounding coil.

A core enhances the effectiveness of an inductor because it has a lower reluctance than that of air. In other words, magnetic flux will flow much more readily through the core than through air.

Roughly speaking, the *permeability* of a magnetic circuit is the opposite of reluctance; it is a measure of how easily a magnetic field can be induced, and is usually expressed relative to the permeability of air, which is approximately 1. The permeability of different core types is discussed in the following "Values" section.

The core of the coil contains *magnetic domains* that behave as tiny magnets, with north and south poles. In the absence of a polarizing magnetic field, the domains are randomly aligned. As a magnetic field is introduced around them and grows stronger, the domains align themselves with it, increasing the total magnetic force. When the domains are almost all uniformly aligned, the core approaches *magnetic saturation* and ceases adding to the net magnetic field. At this point the current in the inductor is said to be *continuous*.

When power to the coil is disconnected, the domains revert partially to their previous random orientation. Thus the core remains a weak permanent magnet. This effect is known as *hysteresis*, while the weak residual field is known as *remanent magnetism*.

EMF and Back-EMF

When DC current is connected through an inductor, the creation of a magnetic field takes a brief but measurable period of time. The field induces an *EMF* (*electro-motive force*) in the wire. Since this force opposes the supplied current, it is referred to as *back-EMF*. It lasts only so long as the field is increasing to its full strength. After the field reaches a steady state, current flows through the coil normally.

This transient resistive effect is caused by the *self-inductance* of the coil, and is opposite to the behavior of a capacitor, which encourages an initial inrush of direct current until it is fully charged, at which point it blocks subsequent current flow.

When high-frequency alternating current attempts to flow through an inductor, if each pulse is too brief to overcome the back-EMF, the coil will block the current. A coil can thus be designed to block some frequencies but not others.

Even a simple electrical circuit that does not contain a coil will still have some self-inductance, simply because the circuit consists of wires, and even a straight length of wire induces a magnetic field when the power is switched on. However, these inductive effects are so small, they can generally be ignored in practical applications.

The transient electrical resistance to alternating current caused by either an inductor or a capacitor is known as *reactance*, although it occurs under opposite electrical conditions, as the coil impedes an initial pulse of DC current and then gradually allows it to pass, while a capacitor allows an initial pulse of DC current and then impedes it.

When a flow of DC current through a coil is switched off, the magnetic field that was created by the coil collapses and releases its stored energy. This can cause a pulse of forward EMF, and like back-EMF, it can interfere with other components in a circuit. Devices such as motors and large relays that contain substantial coils can create problematic spikes of back-EMF and forward-EMF. The forward-EMF that occurs when power to the coil is interrupted is typically dealt with by putting a diode in parallel with the coil, allowing current to circulate through it. This is known as *clamping* the voltage transient. A diode-capacitor combination known as a *snubber* is also commonly used. For a schematic and additional information on this topic, see "Snubber" on page 102.

A schematic to demonstrate EMF and back-EMF is shown in Figure 14-9. The coil can be a 100-foot spool of 26-gauge (or smaller) hookup wire, or magnet wire. It will function more effectively if a piece of iron or steel, such as half-inch galvanized pipe, is inserted through its center. When the button is pressed, current is briefly impeded by the back-EMF created by the coil, and is diverted through D1, making it flash briefly. Then the coil's reactance diminishes, allowing the current to flow through the coil and bypass the LED. When the pushbutton is released, the coil's magnetic field collapses, and the consequent forward-EMF circulates through D2, causing it to flash briefly. Note that the polarity of back-EMF and forward-EMF are opposite, which is why the LEDs in the circuit are oriented with opposite polarities.

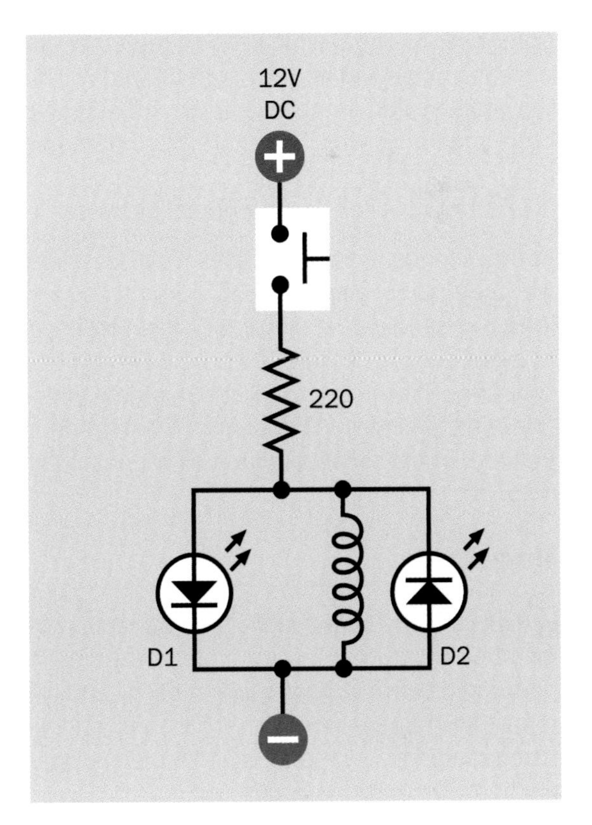

Figure 14-9. *A test circuit to demonstrate the EMF and back-EMF created when DC current starts and stops passing through a coil. See text for details.*

The 220Ω resistor should be rated at 1/4 watt minimum, and the button should not be held down for long, as the electrical resistance of the coil is relatively low. The LEDs ideally should be rated for a minimal forward current of no more than 5mA.

Electrical and Magnetic Polarity

Various mnemonics and images have been created to assist in memorizing the polarity or direction of the magnetic field that will be created by a flow of electricity. The *right-hand rule* suggests that if the fingers of the right hand are curled around a coil in the same direction in which the turns of the coil were wound, and if conventional DC current also flows in this direction, the extended thumb will point in the direction of the principal force that can be created by the magnetic field.

By convention, the magnetic field is oriented from *south* to *north*, which can be remembered since the north end of the magnetic field will be the negative end of the coil (north and negative both beginning with letter N). This mnemonic only works if conventional (positive) current flows through a coil that is wound clockwise.

Another model is the "corkscrew rule" in which we imagine conventional DC current flowing from the handle of a corkscrew, down through its metal section, toward the pointed end. If the corkscrew is turned clockwise, in the same direction as the electricity, the corkscrew will sink into the cork in the same direction as the resulting magnetic force.

Variants

Variants include core materials, core shapes, termination style (for through-hole mounting in perforated board, or for surface-mount), and external finish (some inductors are dipped in insulating material, while others allow their copper magnet wire to be exposed).

In addition there are two functional variants: variable inductors and ferrite beads. Their schematic symbols are shown in Figure 14-10.

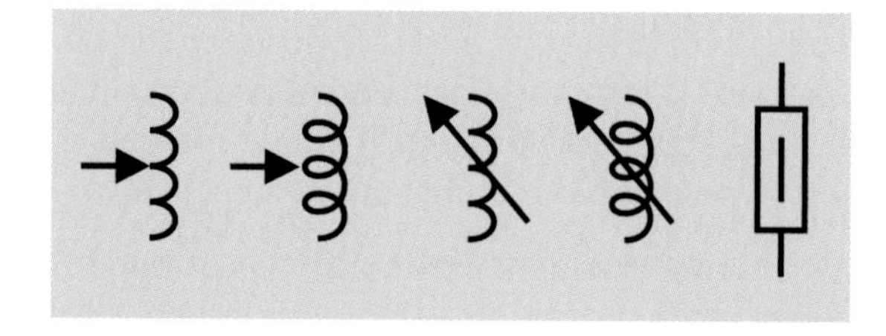

Figure 14-10. *Schematic symbols for a ferrite bead (farthest right) and variable inductors (all other symbols, which are functionally identical).*

Magnetic Cores

A magnetic core may be made from solid iron, plates of iron or steel separated by thin insulating material, powdered iron mixed with a binder, or a ferrite compound derived from nickel, zinc, manganese, or a combination. An iron core has at least 1,000 times the permeability of air, while some ferrites are 10,000 times as permeable.

One major disadvantage of a magnetic core is *hysteresis*, which in this context refers to the tendency of the core to retain some magnetic "memory" as a cycle of alternating current changes from positive to negative. This residual magnetism must be overcome by the next positive pulse of AC. The tendency of the core to retain magnetic polarity is known as its *retentivity*. Iron cores are especially retentive.

Another disadvantage of some magnetic cores is that they may host eddy currents induced by the magnetic field of the coil. These electrical currents tend to circulate through the core, reducing efficiency by generating waste heat, especially if coil currents are high. Forming a core from iron or steel plates, separated by thin layers of insulation, will inhibit these currents. Powdered iron inhibits eddy currents because the particles have limited contact. Ferrites are nonconductive, and are therefore immune to eddy currents. They are widely used.

Hysteresis and eddy currents both incur energy losses with each AC cycle. Therefore, the losses increase linearly as the AC frequency increases. Consequently, inductor cores that suffer either of these problems are not well-suited to high frequencies.

Nonmagnetic Cores

The problems associated with magnetic cores may be avoided by winding the coil around a nonmagnetic core that may be hollow, ceramic, or plastic. A hollow core is referred to as an *air core*. The permeability of ceramic and plastic cores is close to that of air.

An inductor with a nonmagnetic core will be immune to eddy currents and retentivity, but will have to be significantly larger than a magnetic-cored coil with comparable inductance. In the case of a very primitive radio receiver, such as a crystal set, the air-cored coil that selects a radio frequency may be several inches in diameter. A basic circuit diagram for a crystal set (so-called because it uses a diode containing a germanium crystal) is shown in Figure 14-11. The antenna, at top, receives signals broadcast from radio stations. The coil can be tapped (as indicated by the black dots) as a simple way to select different inductance values, blocking all but a narrow range of frequencies. The T-shaped white component at right is a high-impedance earphone. The diode blocks the lower half of the alternating current in a radio signal, and since the signal is *amplitude-modulated*, the earphone responds to variations in intensity in the signal and reproduces the sound encoded in it.

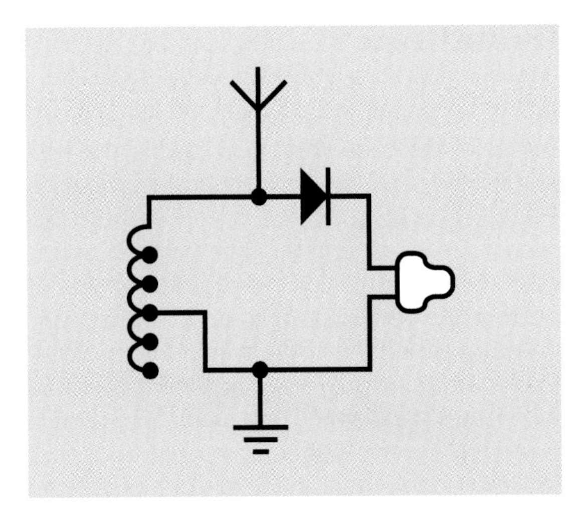

Figure 14-11. *An early and basic application for an inductor is to select radio-station frequencies, as in this schematic for a crystal set. See text for details.*

Variable Inductors

A *variable inductor*, also known as an *adjustable inductor*, is relatively uncommon but can be fabricated by using a magnetic core that penetrates the center of the inductor on an adjustable screw thread. The inductance of the assembly will increase as a larger proportion of the magnetic core penetrates into the open center of the coil. A photograph of a variable inductor is at Figure 14-12.

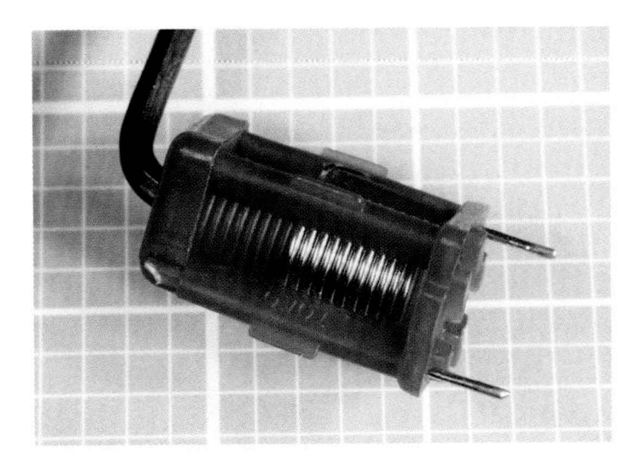

Figure 14-12. *A variable inductor. Its inductance is adjusted via a screw thread that varies the insertion of the core in the coil. In this component the core is turned by inserting a hex wrench, as shown. It is rated from 0.09μH to 0.12μH.*

Ferrite Beads

A *ferrite bead* inverts the design of a typical inductor by running a wire through a hole in the center of the bead, instead of coiling the wire around the core. Two ferrite beads are shown in Figure 14-13. At top, the bead is divided into two sections, each mounted in one-half of a plastic clam shell, which can be closed around a wire. At bottom, the bead must be threaded onto a wire. The purpose is either to limit radio-frequency radiation from a wire by absorbing it into the bead (where it is transformed into heat), or to protect a wire from external sources of radio-frequency radiation. Computer cabling to external devices; lamp dimmers; and some types of motors can be sources of radio frequency.

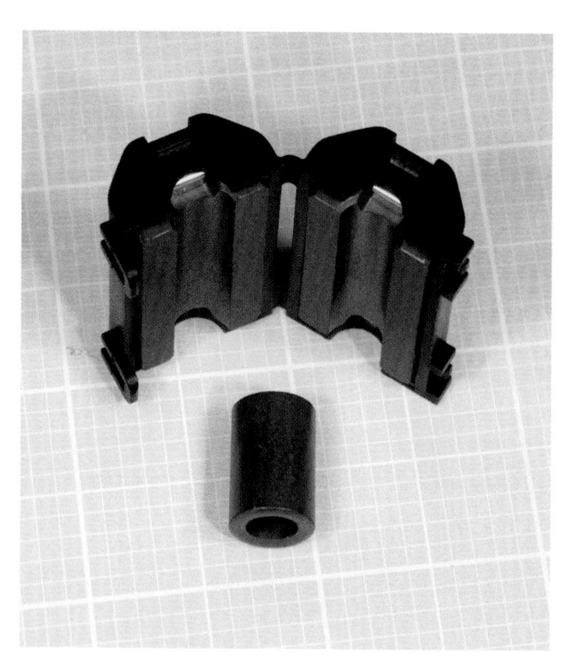

Figure 14-13. *Two examples of ferrite beads. They can inhibit radio-frequency radiation from a wire, or protect the wire from interference.*

Toroidal Cores

The magnetic circuit created by a rod-shaped core must be completed by the lines of force traveling back around from one end of the rod to the other, through the surrounding air. Since air has low permeability, this is a major source of inefficiency. By comparison, a torus (a geometrical shape resembling a donut) completes the entire magnetic circuit inside its core. This significantly increases its efficiency. Also, because its field is better contained, a toroidal inductor needs little or no shielding to protect other components from stray magnetic effects.

Two through-hole toroidal inductors are shown in Figure 14-2. Bottom left: Rated at 345μH. Bottom right: Rated at 15μH. The one at bottom-left has pins beneath it for insertion into a printed circuit board.

Surface-mount inductors often are toroidal to maximize the efficiency of a component that has to function on a very small scale. Examples are shown in Figure 14-14, Figure 14-15, and Figure 14-16.

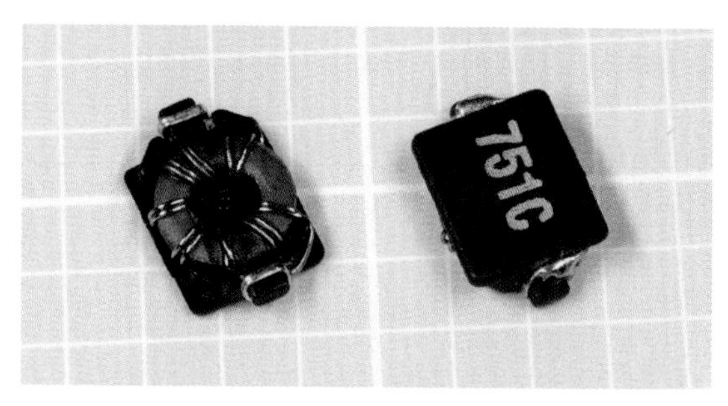

Figure 14-14. *In a typical toroidal inductor, the coil is wrapped around a magnetic core shaped as a torus. This surface-mount component (viewed from the bottom, at left, and from the top, at right) is at the low end of the range of component sizes. It is rated at 750nH.*

Figure 14-15. *A medium-sized surface-mount toroidal inductor (viewed from the bottom, at left, and from the top, at right). It is rated at 25μH.*

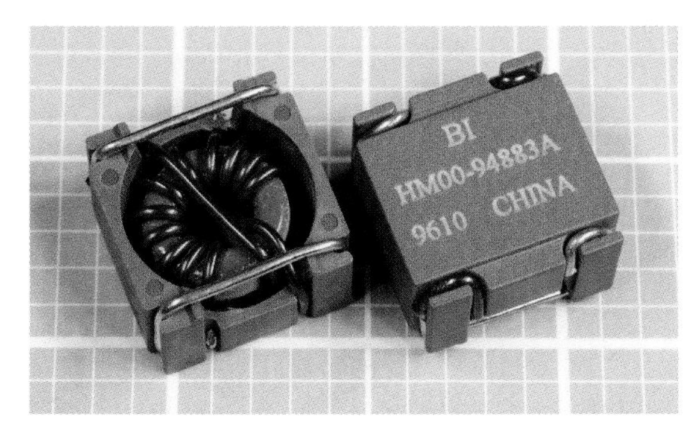

Figure 14-16. *A larger-sized surface-mount toroidal inductor (viewed from the bottom, at left, and from the top, at right). It is rated at 3.8μH.*

A chart showing some types of inductor cores, their commonly available inductances, and their maximum frequencies is shown in Figure 14-17.

Gyrator

A *gyrator* is a small network, sometimes encapsulated in a silicon chip, using **resistors**, a semiconductor, and a **capacitor** to simulate some but not all of the behavior of a coil-based inductor. The semiconductor may be a transistor or a capacitor, depending on the specific circuit. A sample schematic is shown in Figure 14-18. Because no magnetic effects are induced, the gyrator is completely free from the problems of saturation and hysteresis, which affect coils with cores, and also produces no back-EMF. It simply attenuates a signal initially, and then gradually lowers its reactance, thus imitating this aspect of an inductor.

A gyrator may be used where a coil may be unacceptably large (as in a cellular phone) or where signal quality is of paramount importance—for example, in a graphic equalizer or other audio components that perform signal processing at input stages, such as preamplifiers.

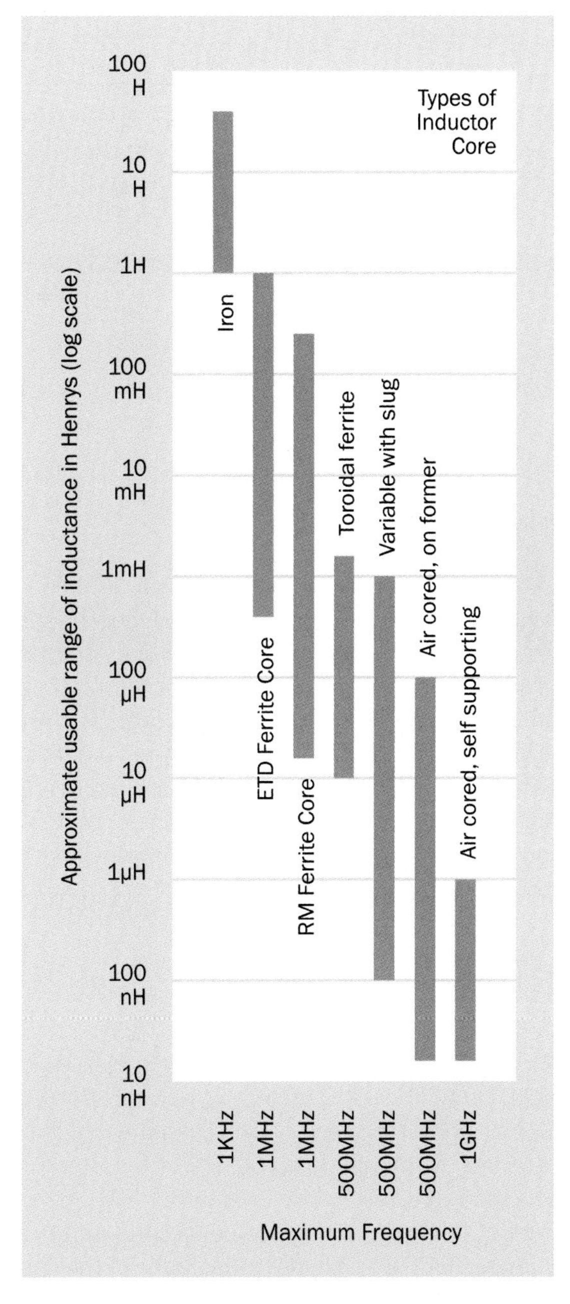

Figure 14-17. *Some commonly used inductor cores and their characteristics. Adapted from "Producing wound components" by R.Clark@surrey.ac.uk.*

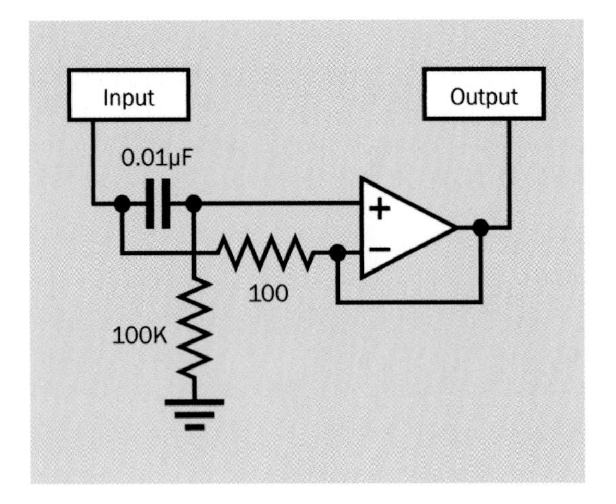

Figure 14-18. *A possible schematic for a coil substitute known as a gyrator, which may be used where a conventional coil would be unacceptably bulky.*

A gyrator does impose some limits on circuit design. While neither side of a real inductor needs to be at ground potential, a gyrator does require a ground connection. However, the performance advantages of gyrators are significant, as they can emulate high inductance without parasitic effects, can be more accurately calibrated (leading to more predictable performance), and do not create magnetic fields that can interfere with other components.

Values

Calculating Inductance

The magnetic inductance of a coil is measured with a unit known as the Henry, named after Joseph Henry, a pioneer in electromagnetism. It is defined by imagining a coil in which current is fluctuating, causing the creation of EMF. If the rate of fluctuation is 1 amp per second and the induced EMF is 1 volt, the inductance of the coil is 1 Henry.

The letter L is commonly used to represent inductance. To derive a useful formula, L will be expressed in microhenrys. If D is the diameter of a coil, N is the number of turns of wire, and W is the width of coil (when the windings are viewed from the side, as shown in Figure 14-19), the precise relationship of the variables is complex but can be reduced to an approximate formula:

$$L = \text{approx } (D^2 * N^2) \ / \ ((18*D) + (40*W))$$

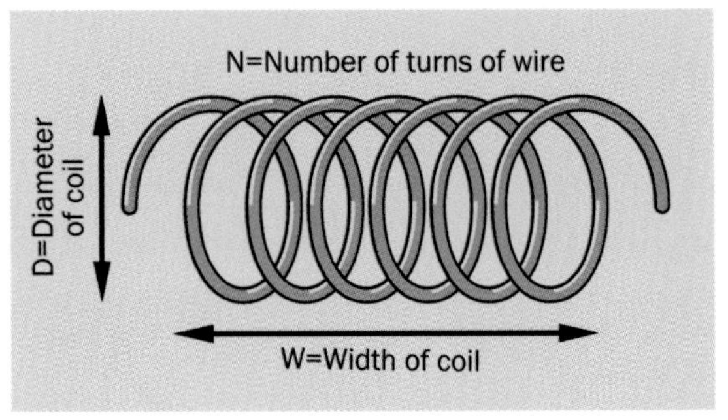

Figure 14-19. *Dimensions of a coil, referenced by a formula to calculate its approximate inductance. See text for details.*

From this, it is clear that inductance tends to increase with coil diameter, and also increases (more significantly) with the square of the number of turns. If the number of turns remains constant, inductance will be higher for a coil that is short and fat than for a coil that is narrow and long.

Because the Henry is a large unit, inductors in electronics circuits typically have their inductances measured in millihenrys (mH), microhenrys (µH), and nanohenrys (nH), where 1H = 1,000mH, 1 mH = 1,000µH, and 1µH = 1,000 nH. This relationship is shown in Figure 14-20.

Calculating Reactance

The reactance of an inductor (that is, its dynamic resistance to alternating current) varies with the frequency of the current. If f is the AC frequency (in Hertz), and L is the inductance (in Henrys), the reactance, X_L in ohms, is given by the formula:

$$X_L = 2 * \pi * f * L$$

From this equation, it's apparent that as the frequency tends toward zero (DC current), or if the inductance tends toward zero (a short piece of straight wire), the reactance will tend toward zero. Conversely, the inductor will impede cur-

nH	μH	mH
1	0.001	0.000001
10	0.01	0.00001
100	0.1	0.0001
1,000	1	0.001
10,000	10	0.01
100,000	100	0.1
1,000,000	1,000	1

Figure 14-20. *Inductance is typically measured in nanohenrys (nH), microhenrys (μH), and millihenrys (mH). Equivalent values in these units are shown here.*

rent increasingly as the frequency and/or the inductance increases.

Calculating Reluctance

The letter S is often used to represent reluctance, while Greek letter μ customarily represents permeability (not to be confused with the use of μ as a multiplication factor of 1/1,000,000, as in μF, meaning "microfarad"). If A is the area of cross-section of the magnetic circuit and L is its length:

```
S = L / μ * A
```

Datasheet Terminology

A typical manufacturer's datasheet should include an *inductance index* for an inductor, expressed in μH per 100 turns of wire (assuming the wire is in a single layer) for inductors with a powdered iron core, and mH per 1,000 turns of wire for inductors with ferrite cores.

The *DCR* is the DC resistance of an inductor, derived purely from the wire diameter and its length.

The *SRF* is the *self-resonant frequency*. An inductor should be chosen so that AC current passing through it will never get close to that frequency.

ISAT (or I_{sat}) is the saturation current, which results in a magnetic core losing its function as a result of magnetic saturation. When this occurs, inductance drops and the charge current rate increases drastically.

Series and Parallel Configurations

Because the inductance of a coil conducting DC current is proportional to the current, the calculations to derive the total inductance of coils in series or in parallel are identical to the calculations used for resistors.

In series, all the coils inevitably pass the same current, and the total inductance is therefore found by summing the individual inductances. When coils are wired in parallel, the current distributes itself according to the inductances; therefore, if L1 is the reluctance of the first coil, L2 is the reluctance of the second coil, and so on, the total reluctance L of the network is found from the formula:

```
1/L = 1/L1 + 1/L2 + 1/L3. . .
```

This is shown in Figure 14-21. In reality, differences between the coils (such as their electrical resistance), and magnetic interaction between the coils, will complicate this simple relationship.

Time Constant

Just as the *time constant* of a capacitor defines the rate at which it accumulates voltage when power is applied through a resistor, the time constant of an inductor defines the rate at which it gradually allows amperage to pass through it, overcoming the EMF generated by the coil. In both cases, the time constant is the number of seconds that the component requires to acquire approximately 63% of the difference between its current value and its maximum value. In the case of an inductor, suppose we assume zero internal resistance in the power source, zero resistance in the coil windings, and an initial current of zero. If L is the inductance of the coil and R is the value of the series resistor, then the time constant—TC—is given in seconds by the formula

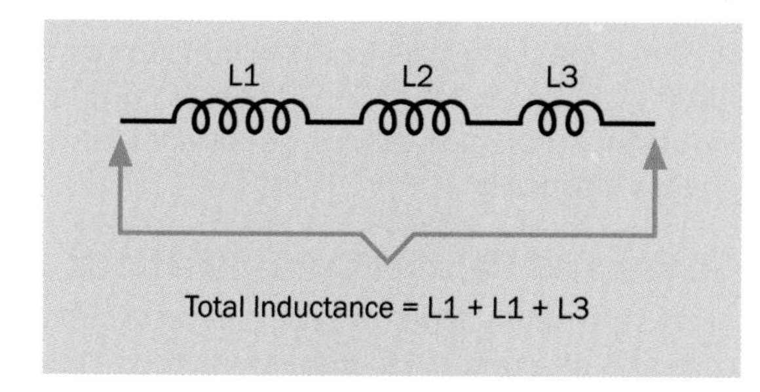

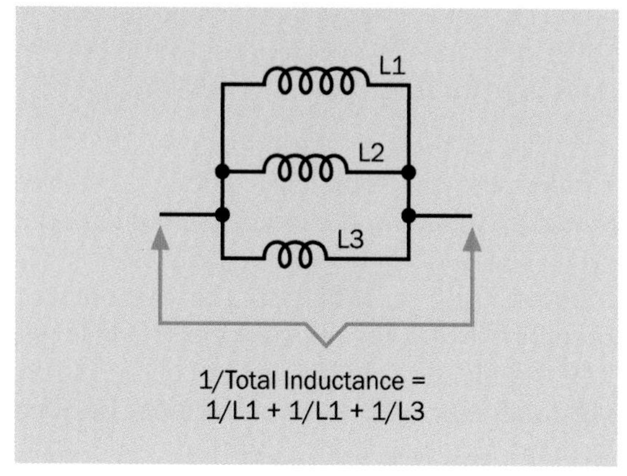

Figure 14-21. *Calculating the total inductance of inductors in parallel (top) and series (bottom).*

```
TC = L / R
```

Therefore a coil of 10 millihenrys (0.01 Henry) in series with a 100-ohm resistor will pass 63% of the full current in 0.0001 seconds, or 1/10 of a millisecond; it will take an equal additional amount of time for the current to rise by another 63% of the remaining difference between its charge and the maximum amperage of the circuit. In theory, the reactance of a coil can never diminish to zero, but in practice, five time constants are considered adequate to allow maximum current flow.

How to Use it

Because the inductance of an inductor peaks as current increases, and then gradually diminishes, an inductor can be used to block or attenuate high frequencies. A circuit that does this is often referred to as a *low-pass filter*. The schematic and a graph suggesting its performance are shown in Figure 14-22. A basic application could be the *crossover network* in a loudspeaker system, where high-frequency signals are blocked from a low-frequency driver and are diverted to a high-frequency driver.

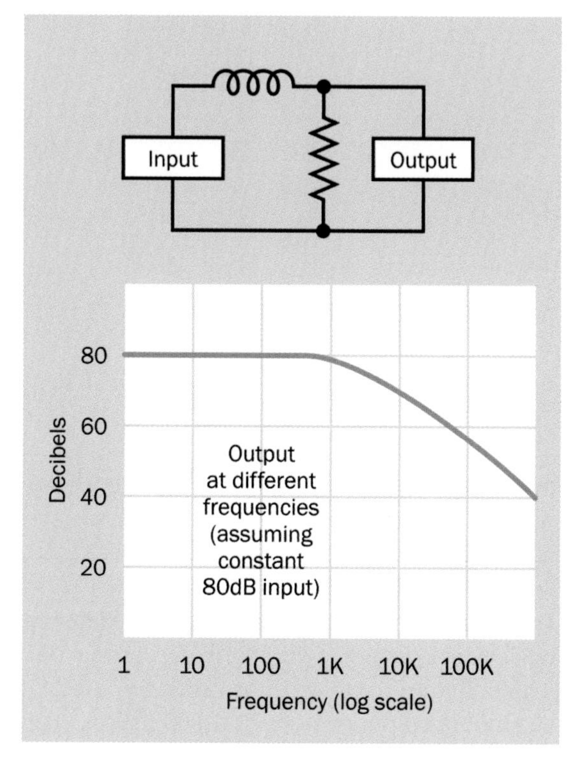

Figure 14-22. *By using the ability of an inductor to block a range of frequencies, a low-pass filter blocks higher frequencies.*

If the location of the inductor is shifted so that it shunts the signal away from the output, the results are reversed, and the circuit becomes a *high-pass filter*. The schematic and a graph suggesting its performance are shown in Figure 14-23.

Note that **capacitors** may also be used to create frequency filters, but because their function is roughly inverse to that of inductors, the placement of a capacitor in a circuit would be opposite to the placement of the inductor. Examples of filter circuits using capacitors are found in the entry for that component in this encyclopedia.

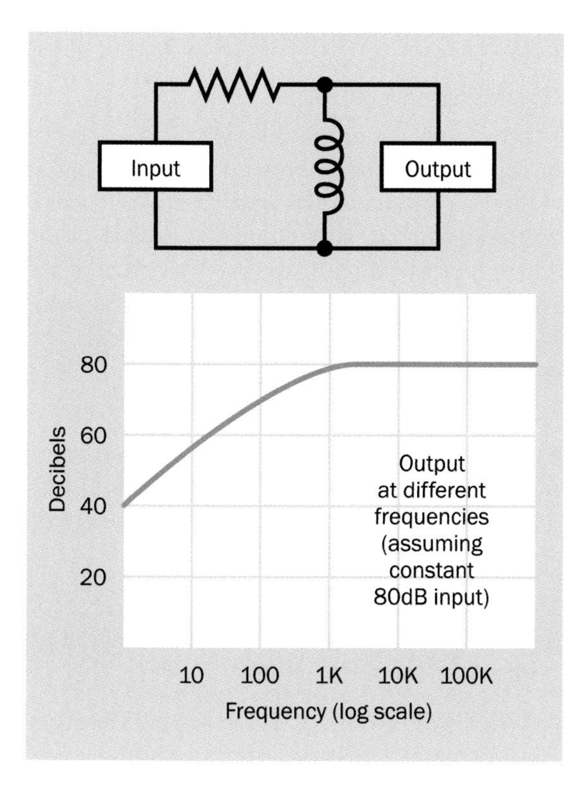

Figure 14-23. *Here the inductor diverts low frequencies away from the output, allowing high frequencies to pass through.*

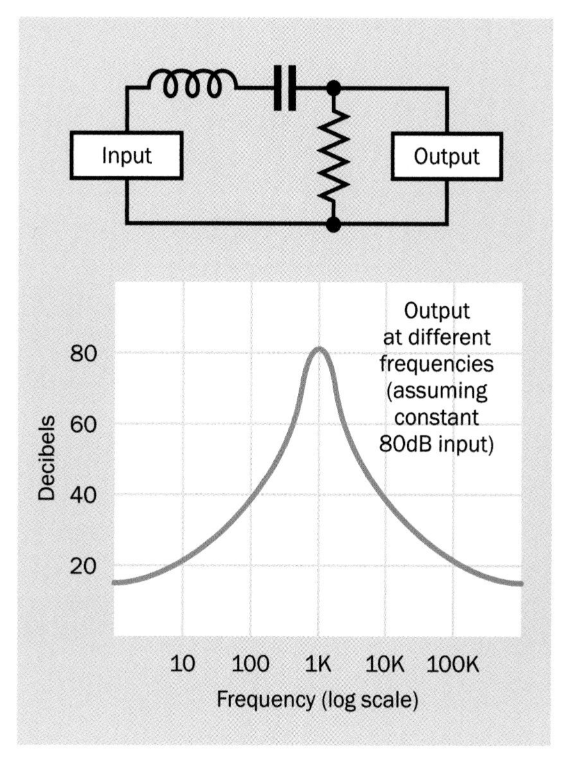

Figure 14-24. *If the values of a capacitor and an inductor are correctly chosen, and the components are placed in series, the inductor blocks high frequencies while the capacitor blocks low frequencies, creating a bandpass filter, in which only a narrow band of frequencies can get through.*

An inductor can be combined with a capacitor to form a *bandpass filter*, as shown in Figure 14-24. In this configuration, the inductor blocks the high frequencies while the capacitor blocks the low frequencies, allowing only a limited band of frequencies to get through.

Once again if the location of the components is shifted to shunt the signal away from the output, the results are reversed, as shown in Figure 14-25. This is known as a *notch filter*.

The performance of these filters will depend on the component values, and in most applications, additional components will be necessary. Sophisticated filter circuits are outside the scope of this encyclopedia.

Inductors are of great importance in **DC-DC converters** and **AC-DC power supplies** where voltage changes are enabled by rapid switching. See the relevant entries of this encyclopedia for additional details.

Generally, as electronic equipment has become increasingly miniaturized, the unavoidable bulk of inductors has limited their application. However they may still be used to tune oscillators, to block sudden spikes in power supplies, and to protect equipment from sudden voltage spikes (they are used, for example, in surge suppressors for computing equipment).

Core Choices

Air-cored inductors have relatively low inductance, because of their low permeability. However, they can be operated at very high frequencies up to the gigahertz range, and can tolerate higher peak currents.

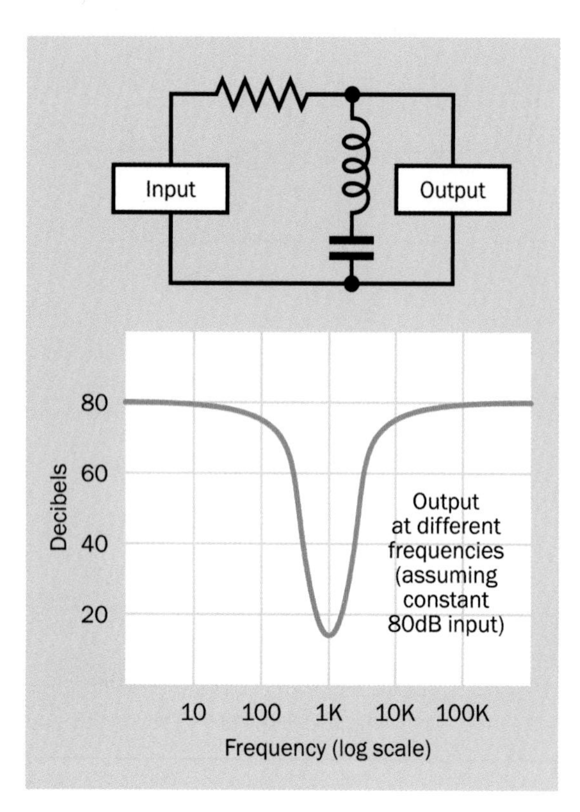

Figure 14-25. *Here the capacitor and inductor block all frequencies except a narrow band, which they divert from the output. The result is a notch filter.*

Inductors with an iron core suffer increasing power losses due to hysteresis and eddy currents as the AC frequency passing through the inductor increases. Consequently, iron-cored inductors are not suitable for frequencies much above 10KHz.

Miniaturization

A low-value inductor can be formed by etching a spiral onto a circuit board, in applications where size must be minimized. They may also be incorporated in integrated circuit chips. However, in small devices such as cellular phones, it is more common to use a coil substitute such as a *gyrator*, as described previously.

What Can Go Wrong

Real-World Defects

The theoretically ideal inductor has no resistance or capacitance and suffers no energy losses. In reality, an inductor possesses both resistance and capacitance, also creates electrical noise, and may pick up electrical noise. It tends to create stray magnetic fields, and generally is more troublesome to deal with than its two cousins, the **resistor** and the **capacitor**.

Parasitic capacitance occurs between adjacent turns of wire. This capacitance becomes more significant at higher frequencies, leading ultimately to a situation where the coil becomes *self resonant*.

The workarounds for these problems involve coil geometries and choices of core material that go beyond the scope of this encyclopedia.

A *gyrator* should be considered as a possible substitute where inductors are troublesome or excessively expensive.

Saturation

Inductance increases as the current passing through a coil increases, but if a magnetic core is used, its contribution to inductance will stop abruptly when the core becomes magnetically *saturated*. In other words, when all of the randomly distributed *magnetic domains* in the core have been induced to align themselves with the pervasive magnetic field, the core cannot become more highly magnetized, and ceases to contribute to the inductance. Note that as a core approaches saturation levels, its hysteresis increases because reversing its magnetization requires greater energy. Antidotes to saturation would include a larger core, a lower current, a smaller number of turns in the coil, and using a core with lower permeability (such as air).

RF Problems

Radio frequencies (*RF*) introduce various problems affecting the efficiency of inductors. The *skin effect* is the tendency of high-frequency AC current to flow primarily on the surface of a strand of wire. The *proximity effect* refers to the tendency of the magnetic fields caused by adjacent wires to introduce eddy currents in the coil. Both of these effects increase the effective resistance of the coil. Various coil geometries have been developed to minimize these effects, but are outside the scope of this encyclopedia. The fundamental lesson is that coils specifically designed for RF are the only ones that should be used with RF.

AC-AC transformer

15

OTHER RELATED COMPONENTS

- **AC-DC power supply** (See Chapter 16)
- **DC-DC converter** (See Chapter 17)
- **DC-AC inverter** (See Chapter 18)

What It Does

A transformer requires an input of *alternating current* (AC). It transforms the input voltage to one or more output voltages that can be higher or lower.

Transformers range in size from tiny impedance-matching units in audio equipment such as microphones, to multi-ton behemoths that supply high voltage through the national power grids. Almost all electronic equipment that is designed to be powered by municipal AC in homes or businesses requires the inclusion of a transformer.

Two small power transformers are shown in Figure 15-1. The one at the rear is rated to provide 36VAC at 0.8A when connected with a source of 125VAC. At front, the miniature transformer is a Radio Shack product designed to provide approximately 12VAC at 300mA, although its voltage will be more than 16VAC when it is not passing current through a load.

Figure 15-1. *Two small power transformers. The one at the rear measures approximately 1" × 2" × 2" and is rated to provide 36VAC at 0.8A. The term "sec" on the smaller unit is an abbreviation for "secondary," referring to the rating for its secondary winding.*

Transformer schematic symbols are shown in Figure 15-2. The different coil styles at left and right are functionally identical. Top: A transformer with a magnetic core—a core that can be magnetized. Bottom: A transformer with an air core. (This type of transformer is rare, as it tends to be less efficient.) The input for the transformer is almost always assumed to be on the left, through the *primary* coil, while the output is on the right, through the *secondary* coil. Often the two coils will show differing numbers of turns to indicate whether the transformer is delivering a reduced voltage (in which case there will be fewer turns in the secondary coil) or an increased voltage (in which case there will be fewer turns in the primary coil).

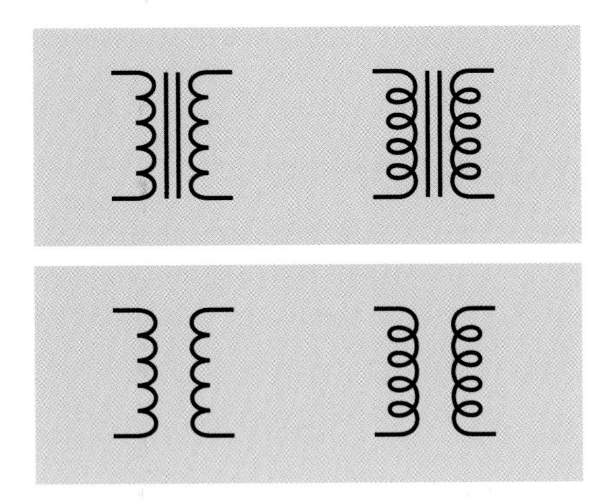

Figure 15-2. *Alternate symbols for a transformer with a ferromagnetic core (top) and air core (bottom). The differing coil symbols at left and right are functionally identical.*

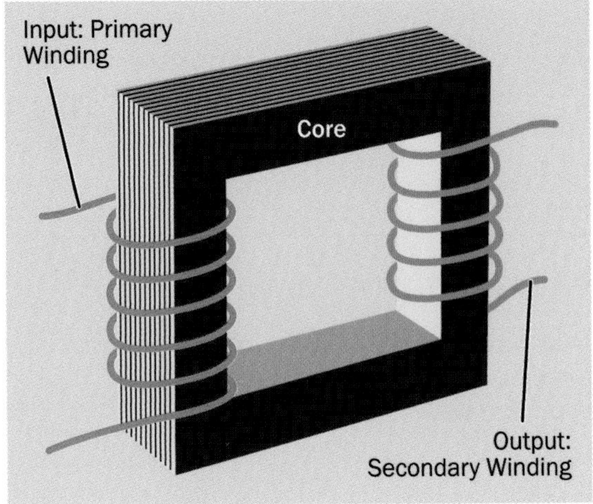

Figure 15-3. *Three basic parts of a transformer, shown in simplified form.*

How It Works

A simplified view of a transformer is shown in Figure 15-3. Alternating current flowing through the primary winding (orange) induces magnetic flux in a laminated core formed from multiple steel plates. The changing flux induces current in the secondary winding (green), which provides the output from the transformer. (In reality, the windings usually consist of thousands of turns of thin magnet wire, also known as enameled wire; and various different core configurations are used.)

The process is known as *mutual induction*. If a load is applied across the secondary winding, it will draw current from the primary winding, even though there is no electrical connection between them.

In an ideal, lossless transformer, the ratio of turns between the two windings determines whether the output voltage is higher, lower, or the same as the input voltage. If V_p and V_s are the voltages across the primary and secondary windings respectively, and N_p and N_s are the number of turns of wire in the primary and secondary windings, their relationship is given by this formula:

$$V_p / V_s = N_p / N_s$$

A simple rule to remember is that fewer turns = lower voltage while more turns = higher voltage.

A *step-up transformer* has a higher voltage at its output than at its input, while a *step-down transformer* has a higher voltage at its input than at its output. See Figure 15-4.

In an ideal, lossless transformer, the power input would be equal to the power output. If V_{in} and V_{out} are the input and output voltages, and I_{in} and I_{out} are the input and output currents, their relationship is given by this formula:

$$V_{in} * I_{in} = V_{out} * I_{out}$$

Therefore, if the transformer doubles the voltage, it allows only half as much current to be drawn from the secondary winding; and if the voltage is cut in half, the available current will double.

Transformers are not 100% efficient, but they can be more than 98% efficient, and relationships between voltage, current, and the number of turns in the windings are reasonably realistic.

When the transformer is not loaded, the primary winding behaves like a simple inductor with reactance that inhibits the flow of current. Therefore a power transformer will consume relatively little electricity if it is left plugged in to an elec-

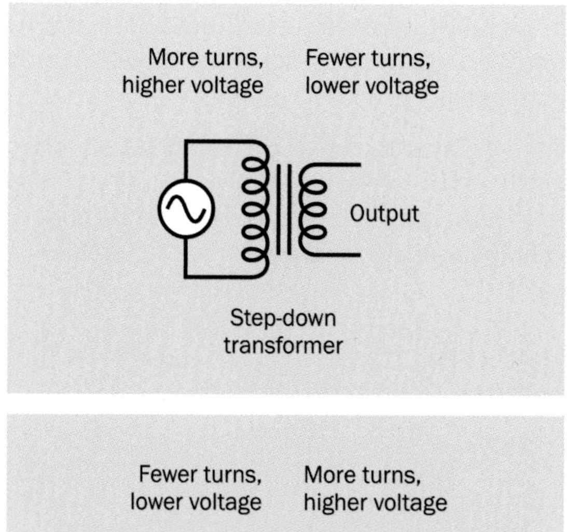

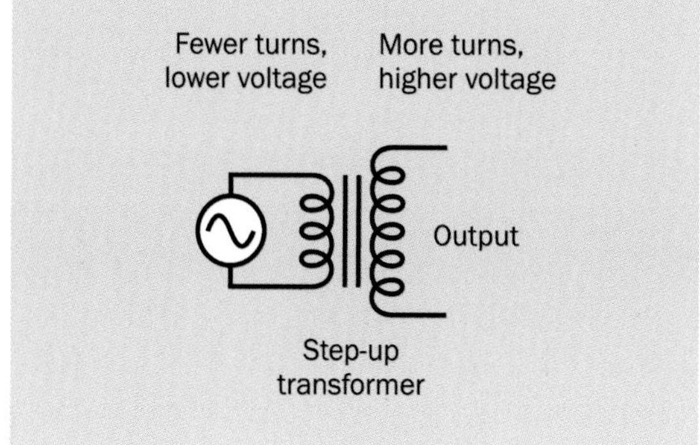

Figure 15-4. *The ratio of input voltage to output voltage is equal to the ratio of primary turns to secondary turns in the transformer windings, assuming a transformer of 100% efficiency.*

trical outlet without any load connected to its output side. The power that it does consume will be wasted as heat.

The Core

The ferromagnetic core is often described as being made of iron, but in reality is more often fabricated from high permeability silicon steel. To reduce losses caused by eddy currents, the core is usually laminated—assembled from a stack of plates separated from each other by thin layers of varnish or a similar insulator. Eddy currents tend to be constrained within the thickness of each plate.

Because a DC voltage would cause magnetic saturation of the core, all transformers must operate with alternating current or pulses of current. The windings and geometry of a transformer are optimized for the frequency range, voltage, and current at which it is designed to operate. Deviating significantly from these values can damage the transformer.

Taps

A *tap* on a transformer is a connection part-way through the primary or (more often) the secondary coil. On the primary side, applying an input between the start of a coil and a tap part-way through the coil will reduce the number of turns to which the voltage is applied, therefore increasing the ratio of output turns to input turns, and increasing the output voltage. On the secondary side, taking an output between the start of a coil and a tap part-way through the coil will reduce the number of turns from which the voltage is taken, therefore decreasing the ratio of output turns to input turns, and decreasing the output voltage. This can be summarized:

- A tap on the primary side can increase output voltage.
- A tap on the secondary side can provide a decreased output voltage.

In international power adapters, a choice of input voltages may be allowed by using a double-throw switch to select either the whole primary winding, or a tapped subsection of the winding. See Figure 15-5. Modern electronics equipment often does not require a voltage adapter, because a **voltage regulator** or **DC-DC converter** inside the equipment will tolerate a wide range of input voltages while providing a relatively constant output voltage.

A transformer's secondary winding is often tapped to provide a choice of output voltages. In fact, most power transformers have at least two outputs, since the cost of adding taps to the secondary winding is relatively small. As an alternative to tapped outputs, two or more separate secondary windings may be used, allowing the outputs to be electrically isolated from each other. See Figure 15-6.

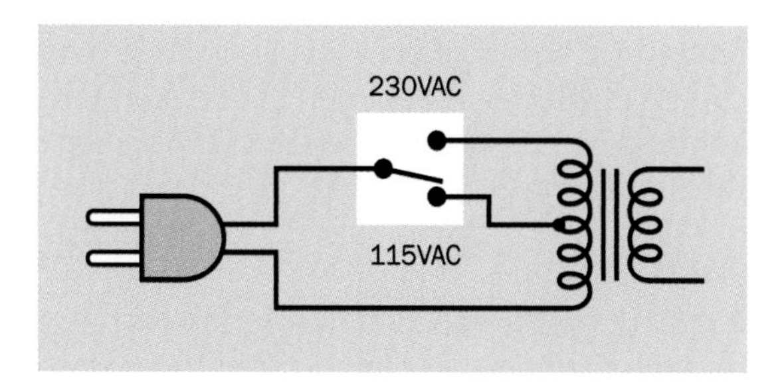

Figure 15-5. *An international power adapter can provide a fixed output voltage by using a double-throw switch to apply 230VAC voltage across a transformer's primary winding, or 115VAC to a tapped midpoint of the primary winding.*

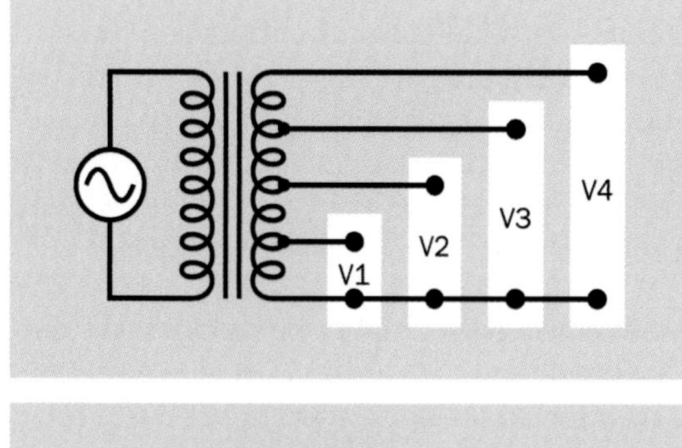

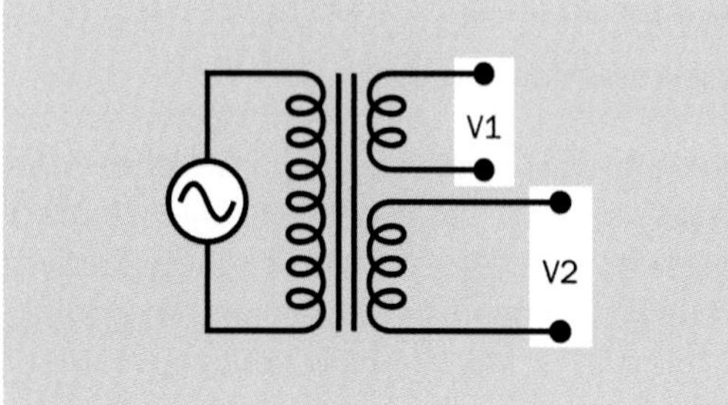

Figure 15-6. *Multiple output voltages may be obtained from a transformer by tapping into the secondary winding (top) or using two or more separate secondary windings (bottom), in which case the outputs will be electrically isolated from each other.*

If the winding on the primary side of a transformer is coiled in the same direction as the winding on the secondary side, the output voltage will be 180 degrees out of phase with the input voltage. In schematics, a dot is often placed at one end of a transformer coil to indicate where the coil begins. If the dots on the primary and secondary sides are at the same ends of the coils, there will be a 180 degree phase difference between input and output. For many applications (especially where the output from a power transformer is going to be converted to DC), this is immaterial.

If there is a center tap on the secondary winding, and it will be referenced as ground, the voltages relative to it, at opposite ends of the secondary winding, will be out of phase. See Figure 15-7.

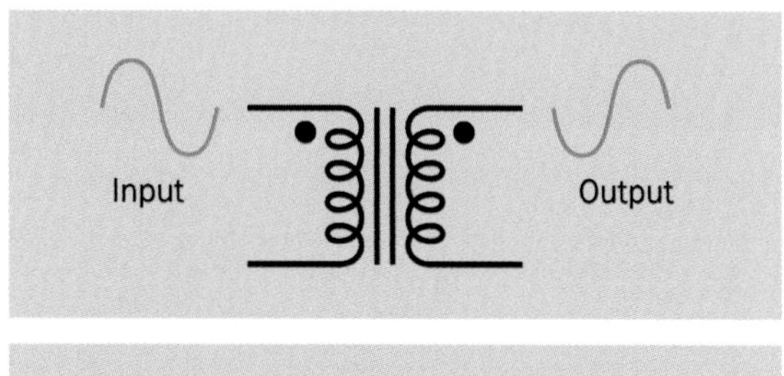

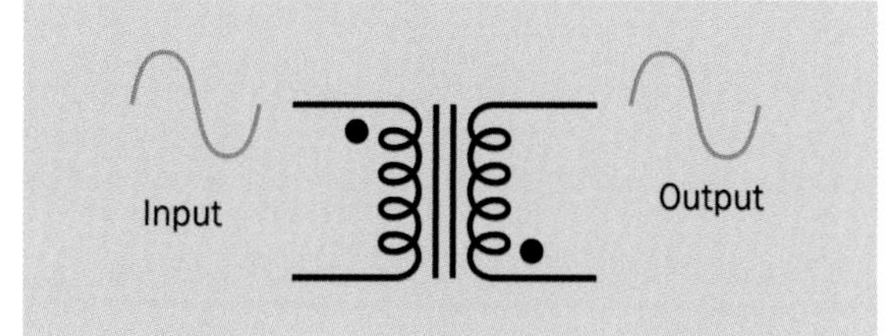

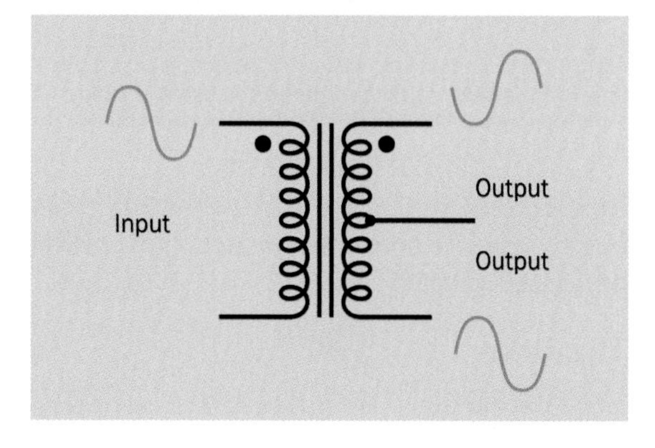

Figure 15-7. *A dot indicates the start of each winding. Where primary and secondary windings are in the same direction, the voltage output will be 180 degrees out of phase with the input. Where the dots indicate windings in opposite directions, the voltage output will have the same phase as the input. Where a center tap on the secondary winding serves as a common ground, the voltages at opposite ends of the secondary winding will be opposite to each other in phase.*

Variants

Core Shapes

The *shell core* is a closed rectangle, as shown in Figure 15-3. This is the most efficient but most costly to manufacture. A C-shaped core is another option (three sides of the rectangle) and an E-I core is popular, consisting of a stack of E-shaped plates with two coils wound around the top and bottom legs of the E, or wound concentrically around the center leg of the E. An additional stack of straight plates is added to close the gaps in the E and form a magnetic circuit.

In Figure 15-8, the small transformer from Figure 15-1 has been sliced open with a band saw and a belt sander to reveal a cross-section of its windings. This clearly shows that its primary and secondary windings are concentric. It also reveals the configuration of its core, which is in the E-I format. In Figure 15-9, the E-I configuration is highlighted to show it more clearly.

Figure 15-8. *The small transformer from the first figure in this entry is shown sliced open to reveal its internal configuration.*

Power Transformer

Typically designed to be bolted onto a chassis or secured inside the case or cabinet housing a piece of electrical equipment with solder tabs or

Figure 15-9. *The "EI" shaped plates that form the core of the transformer are outlined to show their edges.*

connectors allowing wires to connect the transformer to the power cord, on one side, and a circuit board, on the other side. Smaller power transformers such as the one in Figure 15-1 have "through-hole" design with pins allowing them to be inserted directly onto circuit boards.

Plug-in Transformer

Usually sealed in a plastic housing that can be plugged directly into a wall power outlet. They are visually identical to **AC adapters** but have an AC output instead of a DC output.

Isolation Transformer

Also known as a *1:1 transformer* because it has a 1:1 ratio between primary and secondary windings, so that the output voltage will be the same as the input voltage. When electrical equipment is plugged into the isolation transformer, it is separated from the electrical ground of AC power wiring. This reduces risk when working on "live" equipment, as there will be negligible electrical potential between itself and ground. Consequently, touching a grounded object while also touching a live wire in the equipment should not result in potentially lethal current passing through the body.

Autotransformer

This variant uses only one coil that is tapped to provide output voltage. Mutual induction occurs between the sections of the coil. An autotransformer entails a common connection between its input and output, unlike a two-coil transformer, which allows the output to be electrically isolated from the input. See Figure 15-10. Autotransformers are often used for impedance matching in audio circuits, and to provide output voltages that differ only slightly from input voltages.

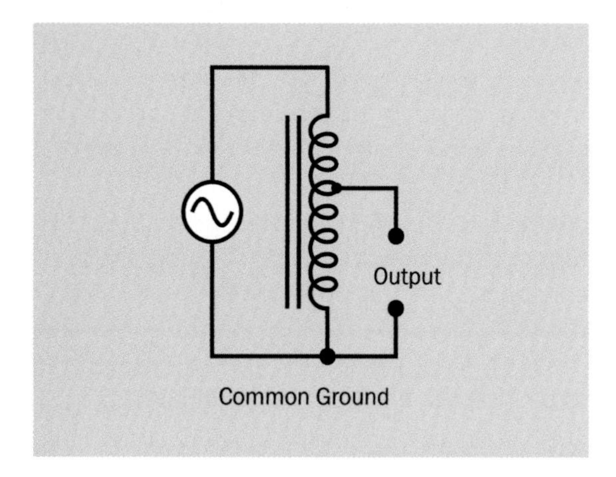

Figure 15-10. *An autotransformer contains only one coil and core. A reduced output voltage can be obtained by tapping into the coil. A common connection prevents the output from being electrically isolated from the input.*

Variable Transformer

A variable transformer, also known as a *variac*, resembles a wire-wound **potentiometer**. Only one winding is used. A *wiper* can be turned to contact the winding at any point, and serves as a movable tap. Like an autotransformer, a variable transformer entails a common connection between input and output.

Audio Transformer

When a signal is transmitted between two stages of a circuit that have different impedance, the signal may be partially reflected or attenuated. (Impedance is measured in ohms but is different from DC electrical resistance because it takes into account reactance and capacitance. It therefore varies with frequency.)

A device of low input impedance will try to draw significant current from a source, and if the source has high output impedance, its voltage will drop significantly as a result. Generally, the input impedance of a device should be at least 10 times the output impedance of the device that is trying to drive it. Passive components (resistors, and/or capacitors, and/or coils) can be used for impedance matching, but in some situations a small transformer is preferable.

If N_p and N_s are the number of turns of wire in the transformer primary and secondary windings, and Z_p is the impedance of a device (such as an audio amplifier) driving the transformer on its primary side, and Z_s is the impedance of a device (such as a loudspeaker) receiving power from the secondary side:

$$N_p / N_s = \sqrt{(Z_p / Z_s)}$$

Suppose that an audio amplifier with rated output impedance of 640Ω is driving a loudspeaker with 8Ω impedance. A matching transformer would be chosen with a ratio of primary turns to secondary turns give by:

$$\sqrt{(640/8)} = \sqrt{80} = \text{approximately } 9:1$$

The two transformers in Figure 15-11 are through-hole components designed for telecommunications purposes, but are capable of passing audio frequencies and can be used for impedance matching in applications such as a preamplifier.

In Figure 15-12, the transformers are designed for audio coupling. The one on the right has impedances of 500 ohms (primary) and 8 ohms (secondary). On the left is a fully encapsulated line matching transformer with a 1:1 turns ratio.

Split-Bobbin Transformer

This variant has primary and secondary coils mounted side by side to minimize capacitive coupling.

Figure 15-11. *Through-hole transformers. See text for details.*

Figure 15-12. *Through-hole transformers. See text for details.*

Surface-Mount Transformer

May be less than 0.2" square and is used for impedance matching, line coupling, and filtering. Two surface-mount transformers are shown in Figure 15-13.

Values

When selecting a power transformer, its power handling capability is the value of primary interest. It is properly expressed by the term VA, de-

Figure 15-13. *Two surface-mount transformers, each measuring less than 0.2" square, typically used in communications equipment and suited for frequencies higher than 5 MHz.*

rived from "volts times amps." VA should not be confused with watts because watts are measured instantaneously in a DC circuit, whereas in an AC circuit, voltage and current are fluctuating constantly. VA is actually the *apparent power*, taking reactance into account.

The relationship between VA and watts will vary depending on the device under consideration. In a worst-case scenario:

```
W = 0.65 VA (approximately)
```

In other words, the averaged power you can draw from a transformer should be no less than two-thirds of its VA value.

Transformer specifications often include input voltage, output voltage, and weight of the component, all of which are self-explanatory. Coupling transformers may also specify input and output impedances.

How to Use it

For most electronic circuits, a power transformer will be followed by a *rectifier* to convert AC to DC, and **capacitors** to smooth fluctuations in the supply. Using a prepackaged **power supply** or *AC adapter* that already contains all the necessary components will be more time-effective and probably more cost-effective than building a power supply from the ground up. See Chapter 16.

What Can Go Wrong

Reversal of Input and Output

Suppose a transformer is designed to provide an output voltage of 10 volts from domestic AC power of 115 volts. If the wrong side of the transformer is connected with 115VAC by mistake, the output will now be more than 1,000 volts—easily enough to cause death, quite apart from destroying components that are connected with it. Reversing the transformer in this way may also destroy it. Extreme caution is advisable when making connections with power transformers. A meter should be used to check output voltage. All devices containing transformers should be fused on the live side and grounded.

Shock Hazard from Common Ground

When working on equipment that uses an autotransformer, the chassis will be connected through the transformer to one side of 115VAC power. So long as a plug is used that prevents reversed polarity, the chassis should be "neutral." However, if an inappropriate power cord is used, or if the power outlet has been wired incorrectly, the chassis can become live. For protection, before working on any device that uses 115VAC power with an autotransformer, plug the device into an isolation transformer, and plug the isolation transformer into the wall outlet.

Accidental DC Input

If DC current is applied to the input side of a transformer, the relatively low resistance of the primary coil will allow high current that can destroy the component. Transformers should only be used with alternating current.

Overload

If a transformer is overloaded, heat will be generated that may be sufficient to destroy the thin layers of insulation between coil windings. Consequently, input voltage can appear unexpectedly on the output side. Transformers with a toroidal (circular) core are especially hazardous in this respect, as their primary and secondary windings usually overlap.

Some (not all) power transformers contain a thermal fuse that melts when it exceeds a temperature threshold. If the fuse is destroyed, the transformer must be discarded.

The consequences of moderate overloading may not be obvious, and can be cumulative over time. Ventilation or heat sinkage should be taken into account when designing equipment around a power transformer.

Incorrect AC Frequency

Single-phase AC power in the United States fluctuates at 60Hz, but Great Britain and some other countries use AC power at 50Hz. Many power transformers are rated to be compatible with either frequency, but if a transformer is specifically designed for 60Hz, it may eventually fail by overheating if it is used with a 50Hz supply. (A 50Hz transformer can be used safely with 60Hz AC.)

AC-DC power supply

16

Also known as an *AC adapter*. When packaged as a palm-sized plastic package that plugs directly into a power outlet, it is occasionally known colloquially as a *wall-wart*.

OTHER RELATED COMPONENTS

- **transformer** (See Chapter 15)
- **DC-DC converter** (See Chapter 17)
- **DC-AC inverter** (See Chapter 18)

What It Does

An AC-DC power supply converts *alternating current* (*AC*) into the *direct current* (*DC*) that most electronic devices require, usually at a lower voltage. Thus, despite its name, a power supply actually requires an external supply of power to operate.

Larger products, such as computers or stereo equipment, generally have a power supply contained within the device, enabling it to plug directly into a wall outlet. Smaller battery-powered devices, such as cellular phones or media players, generally use an external power supply in the form of a small plastic pod or box that plugs into a wall outlet and delivers DC via a wire terminating in a miniature connector. The external type of power supply is often, but not always, referred to as an *AC adapter*.

Although an AC-DC power supply is not a single component, it is often sold as a preassembled modular unit from component suppliers.

Variants

The two primary variants are a *linear regulated power supply* and *switching power supply*.

Linear Regulated Power Supply

A linear regulated power supply converts AC to DC in three stages:

1. A power **transformer** reduces the AC input to lower-voltage AC.
2. A *rectifier* converts the AC to unsmoothed DC. Rectifiers are discussed in the entry on **diodes** in this encyclopedia.
3. A **voltage regulator**, in conjunction with one or more **capacitors**, controls the DC voltage, smooths it, and removes *transients*. The regulator is properly known as a *linear voltage regulator* because it contains one or more **transistors**, which are functioning in linear mode—that is, responding linearly to fluctuations in base current, at less than their saturation level. The linear voltage regulator gives the linear regulated power supply its name.

A simplified schematic of a linear regulated power supply is shown in Figure 16-1.

This type of power supply may be described as *transformer-based*, since its first stage consists of a transformer to drop the AC input voltage before it is rectified.

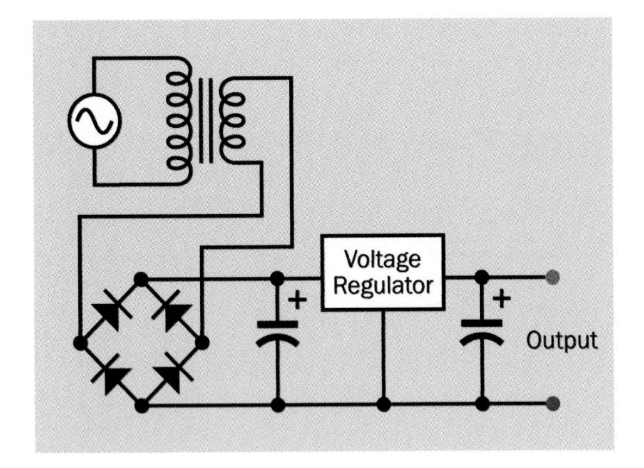

Figure 16-1. *A basic linear regulated power supply.*

Figure 16-2. *A simple transformer-based power supply can be encapsulated in a plastic shell, ready to plug into a power outlet. However, today this format more typically contains a switching power supply, which is usually lighter, smaller, and cheaper.*

Because the rectifier in a power supply generally passes each pulse of AC through a pair of silicon diodes, it will impose a voltage drop of about 1.2V at peak current. A smoothing capacitor will drop the voltage by about 3V as it removes ripple from the current, whereas a voltage regulator typically requires a difference of at least 2V between its input and its output. Bearing in mind also that the AC input voltage may fluctuate below its rated level, the output from the power transformer should be at least 8VAC higher than the ultimate desired DC output. This excess power will be dissipated as heat.

The basic principle of the linear regulated power supply originated in the early days of electronic devices such as radio receivers. A transistorized version of this type of power supply remained in widespread use through the 1990s. Switching power supplies then became an increasingly attractive option as the cost of semiconductors and their assembly decreased, and high-voltage transistors became available, allowing the circuit to run directly from rectified line voltage with no step-down power transformer required.

Some external AC adapters are still transformer-based, but are becoming a minority, easily identified by their relatively greater bulk and weight. An example is shown in Figure 16-2.

Figure 16-3 shows the handful of components inside a cheap, relatively old AC adapter. The output from a power transformer is connected directly to four diodes (the small black cylinders), which are wired as a full-wave rectifier. A single electrolytic capacitor provides some smoothing, but because there is no voltage regulator, the output will vary widely depending on the load. This type of AC adapter is not suitable for powering any sensitive electronic equipment.

Switching Power Supply

Also known as a *switched-mode power supply*, an *SMPS,* or *switcher*, it converts AC to DC in two stages.

1. A *rectifier* changes the AC input to unsmoothed DC, without a power transformer.
2. A DC-DC **converter** switches the DC on and off at a very high frequency using *pulse-width modulation* to reduce its average effective voltage. Often the converter will be the *flyback* type, containing a transformer, but the high-frequency switching allows the transformer to be much smaller than the power transformer required in a linear regulated power supply. See the DC-DC **converter** entry in this encyclopedia for an explanation of the working principles.

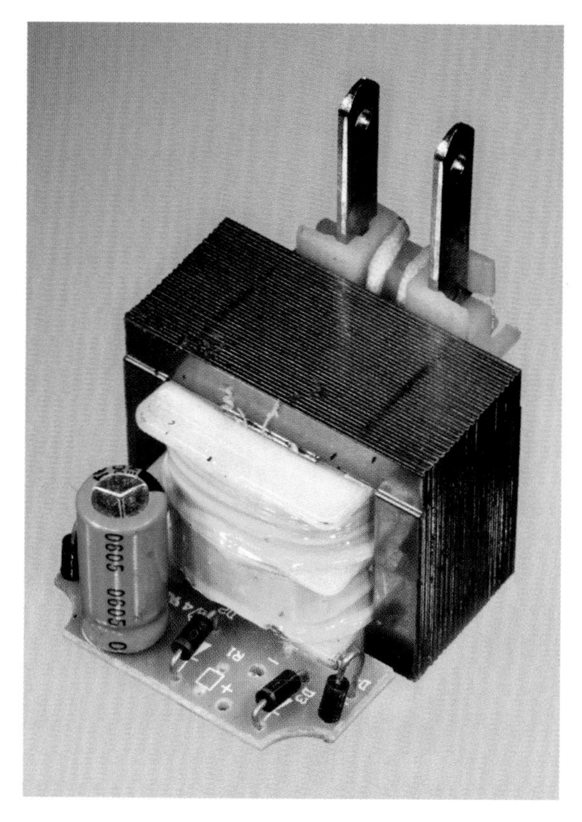

Figure 16-3. *A relatively old, cheap AC adapter contains only the most rudimentary set of components, and does not supply the kind of properly regulated DC power required by electronic equipment.*

A simplified schematic of a switching power supply is shown in Figure 16-4.

The interior of a relatively early switching power supply designed to deliver 12VDC at up to 4A is shown in Figure 16-5. This supply generated considerable waste heat, necessitating well-spaced components and a ventilated enclosure.

The type of small switching power supply that is now almost universally used to power laptop computers is shown in Figure 16-6. Note the smaller enclosure and the higher component count than in the older power supply shown in Figure 16-5. The modern unit also delivers considerably more power, and generates less waste heat. Although this example is rated at 5A, the transformer (hidden under the yellow wrapper at the center of the unit) is smaller than the power transformer that would have been found in an old-style AC adapter delivering just 500mA.

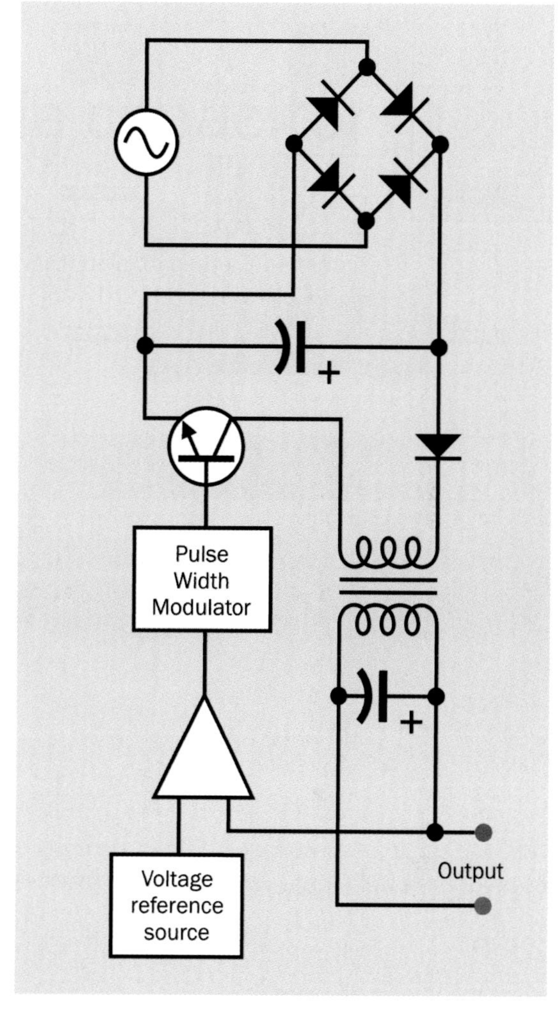

Figure 16-4. *Greatly simplified schematic showing the principal components of a switching power supply. Note the absence of a 115VAC power transformer. The transformer that is inserted subsequently in the circuit functions in conjunction with the high switching frequency, which allows it to be very much smaller, cheaper, and lighter.*

The modern power supply is completely sealed, where earlier versions required ventilation. On the downside, the plastic case of the switching supply requires a metal liner (removed for this photograph) to contain high-frequency electromagnetic radiation.

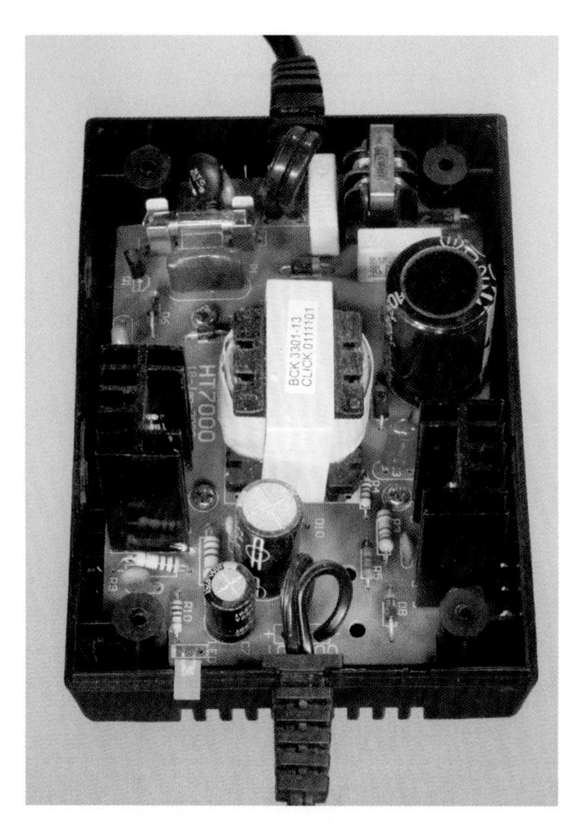

Figure 16-5. *The interior of an early switching power supply.*

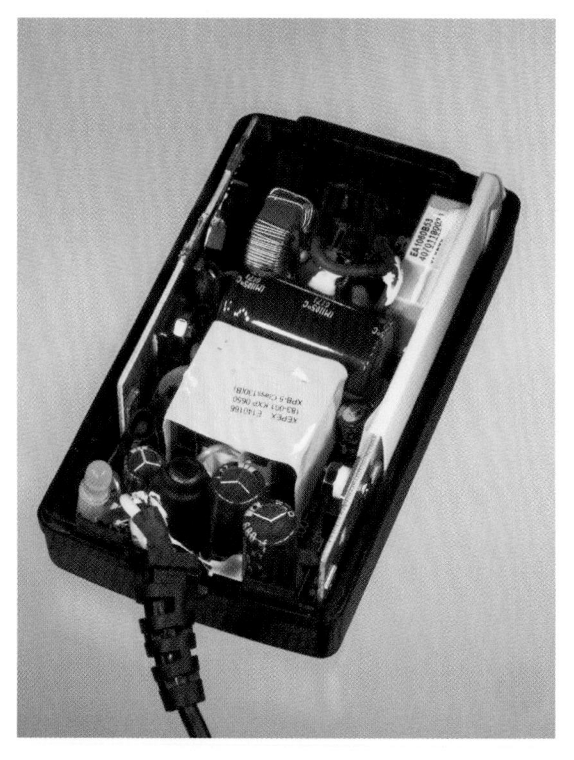

Figure 16-6. *The interior of the type of switching power supply that powers a laptop computer.*

Unregulated Power Supply

Typically this consists of a transformer and rectifying diodes with little or no smoothing or voltage control of the output.

Adjustable Power Supply

This is usually a linear power supply incorporating an adjustable **voltage regulator**. This type of supply has laboratory applications and is found as a benchtop item to power electronics design projects during their development.

Voltage Multiplier

Devices such as photocopiers and laser printers, televisions, cathode-ray tubes, and microwave ovens require voltages significantly higher than those supplied by domestic AC power outlets. A voltage multiplier usually contains a step-up transformer followed by DC conversion components, but detailed consideration is outside the scope of this encyclopedia.

Formats

An *open frame* power supply consists of components on a circuit board, usually mounted on a metal chassis, with no enclosure or fan cooling.

A *covered* power supply is enclosed in a protective perforated metal box with a cooling fan if needed. Power supplies sold for desktop computers are usually in this format.

Power supplies are also available in rack-mount and DIN-rail formats.

How to Use it

Because a switching power supply contains no power transformer, it is lighter and smaller, and may be cheaper than a linear power supply. It is also more efficient and generates less waste heat. These advantages have made switching power supplies the most popular option to provide DC power for electronics devices. However, the high-frequency switching tends to create *electromagnetic interference* (*EMI*), which must be filtered to protect the output of the device and also to minimize the risk of this interference feeding back into AC power wiring. The high-frequency switched power may also generate harmonics, which must be suppressed.

High-quality linear regulated power supplies still find application in laboratory equipment, low-noise signal processing, and other niches where excellent regulation and low-ripple output are necessary. They are relatively heavy, bulky, and inefficient.

See Figure 16-7 for a chart comparing the advantages and disadvantages of linear and switching power supplies.

	Switching power supply	Linear power supply
Component count	High	Medium
Load regulation	0.05% to 0.5%	0.005% to 0.2%
Line regulation	0.05% to 0.2%	0.005% to 0.05%
Ripple (RMS)	10mV to 25mV	0.25mV to 1.5mV
Efficiency	70% to 85%	40% to 60%
EMI	High	Very low
Leakage	High	Low
Physical size	Small	Large
Weight	Light	Heavy
Heat management	Usually fan-cooled	Usually cooled by convection

Figure 16-7. *Comparison of attributes of linear regulated power supplies and switching power supplies. (Adapted from Acopian Technical Company.)*

What Can Go Wrong

High Voltage Shock

One or more capacitors in a power supply may retain a relatively high voltage for some time after the unit has been unplugged. If the power supply is opened for inspection or repairs, caution is necessary when touching components.

Capacitor Failure

If *electrolytic capacitors* fail in a switching power supply (as a result of manufacturing defects, disuse, or age), allowing straight-through conduction of alternating current, the high-frequency switching semiconductor can also fail, allowing input voltage to be coupled unexpectedly to the output. Capacitor failure is also a potential problem in linear power supplies. For additional information on capacitor failure modes, see Chapter 12.

Electrical Noise

If *electrolytic capacitors* are used, their gradual deterioration over time will permit more electrical noise associated with high-frequency switching in a switching power supply.

Peak Inrush

A switching power supply allows an initial inrush or surge of current as its capacitors accumulate their charge. This can affect other components in the circuit, and requires fusing that tolerates brief but large deviations from normal power consumption.

DC-DC converter 17

Often referred to as a *switching regulator*, and sometimes as a *switcher*, not to be confused with a *switching power supply*.

OTHER RELATED COMPONENTS

- **AC-DC power supply** (See Chapter 16)
- **voltage regulator** (See Chapter 19)
- **DC-AC inverter** (See Chapter 18)

What It Does

A DC-DC converter, often referred to simply as a *converter*, receives a DC voltage as its input and converts it to a regulated DC voltage as its output. The output voltage may be higher or lower than the input voltage, may be user-adjustable by adding an external resistor, and may be completely electrically isolated from the input, depending on the type of converter that is used. The overall efficiency is not greatly affected by the difference between input and output voltage, and can exceed 90%, minimizing waste heat and enabling the unit to be extremely compact.

A DC-DC converter is an integrated circuit package that includes a high-speed switching device (almost always, a *MOSFET*) in conjunction with an oscillator circuit, an inductor, and a diode. By comparison, a **linear regulator** is usually based around bipolar transistors. Its input must always be at a higher voltage than its output, and its efficiency will be inversely proportional with the voltage drop that it imposes. See the **voltage regulator** entry in this encyclopedia for additional information.

There is no single symbol to represent a DC-DC converter. Some simplified schematics showing the principles of operation of commonly used converters are referenced under the following Variants section.

A DC-DC converter is also typically found in the output stage of a switching **AC-DC power supply**.

How It Works

An internal oscillator controls a *MOSFET* semiconductor that switches the DC input on and off at a high frequency, usually from 50KHz to 1MHz. Output voltage is adjusted by varying the *duty cycle* of the oscillator—the length of each "on" pulse relative to each "off" interval. This is known as *pulse-width modulation*, or *PWM*. The duty cycle is controlled by sampling the output of the converter and using a comparator to subtract the output voltage from a reference voltage, to establish an error value. This is passed to another comparator, which subtracts the error voltage from an oscillator ramp signal. If the error increases, the oscillator signal is more heavily clipped, thus changing the effective ratio of on/off pulse lengths. A simplified schematic of the PWM circuit is shown in Figure 17-1, which omits other components for clarity. The system of subtracting an error voltage from a ramp oscillator voltage to obtain a pulse-width modulated signal is illustrated in Figure 17-2.

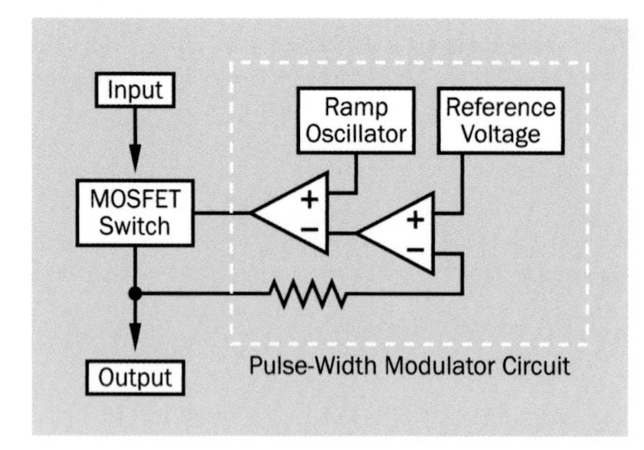

Figure 17-1. *The heart of a DC-DC converter is a MOSFET switch, which operates at a high frequency with pulse-width modulation used to create an adjustable DC output.*

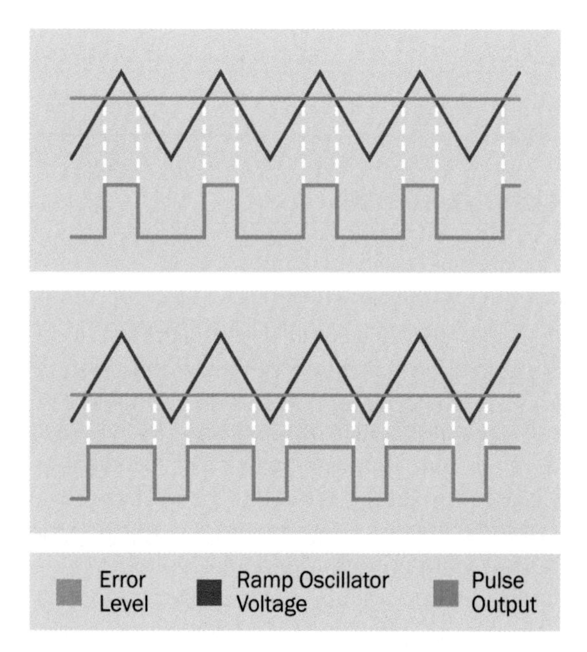

Figure 17-2. *To achieve pulse-width modulation, an error-level voltage is established by comparing the output from the converter with a reference voltage. The error level, shown as an orange line, is then subtracted from the output from a ramp oscillator. The pulse width varies accordingly.*

The key to the efficiency of a DC-DC converter is an inductor, which stores energy in its magnetic field during "on" pulse and releases it in the discharge phase. Thus, the inductor is used as a temporary reservoir and minimizes the ripple current. All converter variants use a coil for this purpose, although its placement varies in relation to the diode and capacitor that complete the basic circuit.

Variants

Four basic switching circuits are used in DC-DC converters and are defined in the coming sections, with a formula to determine the ratio between input voltage (V_{in}) and output voltage (V_{out}) in each case. In these formulae, variable D is the duty cycle in the pulse train generated through an internal MOSFET switch. The duty cycle is the fraction of the total on-off cycle that is occupied by each "on" pulse. In other words, if T_{on} is the duration of an "on" pulse and T_{off} is the "off" time:

$$D = T_{on} / (T_{on} + T_{off})$$

Buck Converter

See Figure 17-3. The output voltage is lower than the input voltage. The input and output share a common ground. For this circuit:

$$V_{out} = V_{in} * D$$

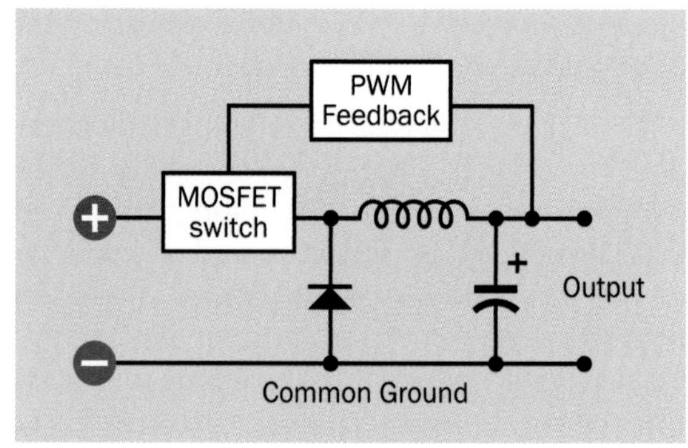

Figure 17-3. *Basic topology of a buck-type DC-DC converter.*

Boost Converter

See Figure 17-4. The output voltage is greater than the input voltage. The input and output share a common ground. For this circuit:

```
Vout = Vin / (1-D)
```

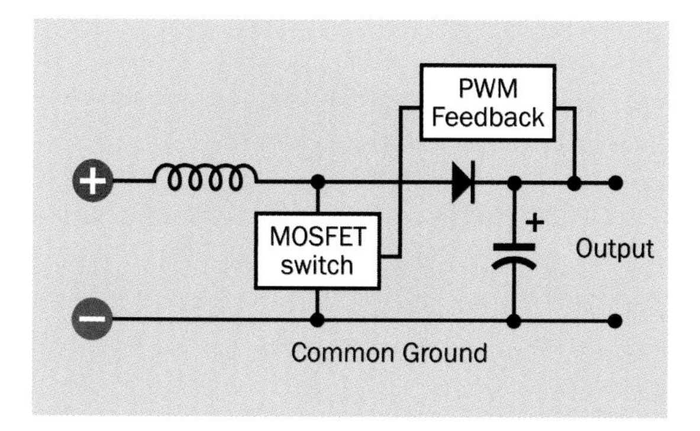

Figure 17-4. *Basic topology of a boost-type DC-DC converter.*

Flyback Converter with Inductor

Commonly known as a *buck-boost converter*. See Figure 17-5. The output voltage can be less than or greater than the input voltage. The input and output share a common ground. For this circuit:

```
Vout = Vin * (D / (1-D))
```

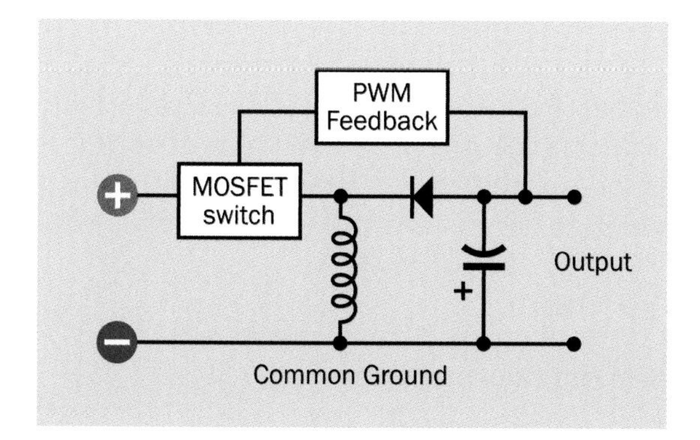

Figure 17-5. *Basic topology of a flyback-type DC-DC converter.*

Flyback Converter with Transformer

See Figure 17-6. The output voltage can be less than or greater than the input voltage. The input and output are isolated from one another. For this circuit:

```
Vout = Vin * (D / (1-D))
```

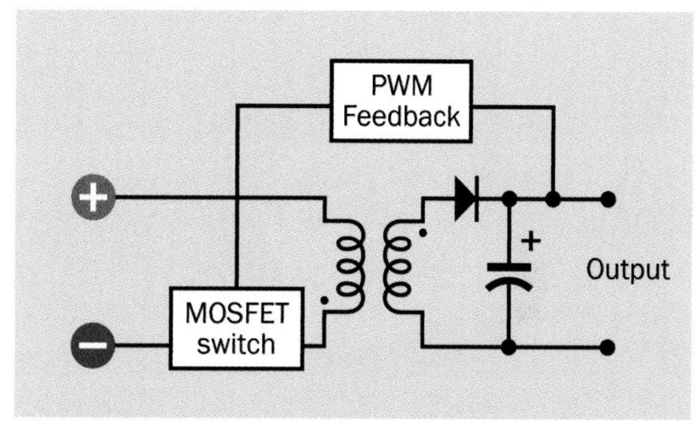

Figure 17-6. *Basic topology of a flyback-type DC-DC converter. (Buck, boost, and flyback topologies adapted from Maxim Integrated Products.)*

Using a transformer in the converter allows multiple outputs with different voltages, supplied through multiple transformer windings.

Formats

A converter may be packaged in a flat rectangular box that requires no additional heat sink and has pins for through-hole insertion into a PC board. Sizes usually range up to to 2″ × 2″. Power handling can range from 5 to 30 watts. Converters of this type are shown in Figure 17-7. (Top: Input range of 9 to 18VDC, fixed output of 5VDC at 3A completely isolated from the input. Typical efficiency of approximately 80%. The case is made of copper, providing good heat dissipation with electrical shielding. Center: Input range of 9 to 18VDC, fixed output of 5VDC at 500mA completely isolated from the input. Typical efficiency of approximately 75%. The manufacturer claims that external capacitors are only needed in critical applications. Bottom: SIP format, fixed input of 12VDC, fixed output of 5VDC at 600mA completely isolated from the input. Typical efficiency of approximately 75%. Requires external capacitors for ripple rejection.)

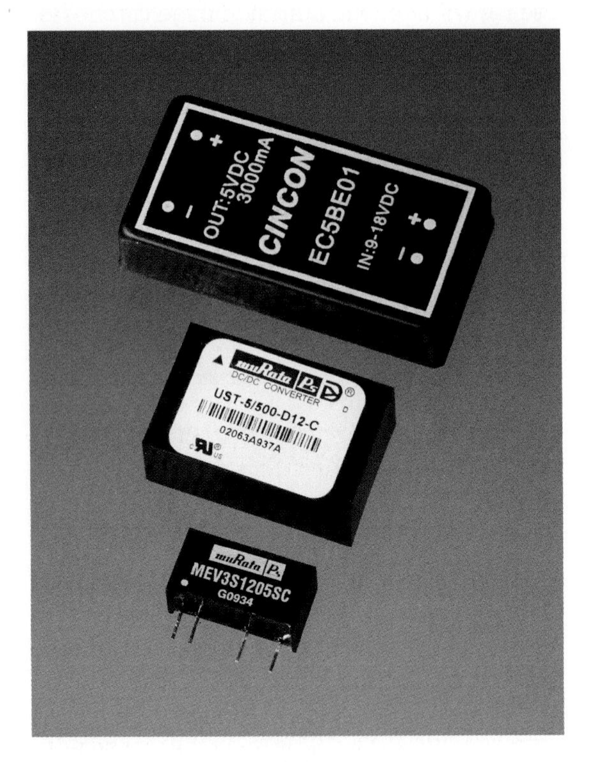

Figure 17-7. *A selection of sealed DC-DC converters.*

Lower-power converters are also available as surface-mount devices.

Some adjustable-output converters are supplied as multiple surface-mounted components preinstalled on a mini-board that has pins for through-hole insertion in a printed circuit board. Their high efficiency enables them to handle a lot of power for their size. In Figure 17-8, the converter accepts a 4.5 to 14VDC input range and has an adjustable output of 0.6 to 6VDC. It is rated at a surprising 10A or 50W and is more than 90% efficient. However, it draws 80mA in a no-load state, causing it to become quite hot. A thermal cutout or automatic shutdown may be used if the converter will not be driving a consistent load.

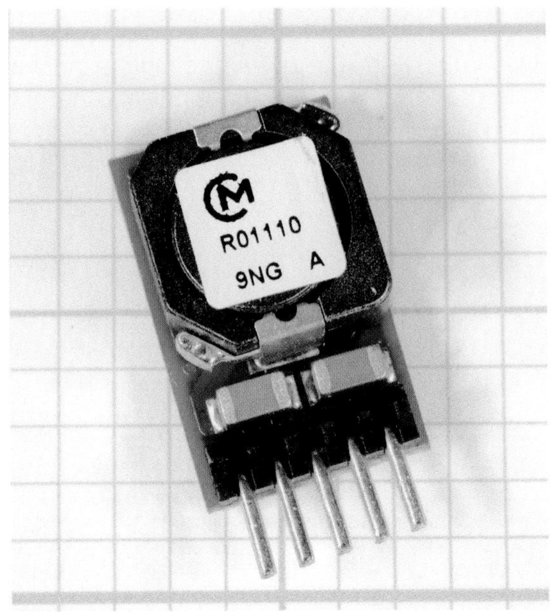

Figure 17-8. *An adjustable DC-DC converter rated for 10A or 50W. The output voltage is determined by adding an external resistor or trimmer potentiometer. External smoothing capacitors are required, as shown in the component's datasheet.*

The miniboard in Figure 17-9 accepts an input voltage from 7 to 36VDC and has an adjustable output ranging from 2.5 to 12.6VDC, at up to 6A. It is non-isolated (has a common negative bus) and claims to be more than 95% efficient at full load.

The miniboard in Figure 17-10 accepts an input voltage from 4.5 to 14VDC and has an adjustable output ranging from 0.6 to 6VDC at up to 20A. It is non-isolated (has a common negative bus) and claims to be more than 90% efficient at full load.

Values

Relevant values include:

Nominal Input Voltage and Frequency

A wide range of input voltages is often acceptable, as the PWM can vary accordingly. Converters often allow equipment to be usable internationally, on any voltage ranging from 100VAC to

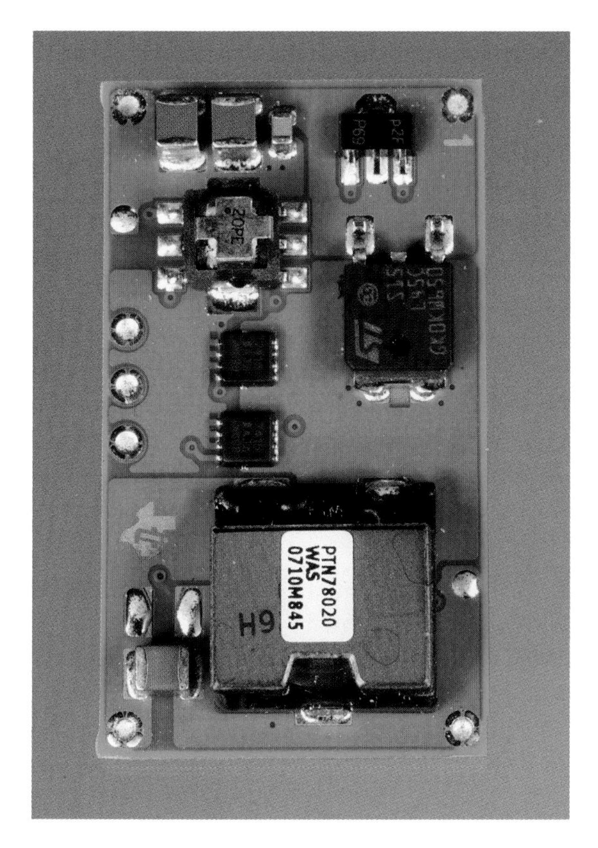

Figure 17-9. Another adjustable DC-DC converter. The output voltage is determined by adding an external resistor or trimmer potentiometer. External smoothing capacitors are required, as shown in the component's datasheet.

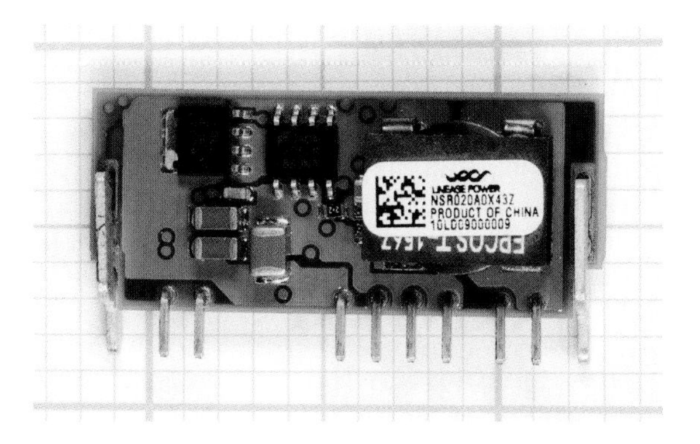

Figure 17-10. Another adjustable DC-DC converter. The output voltage is determined by adding an external resistor or trimmer potentiometer. External smoothing capacitors are required, as shown in the component's datasheet.

250VAC, at a fequency of 50Hz or 60Hz, without any adaptation.

Output Voltage

As previously noted, many converters allow the output voltage to be adjusted by adding an external resistor or potentiometer. Alternatively, there may be multiple fixed output voltages, accessible via different pins on the package. They may also provide a positive voltage and equally opposite negative voltage relative to a ground pin.

Input Current and Output Current

Because input voltage and output voltage are likely to be different, the current alone is not a reliable guide to power handling.

A datasheet should specify input current with no load (open circuit on the output side). This current will have to be entirely dissipated as heat.

Load Regulation

This is usually expressed as a percentage and suggests the extent to which output voltage may be pulled down when the load on a DC-DC converter increases. If V_{nil} is the measured output voltage with no load, and V_{max} is the measured output voltage with the maximum rated load:

$$\text{Load regulation} = 100 * (V_{nil} - V_{max})/V_{max}$$

However, note that some converters are designed with the expectation that they will never be used with zero load across the output. In these cases, V_{nil} will be the voltage at minimum rated load.

Efficiency

This is a measure of how much input current must be dissipated as heat. A converter with a 12-volt input, drawing a maximum 300mA input current, will consume 3.6 watts (3,600mW). If it is 80% efficient, it will have to dissipate roughly 20% of its power as heat, or 720mW.

Ripple and Noise

Sometimes abbreviated R/N, this may be measured in mV or as a percentage. Check the specification carefully to determine whether the ripple-and-noise values require use of external smoothing capacitors. Often, this is the case.

Isolated or Non-Isolated

This crucial piece of information is often found near the top of a datasheet, not in the detailed specifications.

How to Use it

Because a converter creates electrical noise, it should be prevented from affecting other components by adding substantial *bypass capacitors* as close as possible to its input and output pins. For most converters, external capacitors are mandatory, and their *effective series resistance* (ESR) should be as low as possible (see the **capacitor** entry in this encyclopedia for an explanation of ESR). Tantalum capacitors are preferable to electrolytics for this reason, and are also more durable. Some manufacturers recommend placing a tantalum capacitor in parallel with an electrolytic. A small ceramic capacitor, typically 0.1µF, is often recommended in an addition to larger-value capacitors on the output side.

The voltage rating of each capacitor should be twice the voltage at the point in the circuit where it is used. The capacitance value will usually be higher for higher-current converters. Values of 100µF are common, but for high amperage, a value may be as high as 1,000µF.

While datasheets are often inadequate for some types of components, datasheets for DC-DC converters usually include detailed instructions regarding bypass capacitors. Following these instructions is essential. In the relatively rare instances that a datasheet makes no mention of bypass capacitors for a converter, this does not necessarily mean that the capacitors are unnecessary. The manufacturer may assume that they will be used as a matter of course.

Converters are used in a very wide range of devices, supplying power ranging from a few milliamps to tens of amps. At the lower end of the scale, devices such as cellular telephones, portable computers, and tablets contain subcircuits that require different voltages, some of which may be higher than the voltage of a single battery or battery pack that powers the device. A converter can satisfy this requirement. Because a converter can be designed to maintain a fixed output in response to a range of input voltages, it can also compensate for the gradual decline in voltage that occurs during battery usage.

A boost-type converter can be used to double the voltage from a single 1.5V battery in an LED flashlight where 3 volts are required to power the LED. Similarly, a boost-type converter can provide the necessary voltage to run a cold-cathode fluorescent tube that provides backlighting in an LCD computer display.

On a circuit board that is primarily populated with 5VDC components and is fed by a single 5VDC power supply, a converter can be used to supply 12VDC for one special purpose, such as an analog-digital converter or a serial data connection.

If electromechanical **relays** or other inductive loads share a common ground with components ,such as *logic chips* or **microcontrollers**, it may be difficult to protect the sensitive components from voltage spikes. A A flyback converter with a transformer separating the output from the input can allow the "noisy" section of the circuit to be segregated, so long as the converter itself does not introduce noise. Since the electromagnetic interference (EMI) introduced by converters varies widely from one model to another, specifications should be checked carefully.

Very low-power components can pick up EMI from the wires or traces leading into and out of a converter. In this type of circuit, adequate noise suppression may be impossible, and a converter may not be appropriate.

What Can Go Wrong

Electrical Noise in Output

Electrolytic capacitors may be inadequate to smooth the high frequencies used. Multilayer ceramic capacitors or tantalum capacitors may be necessary. Check the manufacturer's datasheet for minimum and maximum values. Also check the datasheet for advice regarding placement of capacitors on the input side as well as across the output.

Excess Heat with No Load

Some converters generate substantial heat while they are powered without a load. The manufacturer's datasheet may not discuss this potential problem very prominently or in any detail. Check the input rating, usually expressed in mA, specified for a no-load condition. All of this current will have to be dissipated as heat, and the very small size of many converters can result in high localized temperatures, especially since many of them allow no provision for a heat sink.

Inaccurate Voltage Output with Low Load

Some converters are designed to operate with at least 10% of full rated load across their output at all times. Below this threshold, output voltage can be grossly inaccurate. Read datasheets carefully for statements such as this: "Lower than 10% loading will result in an increase in output voltage, which may rise to typically double the specified output voltage if the output load falls to less than 5%." Always use a meter to verify the output voltage from a converter at a variety of different loads, and perform this test before installing the converter in a circuit.

DC-AC inverter

18

A power inverter must not be confused with a *logic inverter*, which functions as a digital component in *logic circuits* to invert the state of a low-voltage DC input from high to low or low to high. Logic inverters are discussed in Volume 2.

OTHER RELATED COMPONENTS

- **AC-DC power supply** (See Chapter 16)
- **DC-DC converter** (See Chapter 17)

What It Does

A power inverter is included here as counterpoint to a **power supply** or *AC adapter*, since it has the opposite function. The inverter receives an input of *direct current* (typically 12VDC from a car battery) and delivers an output of *alternating current* (AC) in the range 110VAC-120VAC or 220VAC-240VAC, suitable to power many low-wattage appliances and devices. The interior of a low-cost inverter is shown in Figure 18-1.

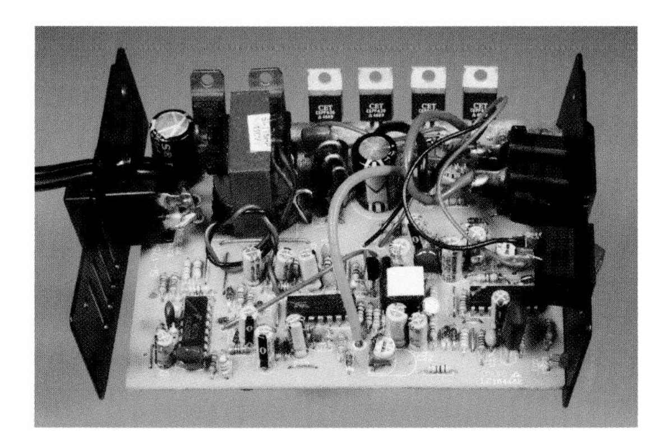

Figure 18-1. *The interior components in a 175-watt inverter.*

How It Works

The first stage of an inverter typically raises a 12VDC input to a higher DC voltage via an internal **DC-DC converter**, then uses a switching circuit to create an approximation of the sinusoidal profile that is characteristic of AC voltage.

Digital switching components naturally tend to create square waves, whose simple appearance conceals the presence of higher frequencies, or *harmonics*, that are ignored by some devices (especially those that convert electricity into heat) but can be troublesome in consumer electronics equipment. A primary objective of inverter design is to adapt or combine square waves to emulate a classic AC sine wave with reasonable fidelity. Generally speaking, the more accurately an inverter emulates a sine wave, the more expensive it tends to be.

The most primitive inverter would create a plain square wave such as that shown in red in Figure 18-2, superimposed on a comparable sine wave (in green). Note that alternating current rated at 115 volts actually peaks at around 163 volts because the number 115 is the approximate *root mean square* (*RMS*) of all the voltage values during a single positive cycle. In other words, if the voltage is sampled x times during a cycle, an

RMS value can be derived by squaring each sample, adding all the samples, dividing by x, and then taking the square root of the result. The RMS value is important as a means to calculate actual power delivered because it can be multiplied by the current to obtain an approximate value in watts.

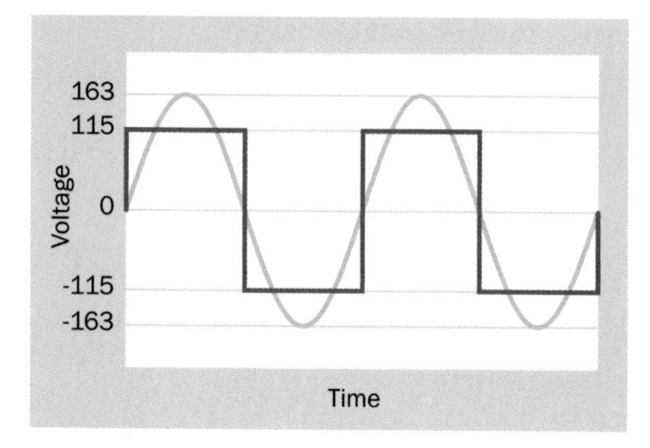

Figure 18-2. *Comparison of an AC voltage sine wave (green) and a square wave of the same frequency (red), both delivering a roughly similar amount of power.*

Variants

As a first step toward a better approximation of a sine wave, gaps of zero voltage can be inserted between square-wave pulses. This "gapped" square wave is shown in Figure 18-3.

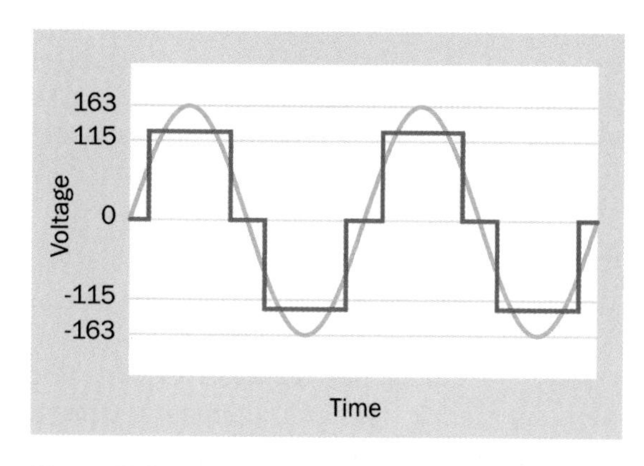

Figure 18-3. *Introducing pauses or gaps of zero voltage between square-wave pulses can produce slightly improved resemblance to a sine wave.*

A further improvement can be achieved if an additional, shorter pulse of higher voltage is added to each primary pulse, as shown in Figure 18-4. Outputs of this kind are referred to as *modified sine wave*, although they are actually square waves modified to emulate a sine wave. Their inaccuracy is expressed as *total harmonic distortion* (*THD*). Some authorities estimate that the THD of gapped square-wave output is around 25%, whereas the addition of shorter square waves reduces this to around 6.5%. This is a topic on which few people agree, but there is no doubt that a "stacked" sequence of square waves provides a closer emulation of a sine wave.

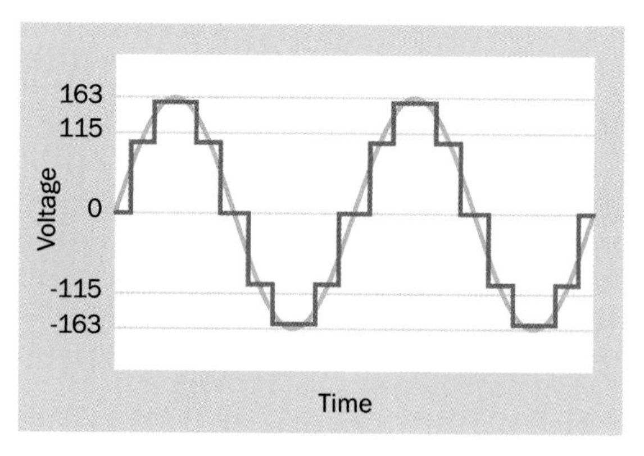

Figure 18-4. *A secondary stream of narrower square-wave pulses can improve the fidelity of an inverter's output.*

A *true sinewave inverter* typically uses *pulse-width modulation* (*PWM*) to achieve THD of less than 1%. It generates a stream of pulses much higher in frequency than that of the AC output, and varies their widths in such a way that their averaged voltage closely approximates the voltage variations in a sine wave. A simplified representation of this principle is shown in Figure 18-5.

Values

Small inverters are typically rated to deliver up to 100 watts and may be fitted with a 12VDC plug for insertion in a vehicle's cigarette lighter. Since a cheap inverter may be only 80% efficient, 100 watts at 135VAC will entail drawing as much as

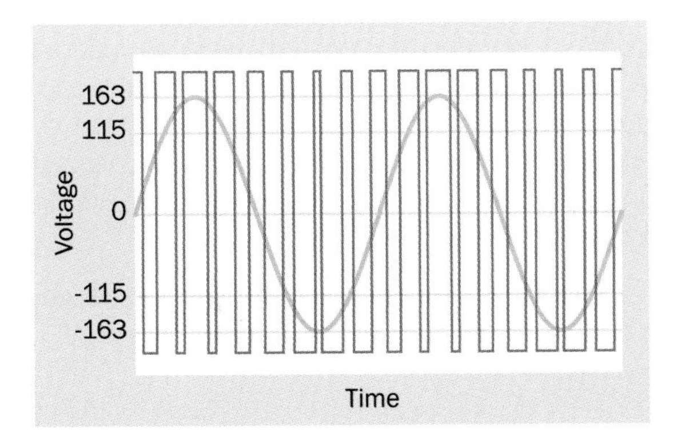

Figure 18-5. *Pulse-width modulation adjusts the widths of pulses delivered at a high frequency. The pulse widths can be averaged to generate voltage that follows a close approximation of a sine wave.*

10 amps at 12VDC. Cigarette lighters are usually fused at 15 or 20 amps, so 100 watts is a reasonable value. Inverters that are rated above 150 watts usually have cables terminating in oversize alligator clips for direct connection to the terminals of a 12V battery.

While the *cold cranking* rating of a car battery may be 100 amps or more, the battery is only designed to deliver that power for up to 30 seconds at a time. Inverters rated for as much as 500 watts will exceed the normal capacity of a single car battery, although if the battery is mounted in a vehicle, it can be supplemented by running the engine so that the alternator shares some of the load. A 500-watt inverter is better supplied by two or more 12-volt car batteries wired in parallel.

How to Use it

Small inverters are typically used in vehicles to run cellphone chargers, music players, or laptop computers. Large inverters are an integral part of off-the-grid solar and wind-powered systems, where battery power must be converted to AC house current. Uninterruptible power supplies contain batteries and inverters capable of running computer equipment for a brief period. Battery-driven electric vehicles with AC motors use inverters with an exceptionally high current rating.

There is a lack of consensus regarding possible harmful effects of powering electronics equipment with a low-cost modified sine wave inverter. Anecdotal evidence suggests that where the equipment uses its own *switching power supply* or uses an *AC adapter* (either mounted internally or as an external plug-in package), the filtering built into the power supply will block harmonics from the inverter.

Other evidence suggests that cheap inverters may have adverse effects on devices containing synchronous motors that run direct from AC. There are reports that fluorescent lighting and photographic electronic flash systems may be unsuitable for use even with modified sine wave inverters. However, differences in product design and component quality make it impossible to generalize. A cheaply made inverter may generate a wave form that is not even a close approximation of a square wave. See Figure 18-6.

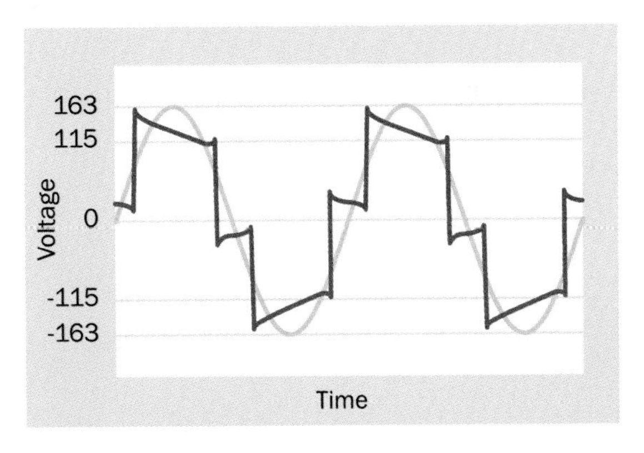

Figure 18-6. *A cheaply made inverter can generate a distorted wave form that is even higher in noise than a pure square wave. This sample is adapted from an actual oscilloscope trace.*

What Can Go Wrong

If multiple batteries are connected in parallel, using suitably heavy-gauge wire to power a large inverter, the batteries must be identical in specification and age, and must all be equally charged to prevent high and potentially dangerous flows of current among the batteries as they attempt to reach an equilibrium among themselves. Interconnections must be firmly clamped to clean battery terminals. For additional information, see the **battery** entry in this encyclopedia.

Problems associated with inverters are likely to be mundane. A 12V wiring to the inverter can overheat if items such as clothes or bedding are left on top of it; a high-wattage fan-cooled inverter can overheat if the fan is obstructed by poor placement or impaired by accumulated dirt; alligator clips may become dislodged from battery terminals; and power surges drawn by inductive loads such as motors may trigger the inverter's *breaker*, especially if they are used in conjunction with other equipment.

As always, high amperage should be treated with caution and respect, regardless of it being delivered at "only 12 volts."

voltage regulator

19

Correctly known as a *linear voltage regulator* to distinguish it from a *switching regulator* or **DC-DC converter**. However, the full term is not generally used, and "voltage regulator" is normally understood to mean a linear voltage regulator.

OTHER RELATED COMPONENTS

- **DC-DC converter** (See Chapter 17)
- **AC-DC power supply** (See Chapter 16)

What It Does

A linear voltage regulator provides a tightly controlled DC output, which it derives from an unregulated or poorly regulated DC input. The DC output remains constant regardless of the load on the regulator (within specified limits). It is a cheap, simple, and extremely robust component.

There is no single schematic symbol for a linear voltage regulator.

The general physical appearance of a commonly used type of regulator, rated for an output of around 1A DC, is shown at Figure 19-1. The LM7805, LM7806, LM7812, and similar regulators in the LM78xx series are encapsulated in this type of package, with pins that are spaced at 0.1" and have functions as shown. Other types of regulator may differ in appearance, or may look identical to this one but have different pin functions. Always check datasheets to be sure.

How It Works

All linear regulators function by taking some feedback from the output, deriving an error value by comparing the output with a reference voltage (most simply provided by a zener diode), and using the error value to control the base of a *pass transistor* that is placed between the input and the output of the regulator. Because the transistor operates below saturation level, its output current varies linearly with the current applied to its base, and this behavior gives the linear regulator is name. Figure 19-2 shows the relationship of these functions in simplified form; Figure 19-3 shows a little more detail, with a *Darlington pair* being used as the pass transistor. The base of the pair is controlled by two other transistors and a comparator that delivers the error voltage. This version of a voltage regulator is known as the *standard type*.

The voltage difference required between the base and emitter of an NPN transistor is a minimum of 0.6V. Because multiple transistors are used inside a standard-type voltage regulator, it requires a minimum total voltage difference, between its input and its output, of 2VDC. This voltage difference is known as the *dropout voltage*. If the voltage difference falls below this minimum, the regulator ceases to deliver a reliable output voltage until the input voltage rises again. *Low dropout regulators* allow a lower voltage difference, but are more expensive and less commonly used. They are described under the following Variants section.

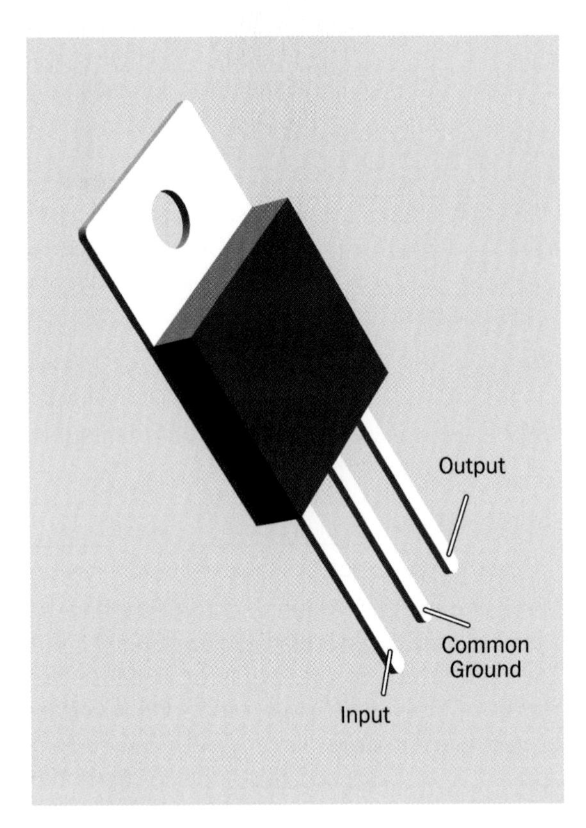

Figure 19-1. *The package design of a commonly used voltage regulator. Others may be significantly different, and the pin functions may vary. Check manufacturer datasheets for details.*

Discrete components could in theory be used to build a voltage regulator, but this ceased to be cost-effective several decades ago. The term is now understood to mean one small integrated package containing the basic circuit augmented with additional, desirable features, such as automatic protection against overload and excessive heat. Instead of burning out if it is overloaded, the component simply shuts down. Most voltage regulators also tolerate accidentally reversed power connection (as when batteries are inserted the wrong way around) and accidentally reversed insertion of the regulator in a circuit board.

Other components can satisfy the requirement to deliver power at a reduced voltage. Most simply, if two resistors in series are placed across a

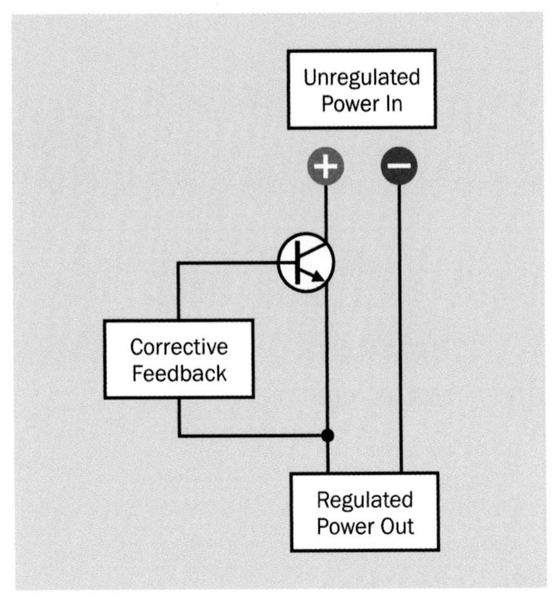

Figure 19-2. *A linear voltage regulator basically consists of a transistor whose base is controlled by corrective feedback derived from the output.*

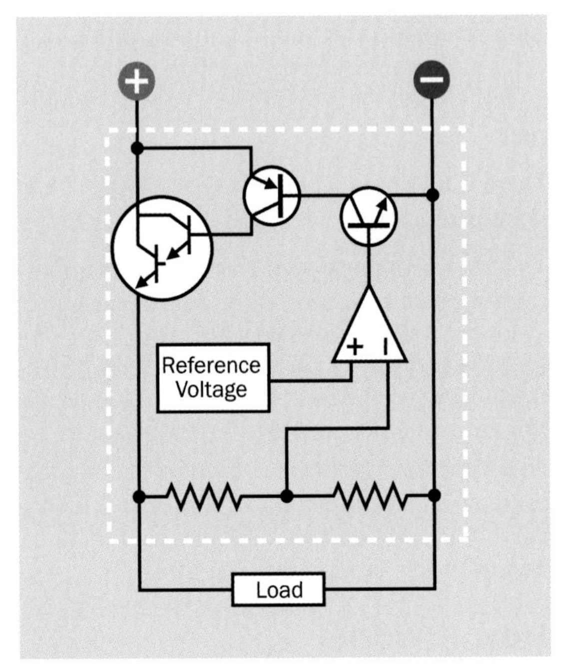

Figure 19-3. *The fundamental internal features of a standard-type voltage regulator, including a Darlington pair, two transistors, a voltage divider, comparator, and reference voltage source, shown inside the dashed white line.*

power source, they form a *voltage divider*, which provides an intermediate voltage at the connection between them. However, this voltage will vary depending on fluctuations in the input voltage and/or load impedance. A voltage regulator is the simplest way to supply a voltage that remains stable regardless of excursions in the input or fluctuations in power consumed by the load.

The disadvantage of a standard-type voltage regulator is that it is inefficient, especially when a relatively high input voltage is used to deliver a relatively low output voltage. If V_{in} is the input voltage, V_{out} is the output voltage, and I_{out} is the output current, the average power loss, P, is given by the formula:

$$P = I_{out} * (V_{in} - V_{out})$$

For example, if the output current is 1A, the input voltage is 9VDC, and the output is 5VDC, 44% of the input power will be wasted, and the component will be only 56% efficient. The wasted power (about 4 watts, in this case) will be dissipated as heat. Even when a standard-type regulator runs at its minimum 2VDC dropout voltage, it must dissipate 1W when delivering 0.5A.

Variants

Packaging

The package for the LM78xx series of regulators, shown in Figure 19-1, incorporates an aluminum plate drilled with a hole so that it can be bolted to a heat sink. Voltage regulators with a lower rated maximum output current (typically, 100mA) do not have the same need for a heat sink, and are available in a package that resembles a small transistor.

Some integrated circuits are available containing two voltage regulators, electrically isolated from each other.

Popular Varieties

In the LM78xx series, the last two digits in the part number specify the output voltage, which is fixed. Thus the LM7805 delivers 5VDC, the LM7806 delivers 6VDC, and so on. For regulators with a fractional voltage output (3.3VDC being common), an additional letter may be inserted in the part number, as in the 78M33.

Many copies of the LM78xx series are made by different manufacturers, the copies being functionally identical, regardless of additional letters that are added to the part number to identify its source or other attributes.

The LM78xx regulators are mostly rated to be accurate within 4%, although actual samples almost always deliver voltages that are more precise than this range suggests.

Adjustable Regulators

While the majority of regulators have a fixed output, some allow the user to set the output by adding one or more resistors. The LM317 is a popular example. Its output voltage can range from 1.25VDC to 37VDC and is set via a resistor and a trimmer potentiometer, as illustrated in Figure 19-4. If R1 is the fixed-value resistor and R2 is the trimmer, as shown in the schematic, the output voltage, V_{out}, is given by the formula

$$V_{out} = 1.25 * (1 + (R2 / R1))$$

Typical values for R1 and R2 would be 240Ω and 5K, respectively. With the trimmer at the middle of its range, V_{out} would be 1.25 * (1 + (2500 / 240)) = approximately 15VDC, requiring an input of at least 17VDC. However, if the trimmer is reduced to 720Ω, the output would be 5VDC. In practice, the value of a trimmer should be chosen so that a mid-range setting provides approximately the desired output. This will enable fine adjustment of the output voltage.

While the versatility of an adjustable regulator is desirable, its overall power dissipation is still proportional to the difference between the input voltage and the output voltage. To minimize heat loss, this difference should not exceed the dropout voltage by a larger amount than is absolutely necessary.

An adjustable regulator may require larger bypass capacitors than a regulator with a fixed output. A manufacturer's recommendations for the LM317 are shown in Figure 19-4.

Negative and Positive Regulators

While most linear voltage regulators are designed for "positive input" (conventional current flow from input to output), some are intended for "negative input." In this variant, the common terminal is positive, and the input and output are negative in relation to it.

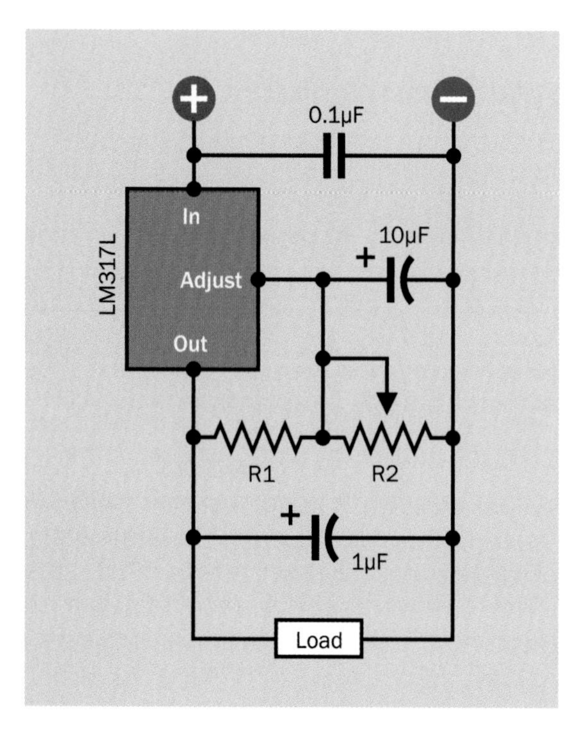

Figure 19-4. *Schematic for the LM317L adjustable voltage regulator, based on a circuit recommended by National Semiconductor, with bypass capacitors added for ripple rejection.*

Low-Dropout Linear Regulators

Low-dropout regulators (sometimes referred to as *LDO* regulators) allow a much lower dropout voltage by using a single PNP or MOSFET transistor. LDO regulators are popularly used in battery-powered devices where efficiency should be maximized and heat dissipation should be minimized. For example, the LM330 is a regulator with a 5VDC output, tolerating a dropout voltage of 0.6V, allowing it to be used with four AAA cells. In an LDO regulator the dropout voltage actually varies with load current and may diminish to as little as one-tenth of its rated value when the output current is minimal.

The majority of low-dropout regulators are sold in surface-mount packages, and are designed for maximum output of 100mA to 500mA. Only a few exceptions exist. They tend to be slightly more expensive than regulators with the typical 2V dropout rating.

Three voltage regulators are shown in Figure 19-5. From left to right, they are rated 5VDC at 1A, 12VDC at 1A, and 5VDC at 7.5A. The two smaller regulators are of the LM78xx series. The larger regulator claims a low maximum dropout voltage of 1.5VDC, and its output voltage can be adjusted with an external potentiometer and resistor.

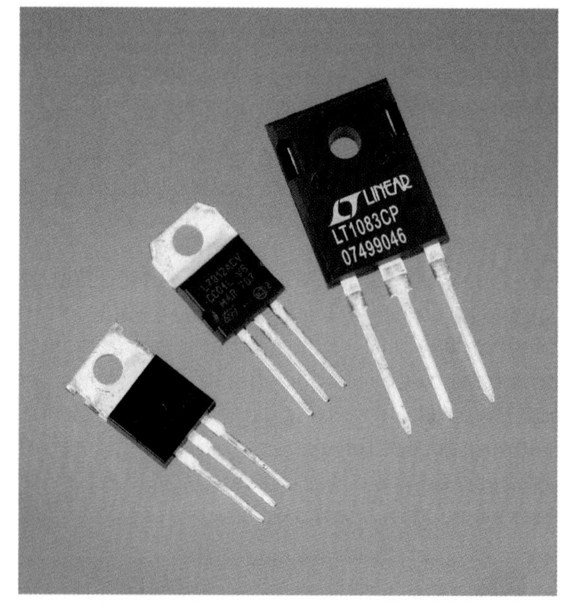

Figure 19-5. *Two voltage regulators from the LM78xx series, and a third high-current, low-dropout, adjustable regulator rated 5VDC (adjustable upward) at 7.5A.*

Quasi-Low-Dropout Linear Regulators

Where a standard regulator uses a Darlington pair as the pass transistor and an LDO uses a single PNP transistor, the so-called Quasi-LDO uses a combination of NPN and PNP transistors and has an intermediate dropout voltage, typically a maximum of 1.5VDC. However, the terms LDO and Quasi-LDO are not used uniformly in the industry. One manufacturer markets Quasi-LDO regulators as LDO regulators, and describes its LDO regulators as *Very Low Dropout* regulators. Consult datasheets to determine the actual specification of the product, regardless of its classification.

Additional Pin Functions

Some voltage regulators include an extra pin, typically known as an *enable pin*, which switches off the device in response to a signal from a microcontroller or logic gate.

Some regulators offer another option, an additional *status pin* that can signal a microcontroller that an error mode exists if the regulator output falls significantly below its rated value.

In battery-powered devices, a low-battery sensor is a desirable feature, since a regulator may simply shut down without warning if the input voltage is insufficient. A few regulators, such as the LP2953, provide a low-battery warning output via an extra pin.

Values

Linear voltage regulators with a single, fixed output are commonly available to supply DC outputs of 3.3, 5, 6, 8, 9, 10, 12, 15, 18, and 24 volts, with a few variants offering fractional values in between. The most commonly used values are 5, 6, 9, 12, and 15 volts. The input voltage may be as high as 35VDC.

Maximum output current is typically 1A or 1.5A, in the traditional three-pin, through-hole, TO-220 format. A surface-mount version is available. Other surface-mount formats have lower power limits.

Accuracy may be expressed as a percentage or as a figure for load regulation in mV. A typical load regulation value would be 50mV, while voltage regulation accuracy ranges from 1% to 4%, depending on the manufacturer and the component. While low-dropout regulators are generally more efficient, they do require more ground-pin current. This is not usually a significant factor.

How to Use it

Some components, such as many old-design CMOS chips or the traditional TTL version of the 555 timer, allow a wide range of acceptable input voltages, but most modern logic chips and microcontrollers must have a properly controlled power supply. Regulators such as the LM7805 are traditionally used to provide this, especially in small and relatively simple devices that draw a moderate amount of current, have a low component count, and are powered via a battery or an **AC adapter**. A fully fledged switching power supply is overkill in this kind of application.

A linear voltage regulator cannot respond instantly to changes in input voltage. Therefore, if the input supply contains voltage spikes, these spikes may pass through the regulator. Bypass capacitors should be applied preventively. A sample schematic showing an LM7805 regulator with bypass capacitors recommended by a manufacturer is shown in Figure 19-6.

In a battery-powered device where standby power is required for long periods and full power is only needed intermittently, the *quiescent current* drawn by a minimally loaded voltage regulator is important. Modern LDO regulators may draw as little as 100μA when they are very lightly loaded. Other types may consume significantly more. Check datasheets to find the most appropriate component for a particular application. Note that DC-DC power **converters** may draw a lot of current when they are lightly loaded, and

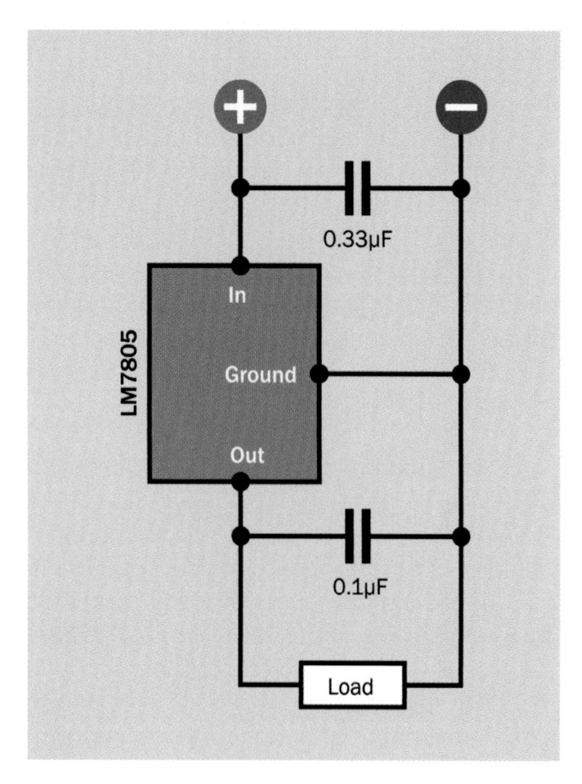

Figure 19-6. *Typical schematic for use of an LM7805 regulator, with capacitor values based on recommendations from Fairchild Semiconductor.*

will dissipate large amounts of heat as a result. An LDO is therefore preferable in this situation.

What Can Go Wrong

Inadequate Heat Management

The ability to "dial up" a wide range of voltages from an adjustable regulator such as the LM317 can be a temptation to use it on a "one size fits all" basis, to deliver any output ranging from 5VDC to 18VDC from a uniform 24VDC input. Assuming 1A output current, the worst-case power dissipation in this scenario would be almost 20W. To achieve reasonable efficiency and maintain waste heat at a manageable level, the input voltage should not exceed the output voltage by much more than the dropout voltage.

Even when a voltage regulator is used correctly, it can generate more heat than was expected if the requirements of a circuit are altered during development. An initial handful of components may draw only 100mA, but as more capabilities are requested and more parts are added (especially relays or LED displays) the power consumption can quickly add up, generating an unexpected amount of waste heat and raising the possibility of a sudden (and mysterious) shutdown if the regulator does not have an adequate heat sink.

Transient Response

When there is a major fluctuation in the demand by the load (for example, if an inductive device is switched on elsewhere in the circuit), the voltage regulator requires a finite time to adjust itself and maintain its specified output voltage. This time lag is known as its *transient response*. If a momentary fluctuation is likely, and other components may be sensitive to it, a larger capacitor should be used between the output of the voltage regulator and ground.

The transient response time may also be insufficient to block sudden, brief spikes in input voltage. This may occur, for example, when a low-cost AC adapter that does not have a properly smoothed output is used as the power source. Additional 1μF bypass capacitors may be added at the input and output of a regulator to provide better protection from power fluctuations.

Misidentified Parts

Many types of linear voltage regulators appear physically identical. Care is needed to distinguish those which have fixed output from those that allow a variable output. When using the LM78xx series, double-check the last pair of digits in the part number, which provide the only guide regarding the output. Using an LM7808 instead of an LM7805 may be sufficient to destroy all the 5VDC chips in a logic circuit. It is advisable to use a meter to check the output of any power supply before connecting it with a circuit.

Misidentified Pins

The LM78xx series of voltage regulators uses a very intuitively obvious and consistent scheme for the functions of its pins: input on the left, ground in the center, and output on the right, when looking at the regulator from the front, with its pins facing downward. Unfortunately the consistency of this scheme can encourage an unthinking habit for making connections. The LM79xx series of negative voltage regulators swaps the identity of the input and ground pins, whereas adjustable regulators use yet another different scheme. Good practice suggests checking a component against the manufacturer's datasheet before connecting it.

Dropout Caused by Low Battery

If a regulator rated to deliver 6VDC has a 2VDC dropout voltage and is powered from a 9V battery, the battery can easily drop below the minimum acceptable 8VDC if it becomes old or depleted. When this happens, the output from the regulator will tend to fall, or may oscillate.

Inaccurate Delivered Voltage

A voltage regulator maintains its output voltage between its output pin and ground pin. Thin traces on a circuit board, or a long run of very small-gauge wiring, can impose some electrical resistance, reducing the actual voltage delivered to a component. Ohm's Law tells us that the voltage drop imposed by a trace (or thin wire) will be proportional to the current flowing through it. For example, if the resistance between the output pin of a voltage regulator and a component is 0.5Ω and the current is 0.1A, the voltage drop will be only 0.05V. But if the current increases to 1A, the voltage drop is now 0.5V. Bearing this in mind, a linear voltage regulator should be positioned close to voltage-sensitive components. In printed circuit designs, the traces that deliver power should not have significant resistance.

When using linear voltage regulators with adjustable output, there may be a temptation to connect adjustment resistor R1 to the positive end of the load, to obtain a "more accurate" delivered voltage. This configuration will not produce the desired result. R1 should always be connected as closely as possible between the output pin and the adjustment pin of the voltage regulator, while R2 should connect between the adjustment pin and the negative end of the load. This is illustrated in Figure 19-7, where the gray wire in each schematic indicates that it possesses significant resistance.

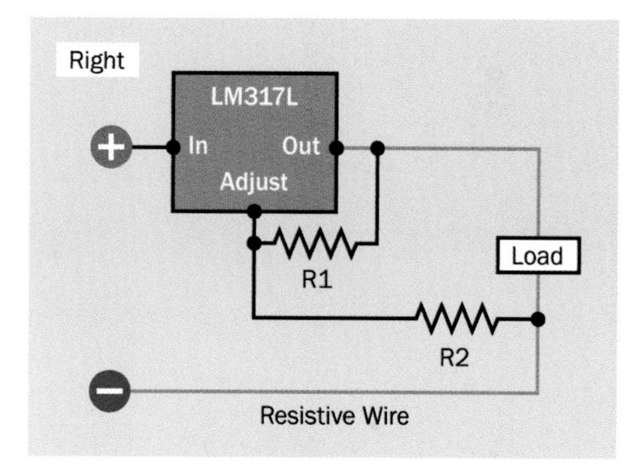

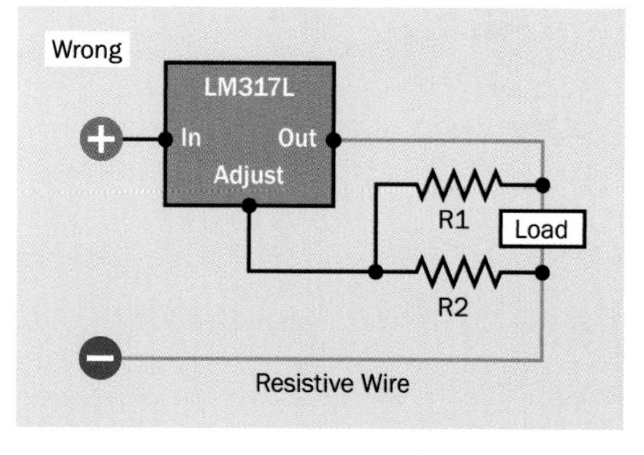

Figure 19-7. *When the connection between an adjustable-output voltage regulator and load components has a significant resistance (shown here as a gray "resistive wire"), R1 should always be connected as closely as possible to the pins of the regulator, as shown in the upper schematic. (Derived from schematics prepared by National Semiconductor.)*

electromagnet

The term **electromagnet** is used here to mean a coil containing a core of ferromagnetic material that does not move relative to the coil. The core is used solely to create a magnetic field that attracts or repels other parts that have appropriate inherent magnetic properties. Where a center component moves in response to the magnetic force created by current through a coil, this is discussed in the **solenoid** entry. By comparison, the **inductor** entry describes a coil that may or may not have a ferromagnetic core, and is used for the specific purpose of creating reactance, or self-inductance, in an electronic circuit, often in association with alternating current and in combination with resistors and/or capacitors. The **inductor** entry contains a basic discussion and explanation of magnetic force created by electricity.

OTHER RELATED COMPONENTS

- **solenoid** (See Chapter 21)
- **electromagnet** (See Chapter 20)
- **DC motor** (See Chapter 22)
- **AC motor** (See Chapter 23)

What It Does

An electromagnet consists of a coil that creates a magnetic field in response to an electric current. The field is channeled and reinforced by a core of magnetic material (that is, material that can be magnetized). Electromagnets are incorporated in motors, generators, loudspeakers, microphones, and industrial-sized applications such as mag-lev trains. On their own, they provide a means for electric current to hold, lift, or move objects in which a magnetic field can be induced.

A very small, basic electromagnet about 1 inch in diameter is shown in Figure 20-1. No specific schematic symbol for an electromagnet exists, and the symbol for an induction coil with a solid core is often used instead, as shown in Figure 14-1 (the center variant of each of the three) in the **inductor** entry of this encyclopedia.

Figure 20-1. *An electromagnet approximately 1 inch in diameter, rated to draw 0.25A at 12VDC.*

How It Works

Electric current flowing through a circle of wire (or a series of connected loops that form a helix or *coil*) will induce a magnetic field through the center. This is illustrated in the **inductor** entry of this encyclopedia, specifically in diagrams Figure 14-3, Figure 14-4, Figure 14-5, and Figure 14-6.

If a stationary piece of ferromagnetic material is placed in the center of the circle or coil, it enhances the magnetic force because the *reluctance* (magnetic resistance) of the material is much lower than the reluctance of air. The combination of the coil and the core is an electromagnet. This is illustrated in Figure 20-2. For a lengthier discussion of this effect, see "Magnetic Core" on page 116.

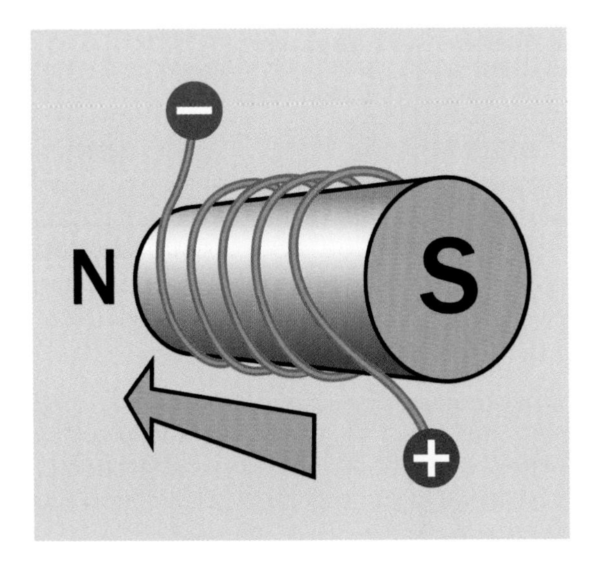

Figure 20-2. *Direct conventional current flowing through a wire coiled around a ferromagnetic rod induces a magnetic force in the rod, conventionally considered to flow from south to north.*

The magnitude of the electromagnetic *flux density* will be proportional to the current flowing through the coil, assuming a DC power source.

Variants

Electromagnet designs vary according to their application. The simplest design consists of a single coil wound around a rod which may terminate in a plate for applications such as lifting scrap metal. This design is relatively inefficient because the magnetic circuit is completed through air surrounding the electromagnet.

A more efficient, traditional design consists of a U-shaped core around which are wound one or two coils. If the U-shaped core is smoothly curved, it resembles a *horse-shoe magnet*, as shown in Figure 20-3. This design has become relatively uncommon, as it is cheaper to make windings across two separate, straight vertical cores and bridge them. However, the horseshoe configuration is extremely efficient, as the coils induce north and south magnetic polarities in the open ends of the U-shaped core, and the magnetic circuit is completed through any object that is attracted toward the open ends and links them. The attracted object is shown as a rectangular plate in Figure 20-3. Because a magnetic circuit will naturally attempt to limit its extent, and because this goal will be achieved when the circuit is completed, the attractive force of the U-shaped magnet is maximized.

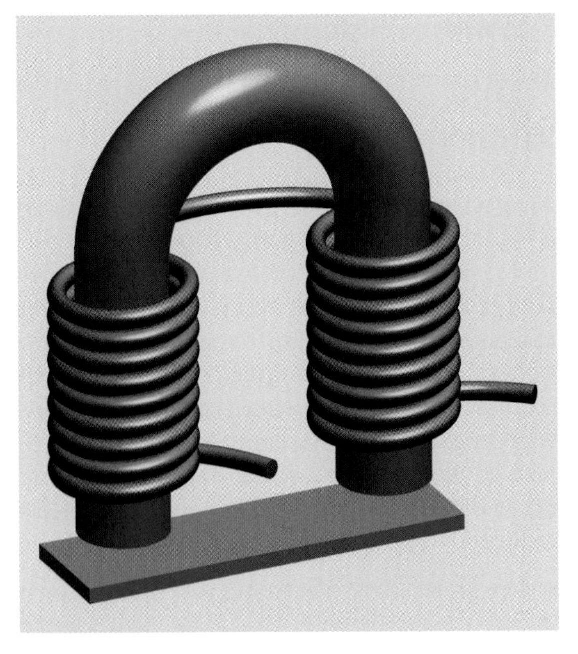

Figure 20-3. *This traditional design for an electromagnet has a pedigree stretching back for more than a century. It maximizes efficiency by completing a magnetic circuit through any object that the magnet attracts.*

An electromagnet powered by direct current naturally produces a consistently polarized, stable magnetic field. When AC current is applied, an electromagnet may still be used to exert an attractive force on a passive object that is not magnetized but is capable of being magnetized. The electromagnet will change its polarity at almost the same frequency as the AC, and will induce equal and opposite fluctuating polarity in the target, causing mutual attraction. The core of the magnet will be composed of plates separated by thin layers of insulation to inhibit the eddy currents induced by the AC, but still an AC-powered electromagnet will be less efficient than a comparable DC-powered electromagnet because it will also suffer from *hysteresis* as power is consumed by repeatedly reversing the polarity of the *magnetic domains* in the core.

Some electromagnets that are described as suitable for AC power actually contain *rectifiers* that convert the AC to DC.

Values

Electromagnets are typically calibrated in terms of their power consumption and retaining force (the weight of an iron target that they can support). The retaining force is usually measured in grams or kilograms.

How to Use it

Electromagnets are used mostly as subassemblies in other components, such as motors and generators, relays, loudspeakers, and disk drives. They have also been used in audio (and video) tape recorders to magnetize ferric oxide on tape, using a magnetic field of varying strength to record an audio signal. In this application, a form of horseshoe magnet with an extremely narrow gap is used, the width of the gap determining the highest frequency that the electromagnet can record, in conjunction with the speed of the tape moving past the head.

The tape recording process can be reversed when the electromagnet "reads" the tape and turns the signal back into a weak alternating current that can be amplified and reproduced through a loudspeaker.

A simple application for an electromagnet is in a traditional-style doorbell, where one or two coils attract a spring-loaded lever, at the top of which is a knob that hits a bell. When the lever is pulled toward the bell, it breaks a contact that supplies power to the electromagnet. This allows the lever to spring back to its original position, which reestablishes the circuit, repeating the process for as long as power is applied to the bell. The bulk and weight of the component parts in this type of doorbell are making it obsolete, as electronic versions containing small loudspeakers become relatively cheaper. However, a **solenoid** may still be used in the type of bell that creates a single chime or pair of chimes.

In any device using a *cathode-ray tube*, electromagnetic coils are used to form a *yoke* around the neck of the tube, to deflect the beam of electrons on its way to the screen. A similar principle is used in electron microscopes. In some cases, electrostatically charged plates are used to achieve the same purpose.

An electromagnet may be used to activate a *reed switch* (the diagram in Figure 9-7 shows such a switch). In this application, the combination of the electromagnet and the switch are functioning as a **relay**.

When an electromagnet is energized by alternating current, it can be used to *degauss* (in other words, to demagnetize) other objects. The AC is either applied with diminishing current, so that the alternating magnetic polarities gradually subside to zero, or the electromagnet is gradually moved away from the target, again reducing the magnetic influence to (virtually) zero. This latter procedure may be used periodically to demagnetize record and replay heads on tape recorders, which otherwise tend to acquire residual magnetism, inducing background hiss on the tape.

Traditional large-scale applications for electromagnets tend to involve lifting and moving heavy objects or scrap metal, such as junked cars. A more modern application is in magnetic resonance imaging (MRI), which has revolutionized some areas of medicine.

Very large-scale applications for electromagnets include particle accelerators, in which multiple magnetic coils are energized sequentially, and fusion-power generators, where high-temperature plasma is contained by a magnetic field.

What Can Go Wrong

Because an electromagnet requires constant power to maintain its magnetic force, yet it is not doing any actual work so long as its target remains stationary (in contact with the core of the magnet), the current running through the coil of the magnet must be dissipated entirely as heat. Further discussion of this issue will be found at "Heat" on page 171 in the **solenoid** section of this encyclopedia.

solenoid

21

The term **solenoid** was historically used to describe any coil without a magnetic core. More recently and more commonly it describes a coil inside of which a cylindrical plunger moves in response to the magnetic field generated by the coil. In this encyclopedia, the term **electromagnet** has its own entry, and describes a coil with a center component of ferromagnetic material that does not move relative to the coil. It is used solely to attract or repel other parts that have inherent magnetic properties. By comparison, the **inductor** entry describes a coil that is used for the specific purpose of creating reactance, or self-inductance, in an electronic circuit, often in association with alternating current and in combination with resistors and/or capacitors. The **inductor** entry contains a basic discussion and explanation of magnetic force created by electricity.

OTHER RELATED COMPONENTS

- **inductor** (See Chapter 14)
- **solenoid** (See Chapter 21)

What It Does

A typical solenoid consists of a hollow coil inside a *frame*, which may be a sealed cylinder or box-shaped with open sides. In the case of a cylinder, its opposite ends may be referred to as *pole faces*.

At least one of the pole faces has a hole through which a *plunger* (also known as an *armature*) is pulled or pushed by the solenoid. Thus, the solenoid is a device for applying a linear mechanical force in response to current passing through it. In most solenoids, current must be maintained in order to maintain the mechanical force.

A small open-frame solenoid is pictured in Figure 21-1. The upper section of the figure shows the three basic parts: frame, compression spring, and plunger. The lower part of the figure shows the parts assembled.

A larger, closed, cynlindrical solenoid is shown in Figure 21-2, with the plunger and spring removed.

A 3D rendering showing a simplified, imaginary, cylindrical solenoid cut in half appears in Figure 21-3. The diagram includes a gray cylindrical shell, often described as the frame; the coil, shown in orange; the plunger, which is pulled into the coil by its magnetic field; and the triangular stop, which limits the plunger's upward travel. The frame of the solenoid exists not merely to protect the coil, but to provide a magnetic circuit, which is completed through the plunger.

The lower end of the plunger is often fitted with a nonmagnetic yoke or perforated plate for connection with other components. Stainless steel can be used for this purpose. The stop may be fitted with a thrust rod (also fabricated from stainless steel) if the solenoid is intended to "push" as well as "pull." Springs to adjust the force of the plunger, or to return it to its initial position when the current through the coil is interrupted, are not shown in the rendering.

Figure 21-1. *A small 12VDC solenoid.*

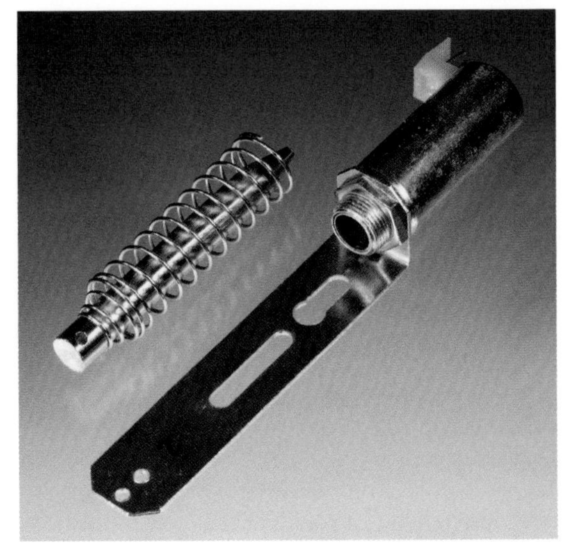

Figure 21-2. *A larger solenoid rated for 24VDC.*

Because there is no standardized schematic symbol for a solenoid, and because this type of component is so widely used in conjunction with valves, any diagram involving solenoids is more likely to emphasize fluid or gas flow with symbols that have been developed for that purpose. In such circuits, a solenoid may be represented simply by a rectangle. However, the symbols shown in Figure 21-4 may occasionally be found.

How It Works

Current flowing through the coil creates a magnetic force. This is explained in the **inductor** entry of this encyclopedia, using diagrams in Figure 14-3, Figure 14-4, Figure 14-5, and Figure 14-6.

If the plunger is fabricated from a material such as soft iron, the coil will induce an equal and opposite magnetic polarity in the plunger. Consequently the plunger will attempt to occupy a position inside the coil where the ends of the plunger are equal distances from the ends of the coil. If a collar is added to the free end of the plunger, this can increase the pulling force on the plunger when it is near the end of its throw because of the additional magnetic pull distributed between the collar and the frame of the solenoid.

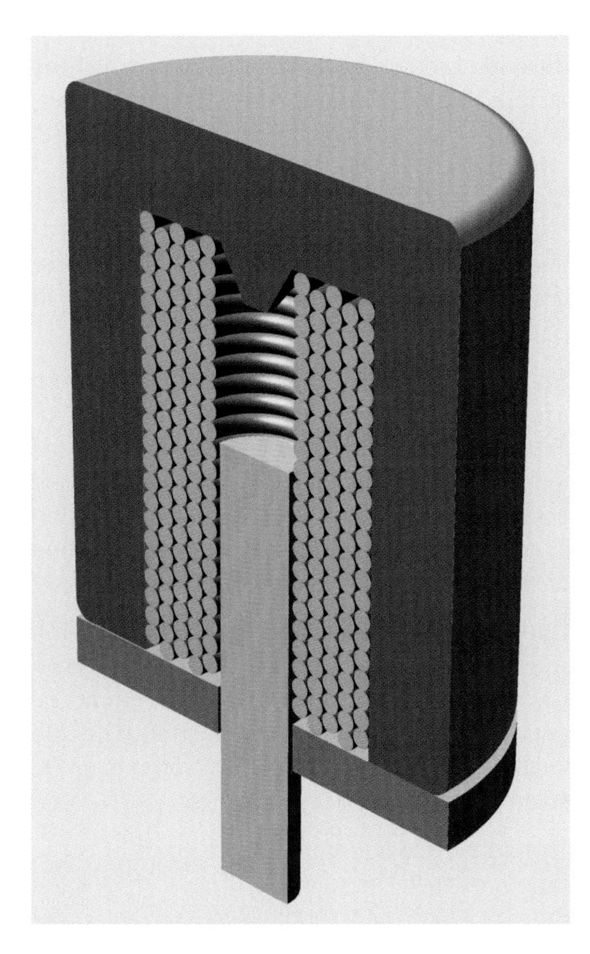

Figure 21-3. *A simplified view of a solenoid cut in half, showing the primary parts.*

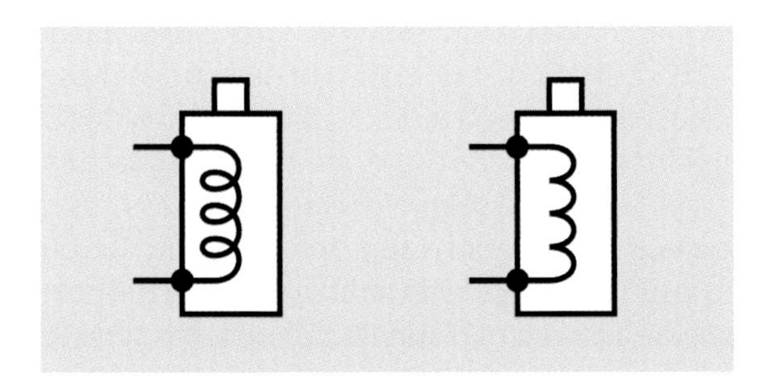

Figure 21-4. *Although no standard schematic symbol for a solenoid exists, these symbols may sometimes be found.*

A spring can be inserted to apply some resistive force to compensate for the increase in pulling force that occurs as a larger proportion of the plunger enters the coil. A spring may also be used to eject the plunger, partially at least, when current to the coil is interrupted.

If the plunger is a permanent magnet, reversing DC current to the coil will reverse the action of the plunger.

A solenoid with a nonmagnetized plunger may be energized by AC current, since polarity reversals in the magnetic field generated by the coil will induce equal and opposite reversals in the polarity of the plunger. However, the force curve of an AC-powered solenoid will be different from the force curve of a DC-powered solenoid. See Figure 21-5. The alternating current is likely to induce humming, buzzing, and vibration.

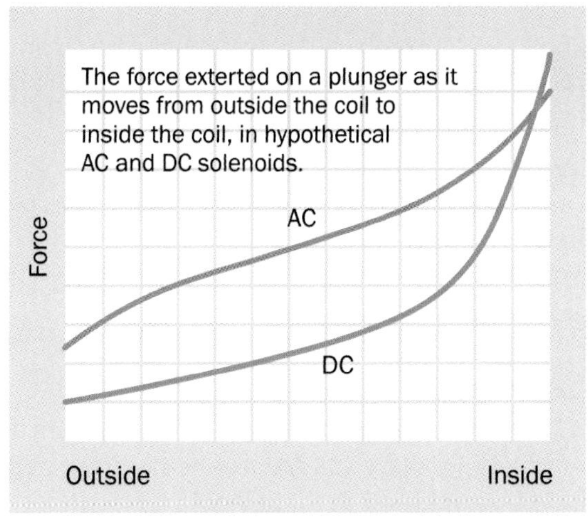

Figure 21-5. *A comparison of the force exerted on a plunger, relative to its position as it enters the coil, in hypothetical AC and DC solenoids.*

The frame of the solenoid increases the magnetic power that the coil can exert by providing a magnetic circuit of much lower *reluctance* than that of air (reluctance being the magnetic equivalent of electrical resistance). For a lengthier discussion of this effect, see "Magnetic Core" on page 116 in the **inductor** entry of this encyclopedia. If current flowing through the coil increases to the point where the frame becomes magnetically *saturated*, the pulling power of the solenoid will level off abruptly.

The heat generated by a solenoid when it is maintained in its energized state may be reduced if the manufacturer includes a series resistor and a switch that functions as a *bypass switch*. The switch is normally closed, but is opened mechanically when the plunger reaches the end of its throw, thus diverting electricity through the series resistor. This itself will generate some heat as a result of the current flowing through it, but by increasing the total resistance of the system, the total heat output will be reduced. The resistor value is chosen to provide the minimum power needed to retain the plunger at the end of its throw.

Variants

The most common variant is tubular, with open-frame as a secondary option. A tubular solenoid has been shown in Figure 21-2.

Additional variants include:

Low Profile

A shorter, fatter solenoid which may be used if a short throw is acceptable.

Latching

A permanent magnet holds the plunger when it reaches the end of its travel, and continues to hold it after power to the solenoid is disconnected. The plunger itself is also a permanent magnet, and is released by running current of reverse polarity through the coil.

Rotary

This variant is similar in principle to a brushless DC motor and causes the armature to rotate through a fixed angle (typically ranging from 25 to 90 degrees) instead of moving linearly. It is used as a mechanical indicator in control panels, although it is being displaced by purely electronic indicators.

Hinged Clapper

Instead of a plunger, a small hinged panel (the "clapper") moves in when the solenoid is active, and springs back when the power is interrupted.

Values

The stroke length, duty cycle, and holding force are the most significant values found in solenoid datasheets.

Holding forces for DC solenoids can range from a few grams to hundreds of kilograms. The holding force will be inversely proportional to the length of the solenoid, if all other variables are equal. The force that the solenoid can exert on its plunger also varies depending on the position of the plunger in the length of its throw.

Duty cycle is of special importance because the solenoid continues to draw power and create heat so long as it is holding the plunger at the end of its throw (assuming the solenoid is not the latching type). The initial current surge in an AC solenoid generates additional heat.

The duty cycle is simply calculated. If T1 is the time for which the solenoid is on and T2 is the time for which the solenoid is off, the duty cycle, D, is derived as a percentage from the formula

```
D = 100 * (T1 / (T1 + T2))
```

Some solenoids are designed to withstand a 100% duty cycle, but many are not, and in those cases, there is a maximum value not only for D but for the peak "on" time, regardless of the duty cycle. Suppose a solenoid is rated for a 25% duty cycle. If the solenoid is appropriately switched on for one second and off for three seconds, the heat will be allowed to dissipate before it has time to reach overload levels. If the solenoid is switched on for one minute and off for three minutes, the duty cycle is still 25%, but the heat that may accumulate during a one-minute "on" cycle may overload the component before the "off" cycle can allow it to dissipate.

Coil Size vs. Power

Because additional windings in a coil will induce a greater magnetic force, a larger solenoid tends to be more powerful than a smaller solenoid. However this means that if a larger and a smaller solenoid are both designed to generate the same force over the same distance, the smaller solenoid will probably draw more current (and will therefore generate more heat) because of its fewer coil windings.

How to Use it

Solenoids are primarily used to operate valves in fluid and gas circuits. Such circuits are found in laboratory and industrial process control, fuel injectors, aircraft systems, military applications, medical devices, and space vehicles. Solenoids may also be used in some electronic locks, in pinball machines, and in robotics.

What Can Go Wrong

Heat

Overheating is the principal concern when using solenoids, especially if the maximum "on" time is exceeded, or the duty cycle is exceeded. If the plunger is prevented from reaching the end of its throw, this can be another cause of overheating.

Because coil resistance increases with heat, a hot solenoid passes less current and therefore develops less power. This effect is more pronounced in a DC solenoid than an AC solenoid.

A manufacturer's force curve should show the solenoid performance at its maximum rated temperature, which is typically around 75 degrees Centigrade, in a hypothetical ambient temperature of 25 degrees Centigrade. Exceeding these values may result in the solenoid failing to perform. As in all coils using magnet wire, there is the risk of excessive heat melting the insulation separating the coil windings, effectively shortening the coil, which will then pass more current, generating more heat.

AC Inrush

When an AC solenoid reaches the end of its travel, the sudden stop of the plunger results in forward *EMF* that generates additional heat. Generally speaking, a longer stroke creates a greater surge. Rapid cycling will therefore exacerbate coil heating.

Unwanted EMF

Like any device containing a coil, a solenoid creates back EMF when power is connected, and forward EMF when the power is disconnected. A protection diode may be necessary to suppress power spikes that can affect other components.

Loose Plunger

The plunger in many solenoids is not anchored or retained inside the frame and may fall out if the solenoid is tilted or subjected to extreme vibration.

DC motor

22

In this section, the term "traditional DC motor" is used to describe the oldest and simplest design, which consists of two *brushes* delivering power via a rotating, sectioned *commutator* to two or more electromagnetic coils mounted on the motor shaft. *Brushless* DC motors (in which DC is actually converted to a pulse train) are also described here because "brushless DC" has become a commonly used phrase, and the motor is powered by direct current, even though this is modified internally via *pulse-width modulation.*

OTHER RELATED COMPONENTS

- **AC motor** (See Chapter 23)
- **stepper motor** (See Chapter 25)
- **servo motor** (See Chapter 24)

What It Does

A traditional DC motor uses direct current to create magnetic force, which turns an output shaft. When the polarity of the DC voltage is reversed, the motor reverses its direction of rotation. Usually, the force created by the motor is equal in either direction.

How It Works

Current passes through two or more coils that are mounted on the motor shaft and rotate with it. This assembly is referred to as the *rotor*. The magnetic force produced by the current is concentrated via cores or poles of soft iron or high-silicon steel, and interacts with fields created by permanent magnets arrayed around the rotor in a fixed assembly known as the *stator*.

Power to the coils is delivered through a pair of *brushes*, often made from a graphite compound. Springs press the brushes against a sleeve that rotates with the shaft and is divided into sections, connected with the coils. The sleeve assembly is known as the *commutator*. As the commutator rotates, its sections apply power from the brushes to the motor coils sequentially, in a simple mechanical switching action.

The most elementary configuration for a traditional DC motor is shown in Figure 22-1.

In reality, small DC motors typically have three or more coils in the rotor, to provide smoother operation. The operation of a three-coil motor is shown in Figure 22-2. The three panels in this figure should be seen as successive snapshots of one motor in which the rotor turns progressively counter-clockwise. The brushes are colored red and blue to indicate positive and negative voltage supply, respectively. The coils are wired in series, with power being applied through the commutator to points between each pair of coils. The direction of current through each coil determines its magnetic polarity, shown as N for north or S for south. When two coils are energized in series without any power applied to their midpoint, each develops a smaller magnetic field than an individually energized coil. This is indicated in the diagram with a smaller white lowercase n and s. When two ends of a coil are at

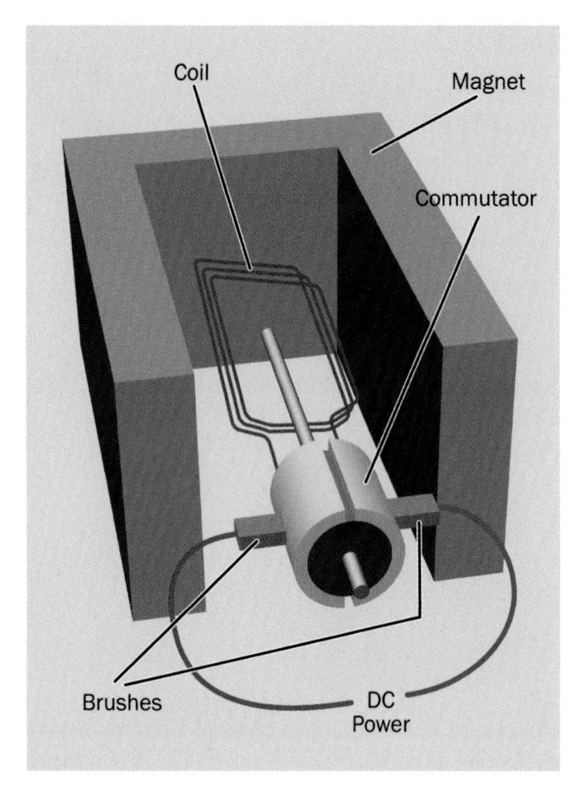

Figure 22-1. *The simplest traditional DC motor contains these parts. The combination of coil, shaft, and commutator is the rotor. The fixed magnetic structure in which it rotates is the stator.*

equal potential, the coil produces no magnetic field at all.

The stator consists of a cylindrical permanent magnet, which has two poles—shown in the figure as two black semicircles separated by a vertical gap for clarity—although in practice the magnet may be made in one piece. Opposite magnetic poles on the rotor and stator attract each other, whereas the same magnetic poles repel each other.

DC motors may be quite compact, as shown in Figure 22-3, where the frame of the motor measures about 0.7" square. They can also be very powerful for their size; the motor that is shown disassembled in Figure 22-4 is from a 12VDC bilge pump rated at 500 gallons per hour. Its output was delivered by the small impeller attached to the rotor at right, and was achieved by using

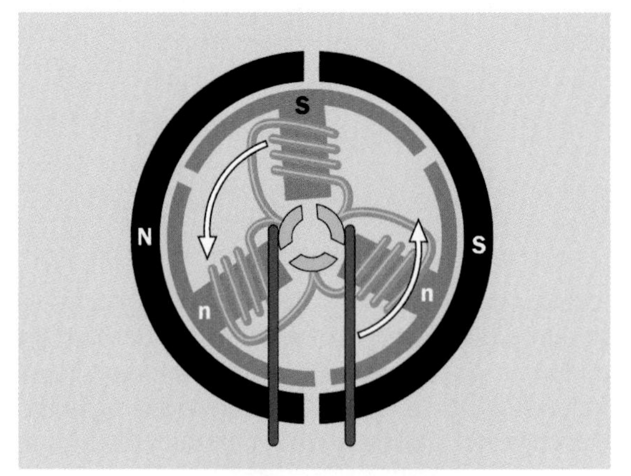

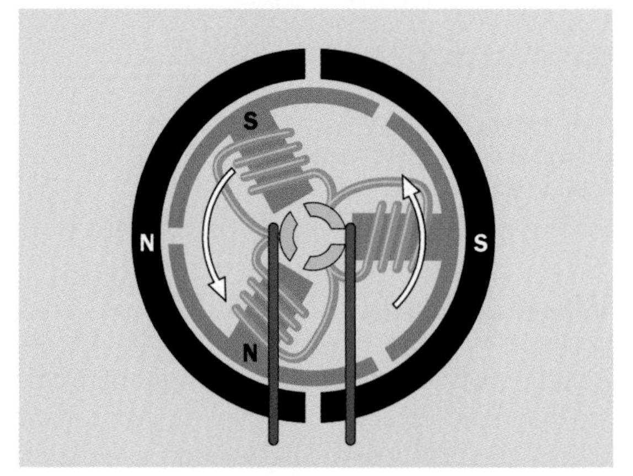

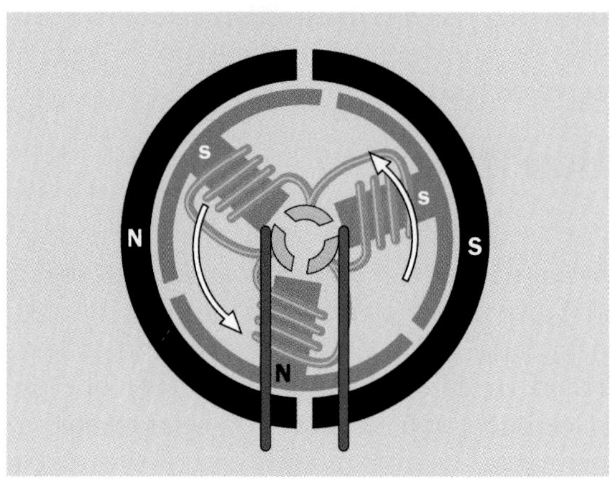

Figure 22-2. *Three sequential views of a typical three-coil DC motor viewed from the end of its shaft (the shaft itself is not shown). Magnetic effects cause the rotor to turn, which switches the current to the coils via the commutator at the center.*

two extremely powerful neodymium magnets, just visible on the inside of the motor's casing (at top-left) in conjunction with five coils on the rotor.

Figure 22-3. *A miniature 1.5VDC motor measuring about 0.7" square.*

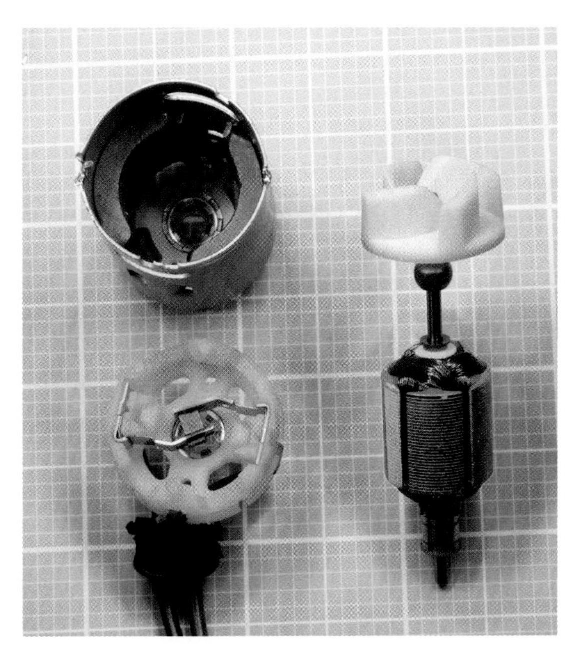

Figure 22-4. *A traditional DC motor removed from its cylindrical casing. The brushes of the motor are attached to the white plastic end piece at bottom-left. Large squares on the graph paper in the background are 1" × 1", divided in 0.1" increments. The motor was used in a small bilge pump.*

Variants

Coil Configurations

The series connection of coils used in Figure 22-2 is known as the *delta configuration*. The alternative is the *wye configuration* (or *Y configuration*, or *star configuration*). Simplified schematics are shown in Figure 22-5. Generally speaking, the delta configuration is best suited to high-speed applications, but provides relatively low torque at low speed. The wye configuration provides higher torque at low speed, but its top speed is limited.

Gearhead Motor

A *gearhead motor* (also often known as a *gear motor*) incorporates a set of reduction gears that increase the torque available from the output shaft while reducing its speed of rotation. This is often desirable as an efficient speed for a traditional DC motor may range from 3,000 to 8,000 RPM, which is too fast for most applications. The gears and the motor are often contained in a single sealed cylindrical package. Two examples are shown in Figure 22-6. A disassembled motor, revealing half of its gear train under the cap and the other half still attached to a separate circular plate, appears in Figure 22-7. When the motor is assembled, the gears engage. As in the case of the bilge-pump motor, the stator magnets are mounted inside the cylindrical casing. Note that the brushes, inside the circular plate of white plastic, have a resistor and capacitor wired to suppress voltage spikes.

Spur gears are widely used for speed reduction. *Planetary gears* (also known as *epicyclic gears*) are a slightly more expensive option. Spur gears such as those in Figure 22-8 may require three or more pairs in series. The total speed reduction is found by multiplying the individual ratios. Thus, if three pairs of gears have ratios of 37 : 13, 31 : 15, and 39 : 17, the total speed reduction ® is obtained by:

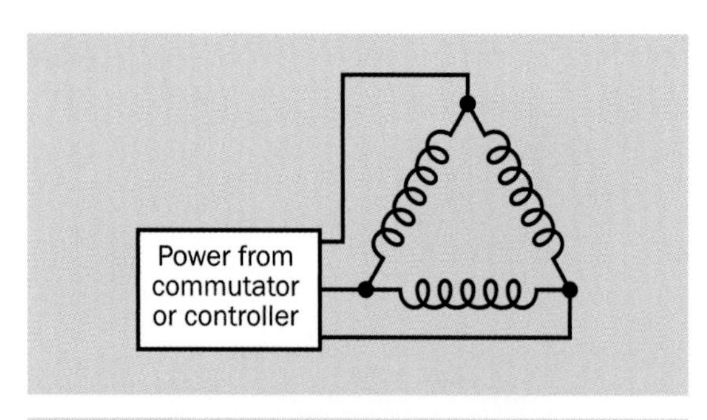

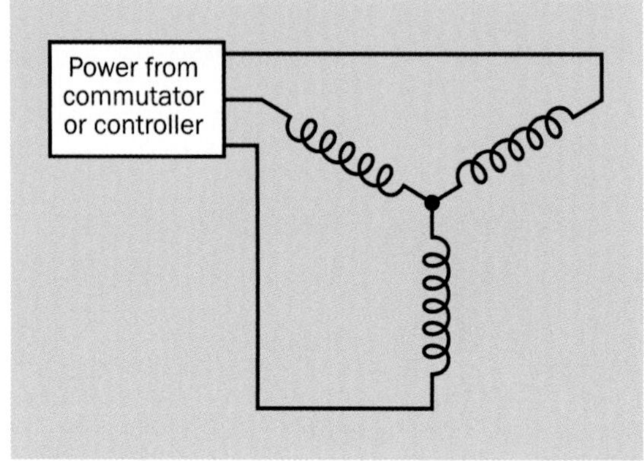

Figure 22-5. *Coils on the rotor of a traditional DC motor may be connected in delta configuration (top) or wye configuration (bottom).*

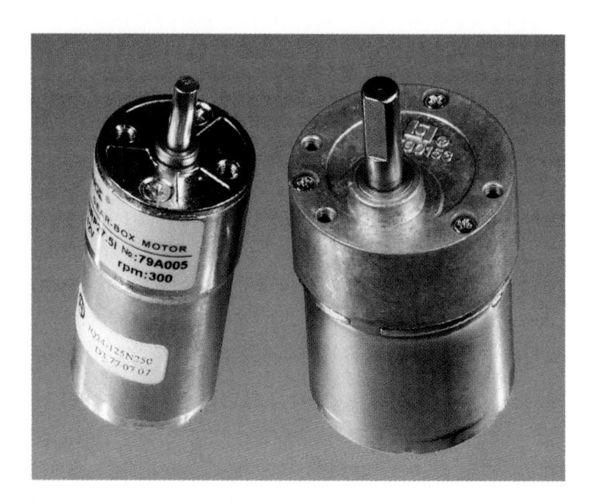

Figure 22-6. *Two typical small gearhead motors.*

```
R = ( 37 * 31 * 39 ) / ( 13 * 15 * 17 )
```

Figure 22-7. *Spur gears from a gearhead motor provide speed reduction and increased torque.*

Therefore:

```
R = 44733 / 3315 = about 13.5 : 1
```

Datasheets almost always express R as an integer. For example, the gear train shown in Figure 22-7 is rated by the manufacturer as having an overall reduction of 50:1. In reality, the reduction can be expected to have a fractional component. This is because if two gears have an integer ratio, their operating life will be shortened, as a manufacturing defect in a tooth in the smaller gear will hit the same spots in the larger gear each time it rotates. For this reason, the numbers of teeth in two spur gears usually do not have any common factors (as in the example above), and if a motor rotates at 500 RPM, a gear ratio stated as 50:1 is very unlikely to produce an output of exactly 10 RPM. Since traditional DC motors are seldom used for applications requiring high precision, this is not usually a significant issue, but it should be kept in mind.

Figure 22-8. *A pair of spur gears.*

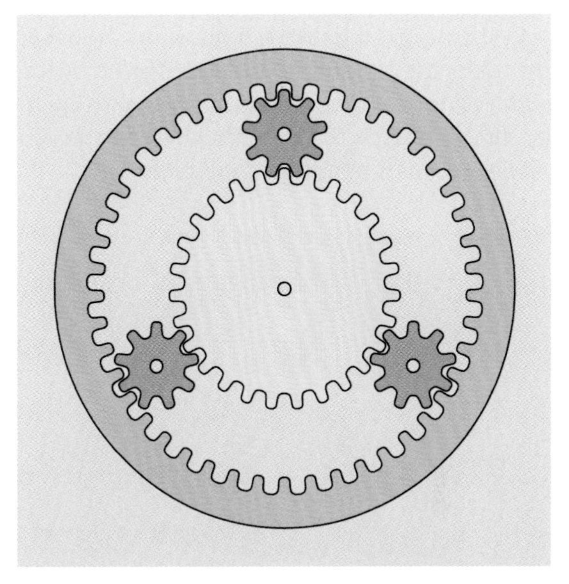

Figure 22-9. *Planetary gears, also known as epicyclic gears, share the torque from a motor among more teeth than simple spur gears.*

Figure 22-9 shows *planetary gears*, also known as *epicyclic gears*. The outer ring gear is properly referred to as the *annulus*, while the *sun gear* is at the center, and the intermediate *planet gears* may be mounted on a *carrier*. The greatest speed reduction will be achieved by driving the sun gear while the annulus is kept in a stationary position and the output is taken from the carrier of the planet gears. If A is the number of teeth in the annulus and S is the number of teeth in the sun gear, the total speed reduction, R, is given by the following formula:

```
R = (S + A) / S
```

Note that in this drive configuration, the number of teeth in each planet gear is irrelevant to the speed reduction. In Figure 22-9, the sun gear has 27 teeth whereas the annulus has 45 teeth. Therefore, the reduction is found by:

```
R = (27 + 45) / 27 = about 2.7 : 1
```

Successive reductions can be achieved by stacking planetary gear sets, using the carrier of one set to drive the sun gear in the next set.

Planetary gears are used primarily if a motor drives a heavy load, as the force is divided among more gear pairs, reducing wear and tear on gear teeth and minimizing the breakdown of lubrication. A planetary gear train may also be more compact than a train of spur gears. These advantages must be evaluated against the higher price and slightly increased friction resulting from the larger number of gears interacting with each other.

Brushless DC Motor

In a *brushless* DC motor, sometimes referred to as a *BLDC motor*, the coils are located in the stator and the permanent magnets are relocated in the rotor. The great advantage of this design is that power can be applied directly to the coils, eliminating the need for brushes, which are the primary source of failure in DC motors as a result of wear and tear. However, since there is no rotating commutator to switch the DC current to the coils, the current must be switched by electronic components, which add to the cost of the motor.

In the *inrunner* configuration the stator surrounds the rotor, whereas in the *outrunner* configuration the stator is located in the center of the motor while the rotor takes the form of a ring or cup that spins around the stator. This is a common design for small cooling fans, where the blades are attached to the outer circumference of a cup that is lined with permanent magnets. An example is shown in Figure 22-10. In this picture, the stator coils are normally hidden from view, being fixed to the fan housing (shown at the top of the picture). Power is controlled by the surface-mount components on the green circular circuit board. The cup attached to the fan blades contains permanent magnets.

Figure 22-10. *A typical brushless DC cooling fan uses stationary coils, with permanent magnets rotating around them.*

The use of a solid-state switching system to energize the coils sequentially is known as *electronic commutation*. *Hall effect* sensors may be used to detect the position of the rotor and feed this information back to the frequency control circuit, so that it stays "one step ahead" of the rotor (when bringing it up to speed) or is synchronized with the rotor (for a constant running speed). The system is comparable to a *reluctance motor* or *synchronous motor*. These variants are described in the **AC motor** section of this encyclopedia.

While traditional DC motors have been commercially available since the late 1800s, brushless DC motors were not introduced until the 1960s, when the availability of solid-state control electronics began to make the motor design economically viable.

Linear Actuator

Linear actuator is a generic term for any device that can exert a pushing or pulling force in a straight line. In industrial applications, actuators may be powered pneumatically or hydraulically, but smaller-scale units are usually driven by a traditional DC motor. These are more properly (but not often) referred to as *electromechanical linear actuators*.

The rotational force of the motor is typically converted to linear motion by using a threaded motor shaft in conjunction with a nut or collar. The unit is often mounted in an enclosure containing *limit switches* that stop the motor automatically at the limits of travel. For an explanation of limit switches, see "Limit Switches" on page 46 in the **switch** entry in this encyclopedia.

Values

A manufacturer's datasheet should list the maximum operating voltage and typical current consumption when a motor is moderately loaded, along with the *stall current* that a motor draws when it is so heavily loaded that it stops turning. If stall current is not listed, it can be determined empirically by inserting an ammeter (or multimeter set to measure amperes) in series with the motor and applying a braking force until the motor stops. Motors should generally be protected with slow-blowing **fuses** to allow for the power fluctuations that occur when the motor starts running or experiences a change in load.

In addition, the *torque* that a motor can deliver should be specified. In the United States, torque is often expressed in *pound-feet* (or *ounce-inches* for smaller motors). Torque can be visualized by imagining an arm pivoted at one end with a weight hung on the other end. The torque exerted at the pivot is found by multiplying the weight by the length of the arm.

In the metric system, torque can be expressed as gram-centimeters, Newton-meters, or dyne-meters. A Newton is 100,000 dynes. A dyne is defined as the force required to accelerate a mass of 1 gram, increasing its velocity by 1 centimeter per second each second. 1 Newton-meter is equivalent to approximately 0.738 pound-feet.

The speed of a traditional DC motor can be adjusted by varying the voltage to it. However, if the voltage drops below 50% of the rated value, the motor may simply stop.

The power delivered by a motor is defined as its speed multiplied by its torque at that speed. The greatest power will be delivered when the motor is running at half its *unloaded speed* while delivering half the *stall torque*. However, running a motor under these conditions will usually create unacceptable amounts of heat, and will shorten its life.

Small DC motors should be run at 70% to 90% of their unloaded speed, and at 10% to 30% of the stall torque. This is also the range at which the motor is most efficient.

Ideally, DC motors that are used with reduction gearing should be driven with less than their rated voltage. This will prolong the life of the motor.

When choosing a motor, it is also important to consider the *axial loading* (the weight or force that will be imposed along the axis or shaft of the motor) and *radial loading* (the weight or force that will be imposed perpendicularly to the axis). Maximum values should be found in motor datasheets.

In the hobby field, motors for model aircraft are typically rated in watts-per-pound of motor weight (abbreviated w/lb). Values range from 50 to 250 w/lb, with higher values enabling better performance.

Relationships between torque, speed, voltage, and amperage in a traditional DC motor can be described easily, assuming a hypothetical motor that is 100% efficient:

If the amperage is constant, the torque will also be constant, regardless of the motor speed.

If the load applied to the motor remains constant (thus forcing the motor to apply a constant torque), the speed of the motor will be determined by the voltage applied to it.

If the voltage to the motor remains constant, the torque will be inversely proportional with the speed.

How to Use it

A traditional DC motor has the advantages of cheapness and simplicity, but is only suitable for intermittent use, as its brushes and commutator will tend to limit its lifetime. Its running speed will be approximate, making it unsuitable for precise applications.

As the cost of control electronics has diminished, brushless DC motors have replaced traditional DC motors. Their longevity and controllability provide obvious advantages in applications such as hard disk drives, variable-speed computer fans, CD players, and some workshop tools. Their wide variety of available sizes, and good power-to-weight ratio, have encouraged their adoption in toys and small vehicles, ranging from remote-controlled model cars, airplanes, and helicopters to personal transportation devices such as the Segway. They are also used in direct-drive audio turntables.

Where an application requires the rotation of a motor shaft to be converted to linear motion, a prepackaged *linear actuator* is usually more reliable and simpler than building a crank and connecting rod, or cam follower, from scratch. Large linear actuators are used in industrial automation, while smaller units are popular with robotics hobbyists and can also be used to control small systems in the home, such as a remote-controlled access door to a home entertainment center.

Speed Control

A *rheostat* or **potentiometer** may be placed in series with a traditional DC motor to adjust its speed, but will be inefficient, as it will achieve a voltage drop by generating heat. Any rheostat must be rated appropriately, and should probably be wire-wound. The voltage drop between the wiper and the input terminal of the rheostat should be measured under a variety of operating conditions, along with the amperage in the circuit, to verify that the wattage rating is appropriate.

Pulse-width modulation (*PWM*) is preferable as a means of speed control for a traditional DC motor. A circuit that serves this purpose is sometimes referred to as a *chopper*, as it chops a steady flow of current into discrete pulses. Usually the pulses have constant frequency while varying in duration. The pulse width determines the average delivered power, and the frequency is sufficiently high that it does not affect smoothness of operation of the motor.

A **programmable unijunction transistor** or PUT can be used to generate a train of pulses, adjustable with a potentiometer attached to its emitter. Output from the transistor goes to a *silicon-controlled rectifier* (*SCR*), which is placed in series with the motor, or can be connected directly to the motor if the motor is small. See Figure 27-7.

Alternatively, a *555 timer* can be used to create the pulse train, controlling a *MOSFET* in series with the motor.

A **microcontroller** can also be used as a pulse source. Many microcontrollers have PWM capability built in. The microcontroller will require its own regulated power supply (typically 5VDC, 3.3VDC, or sometimes less) and a switching component such as an *insulated-gate bipolar transistor* (*IGBT*) to deliver sufficient power to the motor and to handle the flyback voltage. These components will all add to the cost of the system, but many modern devices incorporate microcontrollers anyway, merely to process user input. Another advantage of using a microcontroller is that its output can be varied by rewriting software, for example if a motor is replaced with a new version that has different characteristics, or if requirements change for other reasons. Additionally, a microcontroller enables sophisticated features such as pre-programmed speed sequences, stored memory of user preferences, and/or responses to conditions such as excessive current consumption or heat in the motor.

A PWM schematic using a microcontroller and IGBT is shown in Figure 22-11.

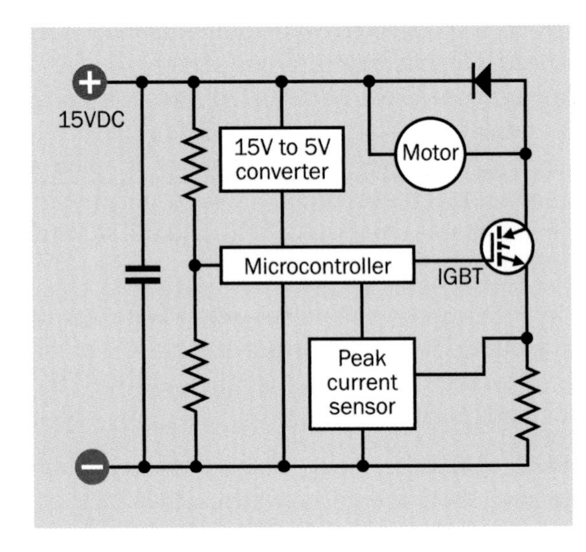

Figure 22-11. *A sample schematic for control of a DC motor via pulse-width modulation, using a microcontroller and an insulated-gate bipolar transistor.*

Direction Control

The *H bridge* is a very early system for reversing the direction of a DC motor simply by swapping the polarity of its power supply. This is shown in Figure 22-12. The switches diagonally opposite each other are closed, leaving the other two switches open; and then to reverse the motor, the switch states are reversed. This is obviously a primitive scheme, but the term "H bridge" is still used when prepackaged in a single chip such as the LMD18200 H bridge motor controller from National Semiconductor.

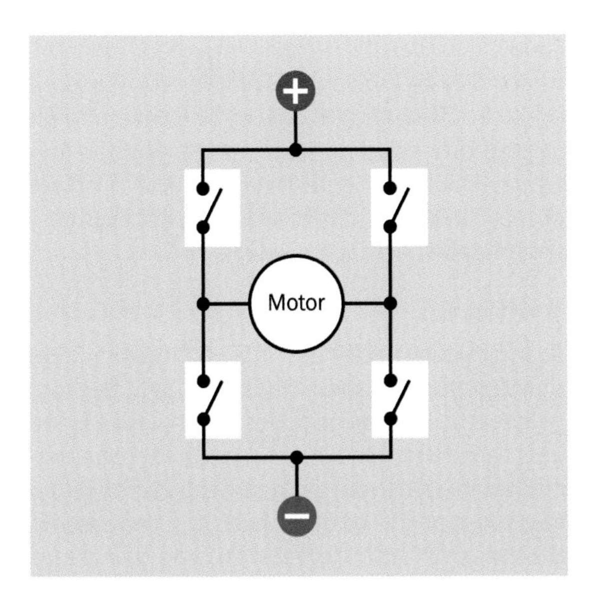

Figure 22-12. *A DC motor can be reversed by this very basic circuit, known as an H bridge, by opening and closing pairs of switches that are diagonally opposite each other.*

A double-throw, double-pole **switch** or **relay** can achieve the same purpose, as shown in Figure 22-13.

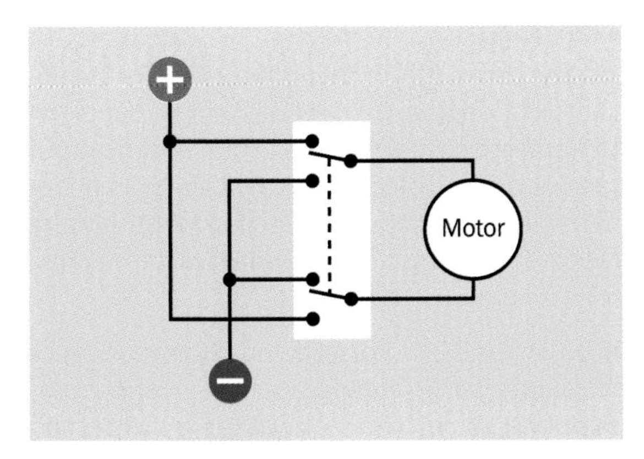

Figure 22-13. *A DPDT switch or relay can reverse the direction of a traditional DC motor simply by swapping the polarity of the power supply.*

Limit Switches

When a traditional DC motor is used reversibly within a restricted range of motion, it can be fitted with *limit switches* to prevent the motor from stalling and burning out at either end of its permitted travel. Limit switches are explained in "Limit Switches" on page 46 in the **switch** entry in this encyclopedia.

What Can Go Wrong

Brushes and Commutator

The primary cause of failure in DC motors is abrasion of the brushes and wear and tear, oxidation, and/or accumulation of dirt on the commutator. Some motors are designed to allow replacement of the brushes; sealed motors and gearhead motors generally are not. High current and high speed will both tend to accelerate wear in the areas where the brushes meet the commutator.

Electrical Noise

The intermittent contact between the brushes and sections of the commutator of a traditional DC motor can induce voltage spikes that may travel back into the power supply for the motor and cause seemingly random effects in other components. Sparking in the commutator can be a significant source of *electromagnetic interference* (*EMI*), especially where cheap or poorly fitted brushes are used. Even if the commutator is running cleanly, the rapid creation of a magnetic field in a motor winding, following by collapse of the field, can create spikes that feed back into the power supply.

Wires that power a motor should be in twisted-pair configuration, so that their radiated EMI tends to cancel itself out. They should be routed away from data lines or encoder outputs, and may be shielded if necessary. Data lines from sensors in brushless motors may also be shielded.

Installing a capacitor across the motor terminals can significantly reduce EMI. Some motors have capacitors preinstalled. If the motor is in a sealed casing, it may have to be disassembled to reveal whether a capacitor is present.

Heat effects

Since all motors in the real world are less than 100% efficient, some power is lost by the motor during normal operation, and will be dissipated as heat. The resistance of motor windings, and consequently the magnetic force that they generate, will decrease as the temperature rises. The motor becomes less efficient, and will try to draw more current, worsening the situation. A manufacturer's rating for maximum temperature should be taken seriously.

The insulation of the coil windings is usually the most vulnerable part of a motor if excess heat persists. Short circuits between adjacent coils as a result of insulation breakdown will degrade the performance of the motor while increasing its power consumption, which will create even more heat.

Where motor casings have protruding ridges, these cooling fins should be exposed to ambient air.

Frequent starting, stopping, and reversing will tend to generate heat as a result of power surges, and will reduce the lifetime of the motor.

Ambient Conditions

A warm, dry environment will tend to dry out bearing lubricants and graphite brushes. Conversely, a very cold environment will tend to thicken the bearing lubricants. If a motor will be used in unusual environmental conditions, the manufacturer should be consulted.

Wrong Shaft Type or Diameter

Motors have a variety of possible output shaft diameters, some measured in inches and others in millimeters, and shafts may be long or short, or may have a D-shaped cross section or splines to mate with appropriate accessories such as gears, pulleys, or couplings. Careful examination of datasheets and part numbers is necessary to determine compatibility. In the hobby-electronics world, retailers may offer purpose-built discs or arms for specific motor shafts.

Incompatible Motor Mounts

Mounting lugs or flanges may or may not be provided, and may be incompatible with the application for which the motor is intended. The same motor may be available with a variety of mount options, differentiated only by one letter or digit in the motor's part number. A mount option that was available in the past may become obsolete or may simply be out of stock. Again, examination of datasheets is necessary.

Backlash

Backlash is the looseness or "slack" in a gear train that results from small gaps between meshing gear teeth. Because backlash is cumulative when gears are assembled in series, it can become significant in a slow-output gearhead motor. When measured at the output shaft, it is generally in the range of 1 to 7 degrees, tending to increase as the load increases. If a geared motor is used as a positioning device, and is fitted with an encoder to count rotations of the motor shaft, control electronics may cause the motor to hunt to and fro in an attempt to overcome the hysteresis allowed by the backlash. A **stepper motor** or **servo motor** is probably better suited to this kind of application.

Bearings

When using a motor that is not rated for significant axial loading, the bearings may be damaged by applying excessive force to push-fit an output gear or pulley onto the motor shaft. Even minor damage to bearings can cause significant noise (see the following section) and a reduced lifetime for the component.

In brushless DC motors, the most common cause of failure is the deterioration of bearings. Attempting to revive the bearings by unsealing them and adding lubricant is usually not worth the trouble.

Audible Noise

While electric motors are not generally thought of as being noisy devices, an enclosure can act as a sounding board, and bearing noise is likely to increase over time. Ball bearings become noisy over time, and gears are inherently noisy.

If a device will contain multiple motors, or will be used in close proximity to people who are likely to be sensitive to noise (for example, in a medical environment), care should be taken to insure that motor shafts are properly balanced, while the motors may be mounted on rubber bushings or in sleeves that will absorb vibration.

AC motor

23

The distinction between AC and DC motors has become blurred as controllers for DC motors make increasing use of *pulse-width modulation*, which can be viewed as a form of alternating current. All motors that consume DC power are referenced in the **DC motor** section of this encyclopedia, regardless of whether they modulate the power internally. **Stepper motors** and **servo motors** are considered as special cases, each with its own entry. AC motors, described here, are those that consume alternating current, usually in the form of a sine wave with a fixed frequency.

OTHER RELATED COMPONENTS

- **DC motor** (see Chapter 22)
- **stepper motor** (see Chapter 25)
- **servo motor** (see Chapter 24)

What It Does

An AC motor uses a power supply of alternating current to generate a fluctuating magnetic field that turns a shaft.

How It Works

The motor consists primarily of two parts: the *stator*, which remains stationary, and the *rotor*, which rotates inside the stator. Alternating current energizes one or more coils in the stator, creating fluctuating magnetic fields that interact with the rotor. A simplified representation is shown in Figure 23-1, where the coils create magnetic forces indicated by the green arrows, N representing North and S representing South.

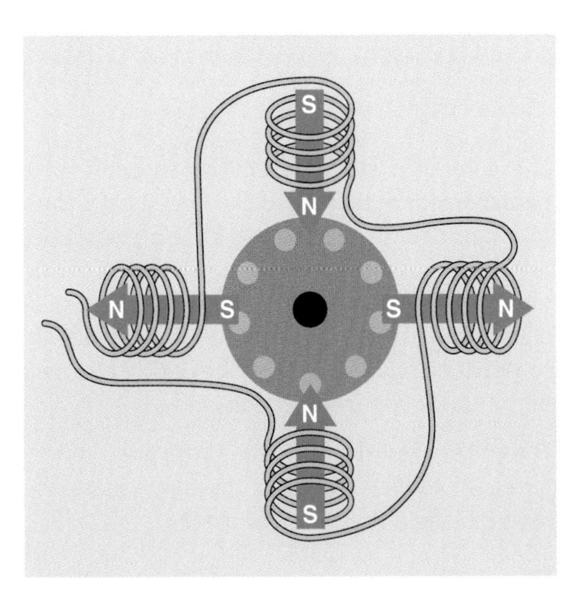

Figure 23-1. *A simplified representation of a basic AC motor. The green arrows indicate magnetic force.*

Stator Design

Plug-in electric fans typically use AC motors. The stator from a large electric fan is shown in Figure 23-2, where the large diameter of each coil maximizes its magnetic effect. The stator from a smaller electric fan is shown in Figure 23-3, in which only one coil is used. (The coil is wrapped in black electrical tape.)

Figure 23-2. *The stator from a large electric fan. Each coil of copper wire is centered on a lug pointing inward to the hole at the center, where the rotor would normally be mounted. The coils overlap because their diameter is maximized to increase their magnetic effect. Each coil is tapped to allow speed selection in steps via an external rotary switch.*

The core of a stator resembles the core of a **transformer** in that it usually consists of a stack of wafers of high-silicon steel (or sometimes aluminum or cast iron). The layers are insulated from one another by thin layers of shellac (or a similar compound) to prevent eddy currents that would otherwise circulate through the entire thickness of the stator, reducing its efficiency.

The coil(s) wound around the stator are often referred to as *field windings*, as they create the magnetic field that runs the motor.

Rotor Design

In most AC motors, the rotor does not contain any coils and does not make any electrical connection with the rest of the motor. It is powered entirely by induced magnetic effects, causing this type of motor to be known generally as an *induction motor*.

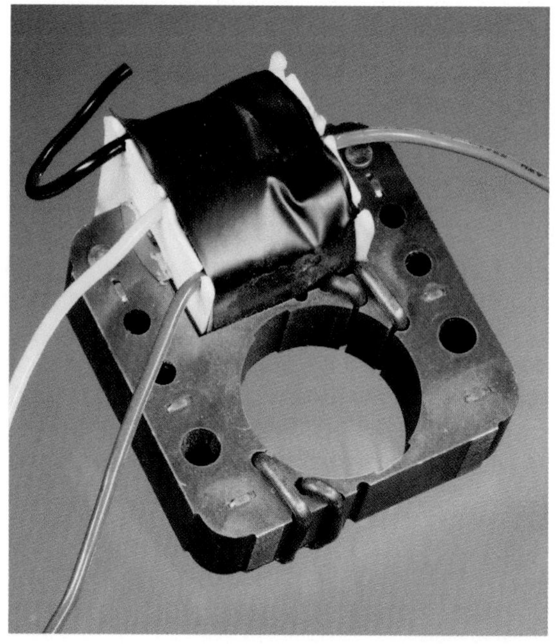

Figure 23-3. *This stator from a smaller electric fan uses only a single coil, wrapped in black tape. It is sufficient to induce a magnetic field but is generally less efficient than a motor using multiple coils.*

As the AC voltage changes from positive to negative, the magnetic force induced in the stator collapses and a new field of opposite polarity is created. Because the stator is designed to create an asymmetrical field, it induces a rotating magnetic field in the rotor. The concept of a rotating magnetic field is fundamental in AC motors.

Like the stator, the rotor is fabricated from wafers of high-silicon steel; embedded in the wafers are nonmagnetic rods, usually fabricated from aluminum but sometimes from copper, oriented approximately parallel to the axis of rotation. The rods are shorted together by a ring at each end of the rotor, forming a conductive "cage," which explains why this device is often referred to colloquially as a *squirrel cage* motor.

Figure 23-4 shows the configuration of a rotor cage with the surrounding steel wafers removed for clarity. In reality, the rods in the cage are almost always angled slightly, as shown in Figure 23-5, to promote smooth running and reduce *cogging*, or fluctuations in torque, which would otherwise occur.

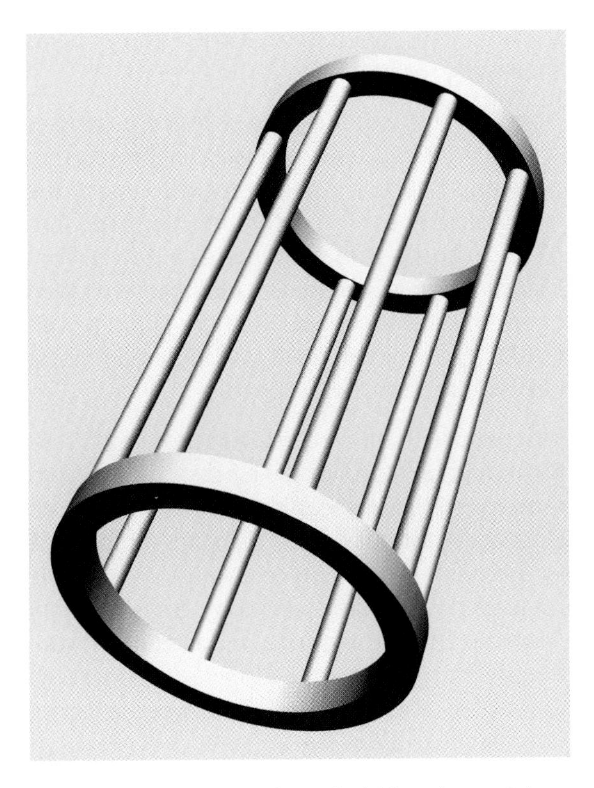

Figure 23-4. *The rotor of a typical AC motor contains a cage of aluminum (or sometimes copper) in which eddy currents occur, as a result of the rotating magnetic field inside the steel body of the rotor (which is omitted here for clarity). These currents cause their own magnetic fields, which interact with the fields generated by coils in the stator.*

In Figure 23-6, the steel wafers of a rotor are shown, with channels to accommodate an angled aluminum cage. Figure 23-7 shows a cross-section of a rotor with the cage elements in pale red and the steel wafer in gray.

The actual rotor from an induction motor is shown in Figure 23-8. This rotor was removed from the stator shown in Figure 23-3. The bearings at either end of the rotor were bolted to the stator until disassembly.

Although the cage is nonmagnetic, it is electrically conductive. Therefore the rotating magnetic field that is induced in the steel part of the rotor generates substantial secondary electric current

Figure 23-5. *To promote smooth running of the motor, the longitudinal elements of the cage are typically angled, as suggested in this rendering.*

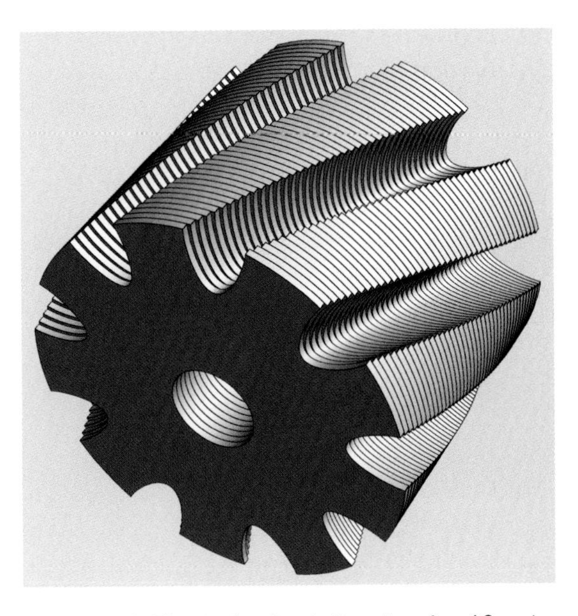

Figure 23-6. *The steel wafers in the rotor of an AC motor are typically offset as shown here. The channels are to accommodate a cage of aluminum or copper conductors.*

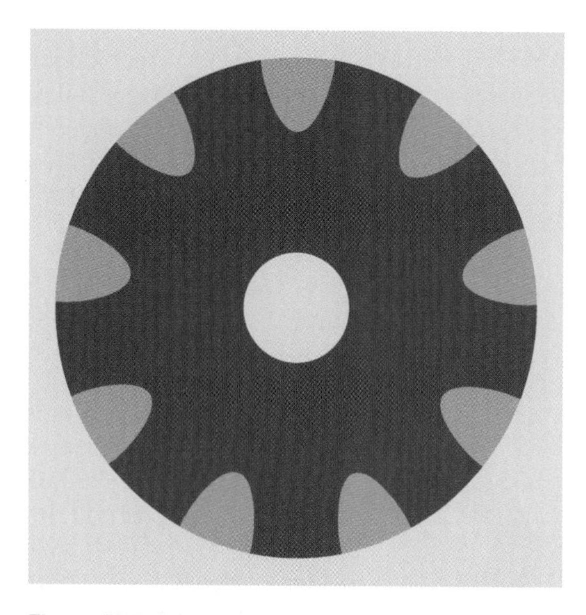

Figure 23-7. *Cross-section of a rotor with steel shown in gray and embedded elements of an aluminum cage shown in pale red.*

Figure 23-8. *The rotor from a small fan motor. The aluminum cage and its end pieces are the pale gray sections, steel plates are the darker sections.*

in the cage, so long as the magnetic field inside the rotor is turning faster than the rotor itself. The current in the longitudinal elements of the cage creates its own magnetic field, which interacts with the fields created by coils in the stator. Attraction and repulsion between these fields causes the rotor to turn.

Note that if the turning speed of the rotor rises to match the frequency of the alternating current powering the coils in the stator, the cage in the rotor is no longer turning through magnetic lines of force, and ceases to derive any power from them. In an ideal, frictionless motor, its unloaded operating speed would be in equilibrium with the AC frequency. In reality, an induction motor never quite attains that speed.

When power is applied while the rotor is at rest, the induction motor draws a heavy surge of current, much like a short-circuited **transformer**. Electrically, the coils in the stator are comparable to the primary winding of a transformer, while the cage in the rotor resembles the secondary winding. The turning force induced in the stationary rotor is known as *locked-rotor torque*. As the motor picks up speed, its power consumption diminishes. See Figure 23-9.

When the motor is running and a mechanical load is applied to it, the motor speed will drop. As the speed diminishes, the cage of conductors embedded in the rotor will derive more power, as they are turning more slowly than the rotating magnetic field. The speed of rotation of the field is determined by the frequency of the AC power, and is therefore constant. The difference in rotational speed between the magnetic field and the rotor is known as *slip*. Higher levels of slip induce greater power, and therefore the induction motor will automatically find an equilibrium for any load within its designed range.

When running under full load, a small induction motor may have a slip value from 4 to 6 percent. In larger motors, this value will be lower.

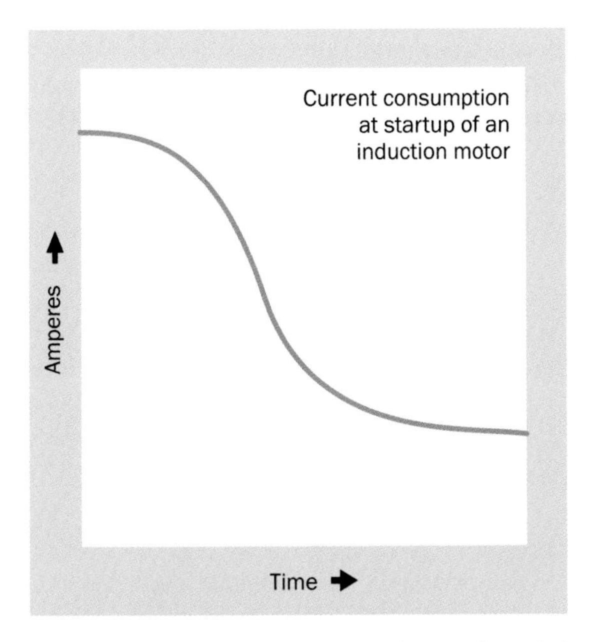

Figure 23-9. *An approximated graph showing the typical current consumption of an AC induction motor as it starts from rest and gains speed over a period of time.*

Variants

Variants of the generic induction motor described above are generally designed to take advantage of either *single-phase* or *three-phase* alternating current.

A *synchronous motor* is a variant in which the rotor maintains a constant speed of rotation regardless of small fluctuations in load.

Some AC motors incorporate a *commutator*, which allows an external connection to coils mounted on the rotor, and can enable variable speed control.

A *linear motor* may consist of two rows of coils, energized by a sequence of pulses that can move a permanent magnet or electromagnet between the coils. Alternatively, the linear motor's coils may move as a result of magnetic interaction with a segmented fixed rail. Detailed description of linear motors is outside the scope of this encyclopedia.

Single-Phase Induction Motor

The majority of induction motors run on single-phase alternating current (typically, from domestic wall outlets). This type of motor is not innately self-starting because the stator coils and rotor are symmetrical. This tends to result in vibration rather than rotation.

To initiate rotation, the stator design is modified so that it induces an asymmetrical magnetic field, which is more powerful in one direction than the other. The simplest way to achieve this is by adding one or more *shorting coils* to the stator. Each shorting coil is often just a circle of heavy-gauge copper wire. This ploy reduces the efficiency of the motor and impairs its starting torque, and is generally used in small devices such as electric fans, where low-end torque is unimportant. Because the shorting coil obstructs some of the magnetic field, this configuration is often known as a *shaded pole* motor.

Copper shorting coils are visible in the fan motor shown in Figure 23-3.

A **capacitor** is a higher-cost but more efficient alternative to a shorting coil. If power is supplied to one or more of the stator coils through a capacitor, it will create a phase difference between these coils and the others in the motor, inducing an asymmetrical magnetic field. When the motor reaches approximately 80% of its designed running speed, a centrifugal switch may be included to take the capacitor out of the circuit, since it is no longer necessary. Switching out the capacitor and substituting a direct connection to the stator coils will improve the efficiency of the motor.

A third option to initiate rotation is to add a second winding in the stator, using fewer turns of smaller-gauge wire, which have a higher resistance than the main winding. Consequently the magnetic field will be angled to encourage the motor to start turning. This configuration is known as a *split-phase* induction motor, in which the starter winding is often referred to as the *auxilliary winding* and consists of about 30% of the total stator windings in the motor. Here again, a centrifugal switch can be incorporated, to eliminate the secondary winding from the circuit when the motor has reached 75 to 80 percent of its designed running speed.

The relationship between motor speed and torque of the three types of motors described above is shown in Figure 23-10. These curves are simplified and do not show the effect that would be produced by introducing a centrifugal switch.

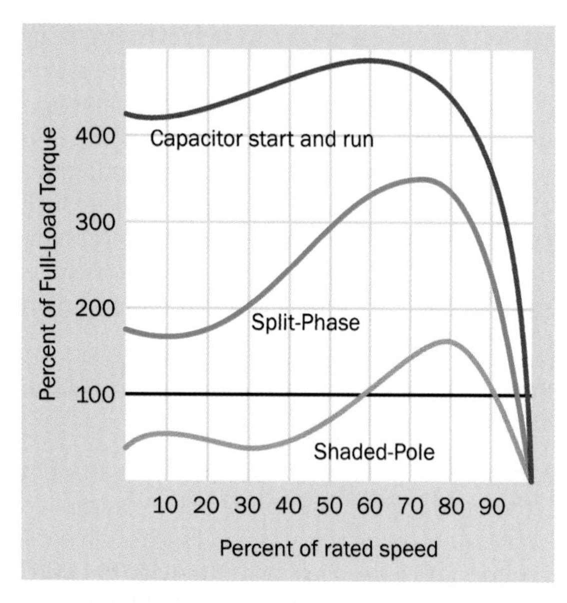

Figure 23-10. *Approximate curves showing the relationship between speed and torque for three types of single-phase induction motor. (Graph derived from AC Induction Motor Fundamentals published by Microchip Technology Inc.)*

Three-Phase Induction Motor

Larger induction motors are often *three-phase* devices. Three-phase AC (which is by far the most common form of *polyphase* AC) is delivered by a power utility company or generator via three wires, each of which carries alternating current with a phase difference of 120 degrees relative to the other two, usually for industrial applications. A common configuration of stator coils for a three-phase motor is shown in Figure 23-11. Since the three wires take it in turns to deliver their peak voltage, they are ideally suited to turn the stator of a motor via induction, and no shorting coil or capacitor is needed for startup. Heavy-duty 3-phase induction motors are extremely reliable, being brushless and generally maintenance-free.

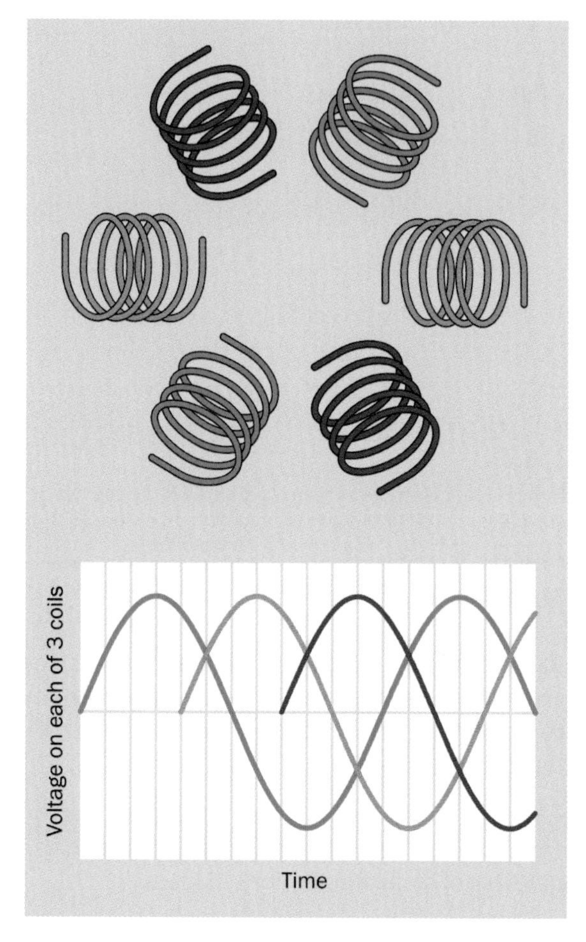

Figure 23-11. *The graph shows voltage delivered via three wires constituting a three-phase power supply. (The curve colors are arbitrary.) A three-phase motor contains a multiple of three coils—often six, as shown here diagrammatically. The three wires of the power supply are connected directly to the coils, which induce a rotating magnetic field.*

Synchronous Motor

A synchronous motor is a form of induction motor that is designed to reach and maintain equilibrium when the rotor is turning in perfect synchronization with the AC power supply. The speed of the motor will depend on the number of poles (magnetic coils) in the stator, and the number of phases in the power supply. If R is the RPM of a synchronous motor, f is the frequency of the AC current in Hz, and p is the number of poles per phase:

```
R = (120 * f) / p
```

This formula assumes 60Hz AC current. In nations where 50Hz AC is used, the number 120 should be replaced with the number 100.

Two basic types of synchronous motors exist: *direct current excited*, which require external power to start turning, and *non-excited*, which are self-starting. Since non-excited synchronous motors are more common in electronic applications, this encyclopedia will not deal with direct current excited variants.

A *hysteresis motor* is a synchronous motor containing a solid rotor cast from cobalt steel, which has high coercivity, meaning that once it is magnetized, a substantial field is required to reverse the magnetic polarity. Consequently the polarity of the rotor lags behind the constantly changing polarity of the stator, creating an attracting force that turns the rotor. Because the lag angle is independent of motor speed, this motor delivers constant torque from startup.

Reluctance Motor

Reluctance is the magnetic equivalent to electrical resistance. If a piece of iron is free to move in a magnetic field, it will tend to align itself with the field to reduce the reluctance of the magnetic circuit. This principle was used in very early reluctance motors designed to work from AC and has been revived as electronics to control variable frequency drives have become cheaper.

The simplest reluctance motor consists of a soft iron rotor with projecting lugs, rotating within a stator that is magnetically energized with its own set of inwardly projecting poles. The rotor tends to turn until its lugs are aligned with the poles of the stator, thus minimizing the reluctance.

A basic reluctance motor design is shown in Figure 25-2. It is located in the **stepper motor** section of this encyclopedia, as stepper motors are a primary application of the reluctance principle.

Although a reluctance motor can be used with polyphase fixed-frequency AC power, a variable frequency drive greatly enhances its usefulness. The timing of the frequency is adjusted by the speed of the motor, which is detected by a sensor. Thus the energizing pulses can remain "one step ahead" of the rotor. Since the rotor is not a magnet, it generates no back-EMF, allowing it to reach very high speeds.

The simplicity of the motor itself is a compensating factor for the cost of the electronics, as it requires no commutator, brushes, permanent magnets, or rotor windings. Characteristics of reluctance motors include:

- Cheap parts, easily manufactured, and high reliability.
- Compact size and low weight.
- Efficiencies greater than 90% possible.
- Capable of high start-up torque and high speed operation.

Disadvantages include noise, cogging, and tight manufacturing tolerances, as the air gap between the rotor and stator must be minimized.

A reluctance motor can function synchronously, if it is designed for that purpose.

Variable Frequency Drive

A basic induction motor suffers from significant problems. The surge of power that it draws when starting from rest can pull down the supply voltage enough to affect other devices that share the AC power supply. (Hence, the brief dimming of lights that may occur when the compressor in an air conditioner or refrigerator starts running.) While the motor is turning, it can introduce electrical noise, which feeds back into the power supply, once again causing potential problems for other devices. In addition, the narrow range of speed of an AC induction motor is a great disadvantage.

The advent of cheap solid-state technology encouraged the development of *variable-frequency power supplies* for induction motors. Because the impedance of the motor will diminish as the frequency diminishes, the current drawn by the motor will tend to increase. To prevent this, a variable frequency supply also varies the voltage that it delivers.

Wound-Rotor AC Induction Motor

The stator of this variant is basically the same as that of a single-phase induction motor, but the rotor contains its own set of coils. These are electrically accessible via a *commutator* and *brushes*, as in a **DC motor**. Because the maximum torque (also known as *pull-out* torque) will be proportional to the electrical resistance of the coils in the rotor, the characteristics of the motor can be adjusted by adding or removing resistance externally, via the commutator. A higher resistance will enable greater torque at low speed when the slip between the rotor speed and rotation of the magnetic field induced by the stator is greatest. This is especially useful in corded power tools such as electric drills, where high torque at low speed is desirable, yet the motor can accelerate to full speed quickly when the external resistance is reduced. Typically, the resistance is adjusted via the trigger of the drill.

Figure 23-12. *A motor in a corded electric drill uses coils in a brushed rotor to enable variable speed output. In most AC motors, the speed is not adjustable and the rotor does not make any electrical connection with the rest of the motor.*

Figure 23-12 shows a wound-rotor AC induction motor. The disadvantage of this configuration is the brushes that supply power to the rotor will eventually require maintenance. Much larger wound-rotor motors are also used in industrial applications such as printing presses and elevators, where the need for variable speed makes a simple three-phase motor unsuitable.

Universal Motor

A wound-rotor motor may also be described as a *universal motor* if its rotor and stator coils are connected in series. This configuration allows it to be powered by either AC or DC.

DC supplied to the rotor and the stator will cause mutual magnetic repulsion. When the rotor turns, the brushes touching the split commutator reverse the polarity of voltage in the rotor coils, and the process repeats. This configuration is very similar to that of a conventional DC motor, except that the stator in a universal motor uses electromagnets instead of the permanent magnets that are characteristic of a **DC motor**.

When powered by AC, the series connection between stator and rotor coils insures that each pulse to the stator will be duplicated in the rotor, causing mutual repulsion. The addition of a shorting coil in the stator provides the necessary asymmetry in the magnetic field to make the motor start turning.

Universal motors are not limited by AC frequency, and are capable of extremely high-speed operation. They have high starting torque, are compact, and are cheap to manufacture. Applications include food blenders, vacuum cleaners, and hair dryers. In a workshop, they are found in routers and miniature power tools such as the Dremel series.

Because a universal motor requires commutator and brushes, it is only suitable for intermittent use.

Inverted AC Motors

Some modern domestic appliances may seem to contain an AC motor, but in fact the AC current is *rectified* to DC and is then processed with pulse-width modulation to allow variable speed control. The motor is really a **DC motor**; see the entry on this type of motor for additional information.

Values

Because a basic AC induction motor is governed by the frequency of the power supply, the speed of a typical four-pole motor is limited to less than 1,800 RPM (1,500 RPM in nations where 50Hz AC is the norm).

Variable-frequency, universal, and wound-rotor motors overcome this limitation, and can reach speeds of 10,000 to 30,000 RPM. Synchronous motors typically run at 1,800 or 1,200 RPM, depending on the number of poles in the motor. (They run at 1,500 or 1,000 RPM in locations where the frequency of AC is 50Hz rather than 60Hz).

For a discussion of the torque that can be created by a motor, see "Values" on page 178 in the **DC motor** entry in this encyclopedia.

How to Use It

Old-fashioned record players (where a turntable supports a vinyl disc that must rotate at a fixed speed) and electric clocks (of the analogue type) were major applications for synchronous motors, which used the frequency of the AC power supply to control motor speed. These applications have been superceded by CD players (usually powered by brushless DC motors) and digital clocks (which use crystal oscillators).

Many home appliances continue to use AC-powered induction motors. Small cooling fans for use in electronic equipment are sometimes AC-powered, reducing the current that must be provided by the DC power supply. An induction motor generally tends to be heavier and less efficient than other types, and its speed limit imposed by the frequency of the AC power supply is a significant disadvantage.

A simple induction motor cannot provide the sophisticated control that is necessary in modern devices such as CD or DVD players, ink-jet printers, and scanners. A **stepper motor**, **servo motor**, and **DC motors** controlled with *pulse-width modulation* are preferable in these applications.

A *reluctance motor* may find applications in high-speed, high-end equipment including vacuum cleaners, fans, and pumps. Large variable reluctance motors, with high amperage ratings, may be used to power vehicles. Smaller variants are being used for power steering systems and windshield wipers in some automobiles.

What Can Go Wrong

Compared with other devices that have moving parts, the brushless induction motor is one of the most reliable and efficient devices ever invented. However, there are many ways it can be damaged. General problems affecting all types of motors are listed at "Heat effects" on page 182. Issues relating specifically to AC motors are listed below.

Premature Restart

Large industrial three-phase induction motors can be damaged if power is reapplied before the motor has stopped rotating.

Frequent Restart

If a motor is stopped and started repeatedly, the heat that is generated during the initial surge of current is likely to be cumulative.

Undervoltage or Voltage Imbalance

A voltage drop can cause the motor to draw more current than it is rated to handle. If this situation persists, overheating will result. Problems also are caused in a three-phase motor where one phase is not voltage-balanced with the others. The most common cause of this problem is an open circuit-breaker, wiring fault, or blown fuse affecting just one of the three conductors. The motor will try to run using the two conductors that are still providing power, but the result is likely to be destructive.

Stalled Motor

When power is applied to an induction motor, if the motor is prevented from turning, the conductors in the rotor will carry a large current that is entirely dissipated as heat. This current surge will either burn out the motor or blow a fuse or circuit breaker. Care should be taken, in equipment design, to minimize the risk that an induction motor may stall or jam.

Protective Relays

Sophisticated protective relays are available for industrial 3-phase motors, and can guard against all of the faults itemized above. Their details are outside the scope of this encyclopedia.

Excess Torque

As has been previously noted, the torque of an induction motor increases with the slip (speed difference) between the rotation of the magnetic field and the rotation of the rotor. Consequently, if the motor is overloaded and forced to run more slowly, it can deliver more rotational force. This can destroy other parts attached to the motor, such as drive belts.

Internal Breakage

An overloaded induction motor may suffer some cracking or breakage of its rotor. This may be obvious because of reduced power output or vibration, but can also be detected if the motor's power consumption changes significantly.

24 servo motor

Should be referred to as an *RC servo* if it is intended for use in small devices that are remote-controlled and battery powered. However, in practice, the RC acronym is often omitted.

OTHER RELATED COMPONENTS

- **AC motor** (See Chapter 23)
- **DC motor** (See Chapter 22)
- **stepper motor** (See Chapter 25)

What It Does

A servo motor is actually a combination of a motor, reduction gearing, and miniaturized control electronics, usually packaged together inside a very compact sealed plastic case. The motor itself may be AC or DC, and if DC, it may be brushed or brushless. What distinguishes a servo from other types of motor is that it is not designed for continuous rotation. It is a position-seeking device. Its rotational range may be more than 180 degrees but will be significantly less than 360 degrees. Two typical RC servos are shown in Figure 24-1. A side view of a motor is shown in Figure 24-2.

Figure 24-1. *A typical RC servo motor is capable of more than 50 inch-ounces of torque yet can be driven by three or four AA alkaline cells in series, and weighs under 2 ounces.*

The electronics inside the motor enclosure interpret commands from an external controller. The command code specifies the desired turn angle measured as an offset either side of the center position of the motor's range. The motor turns quickly to the specified position and stops there. So long as the command signal continues and power to the motor is sustained, the motor will hold its position and "push back" against any external turning force. In the absence of such a force, while the motor is stationary, it will use very little current.

The electronics inside a typical RC servo motor are shown in Figure 24-3.

How It Works

Servo motors are generally controlled via *pulse-width modulation* (*PWM*).

Figure 24-2. *RC servo motors are mostly similar in size. This is a typical side view.*

An industrial servo typically requires a controller that is an off-the-shelf item sold by the manufacturer of the motor. The encoding scheme of the control signals may be proprietary. A heavy-duty servo may be designed to run from three-phase power at a relatively high voltage, and may be used in applications such as production-line automation.

The remainder of this encyclopedia entry will focus primarily on small RC servos rather than industrial servos.

For small RC servos, the stream of control pulses is at a constant frequency of 20ms, with the positive durations of each pulse being interpreted as a positioning command to the motor, and the gaps between the pulses being disregarded. A typical range of pulse widths for a small motor is 1ms to 2ms, specifying a range of -90 to +90 degrees either side of a center location. Many modern motors are capable of excursions beyond these limits, and can be calibrated to establish the precise relationship between pulse width and turn angle. The motor can then be controlled by a lookup table in microcontroller software, or by using a conversion factor between degree-angle and pulse width.

Figure 24-4 shows the typical range of pulse widths within the fixed 20ms period (a frequency of 50Hz) between the start of one pulse and the start of the next, and the meaning of each pulse width to the servo motor. Intermediate pulse widths are interpreted as instructions to rotate to intermediate positions.

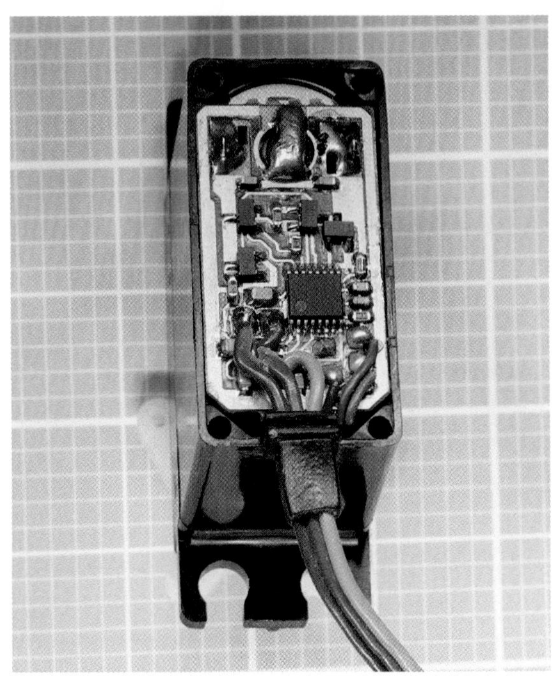

Figure 24-3. *The electronics inside a servo motor decode a stream of pulses that specify the turn angle of the motor.*

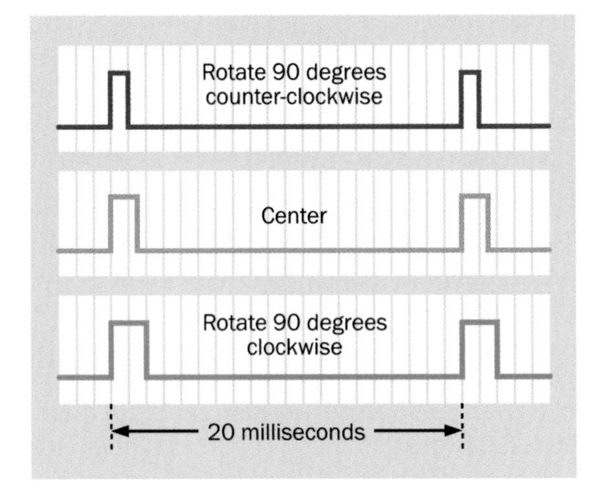

Figure 24-4. *The turn angle of a small RC servo motor is determined by a pulse width from a controller ranging from 1ms to 2ms in duration. The frequency of the pulses is constant at 50Hz.*

Small servo motors require the user to provide a controller that will conform with the above specification. This is often achieved by programming a *microcontroller*, and some microcontroller chips make this especially easy by providing a PWM output specifically tailored to the requirements of an RC servo. Either way, the microcontroller can be directly connected to the servo, enabling an extremely simple and flexible way to manage a positioning device.

Alternatively, a simple pulse generator such as a *555 timer chip* can be used, or controller boards are available from hobbyist supply sources. Some controller boards have USB connections enabling a servo to be governed by computer software.

In Figure 24-5, a schematic illustrates the connection of a 555 timer with an RC servo, with component values to create a constant frequency of about 48Hz (slightly more than 20ms from peak to peak). The 1µF capacitor in the circuit charges through the 2.2K resistor in series with the diode, which bypasses the 28K resistor. This charging time represents the "on" cycle of the timer chip. The capacitor discharges through the 28K resistor, representing the "off" cycle. The 1K potentiometer, in series with the 5K resistors, acts as a voltage divider applied to the control pin of the timer, adjusting the timer's charge and discharge thresholds. Turning the potentiometer will lengthen or shorten the "on" time of each cycle, without changing the frequency. In practice, because capacitors are manufactured with wide tolerances, the frequency of the timer output cannot be guaranteed. Fortunately most servos will tolerate some inaccuracy.

Since the motor shares the power supply of the timer in this circuit, a protection diode and capacitor have been added between the power supply to the motor and negative ground, to suppress noise and back-EMF.

Inside a servo motor's casing, the electronics include a **potentiometer** that turns with the output shaft, to provide feedback confirming the

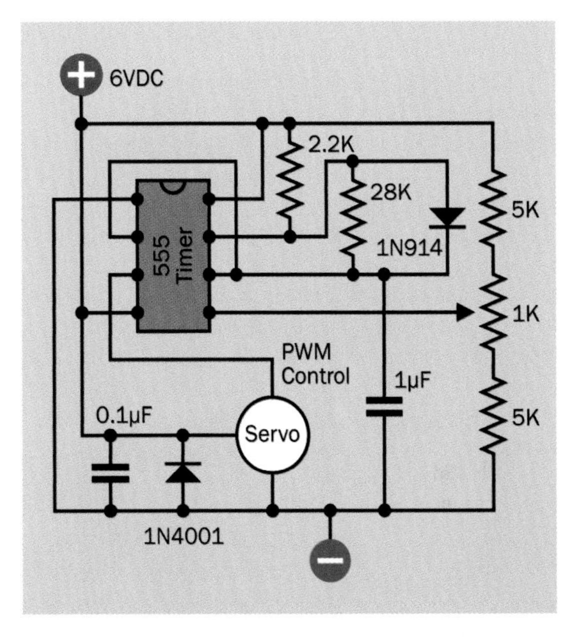

Figure 24-5. *An RC servo can be controlled via a 555 timer with appropriate component values. The potentiometer determines the angular position of the servo.*

motor's position. The limited turning range of the potentiometer determines the turn limits of the motor output shaft.

Variants

Small servos may contain brushed or brushless DC motors. Naturally the brushless motors have greater longevity and create less electrical noise. See the **DC motor** entry in this encyclopedia for a discussion of brushed versus brushless motors.

Servos may use nylon, "Karbonite," or metal reduction gearing. The nylon gears inside a cheaper RC servo are shown in Figure 24-6.

Brushless motors and metal gears add slightly to the price of the motor. Metal gears are stronger than nylon (which can crack under load) but may wear faster, leading to *backlash* and inaccuracy in the gear train. The friction between nylon-to-nylon surfaces is very low, and nylon is certainly adequate and may be preferable if a servo will not be heavily loaded. "Karbonite" is claimed to be five times stronger than nylon and may be a

Figure 24-6. *Nylon gearing inside a servo motor.*

satisfactory compromise. If a gear set experiences a failure (for example, teeth can be stripped as a result of excessive load), manufacturers usually will sell a replacement set to be installed by the user. Installation requires manual dexterity and patience, and some skill.

Servos may have roller bearings or plain sintered bearings, the latter being cheaper but much less durable under side loading.

So-called *digital servos* use faster internal electronics than the older, so-called *analog servos*, and because they sample the incoming pulse stream at a higher frequency, they are more responsive to small, rapid commands from the controller. For this reason they are preferred by hobbyists using servos to control the flight of model airplanes. Externally, the control protocol for digital and analog servos is the same, although a digital servo can be reprogrammed with new code values establishing the limits to its range. A standalone programming unit must be purchased to achieve this.

The most popular manufacturers of small servo motors are Futaba and Hitec. While their control protocols are virtually identical, the motor output shafts differ. The shaft is typically known as the *spline*, and is grooved to fit push-on attachments. The spline of a Futaba motor has 25 grooves, while Hitec uses 24 grooves. Attachments must be appropriate for the brand of motor that has been chosen.

Values

A small servo typically weighs 1 to 2 ounces, has a rotation time of 1 to 2 seconds from one end of its travel to the other, and can exert a surprisingly robust torque of 50 ounce-inches or greater.

Voltage
: Small servos were originally designed to run from 4.8V rechargeable batteries in model aircraft. They can be driven with 5VDC to 6VDC on a routine basis. A few servos are designed for higher voltages.

Amperage
: The datasheets provided by most manufacturers often fail to specify the power that a servo will draw when it is exerting maximum torque (or indeed, any torque at all). Since small servos are often driven by three or four AA alkaline batteries in series, the maximum current draw is unlikely to be much greater than 1 amp. When the motor is energized but not turning, and is not resisting a turning force, its power consumption is negligible. This feature makes servos especially desirable for remote-controlled battery-powered devices.

Some motors that have a turning range exceeding 180 degrees will respond to pulses of less than 1ms or greater than 2ms. A newly acquired motor should be tested with a microcontroller that steps through a wide range of pulse durations, to determine the limits empirically. Pulses that are outside the motor's designed range will generally be ignored and will not cause damage.

The *turn rate* or *transit time* specified in a datasheet is the time a servo takes to rotate through 60 degrees, with no load on the output shaft. A high-torque servo generally achieves its greater turning force by using a higher reduction gear ratio, which tends to result in a longer transit time.

How to Use it

Typical applications for a small servo include rotating the flaps or rudder of a model aircraft, steering a model boat, model car, or wheeled robot, and turning robotic arms.

A servo generally has three wires, colored red (power supply), black or brown (ground), and orange, yellow, or white (for the pulse train from the controller). The ground wire to the motor must be common with the ground of the controller, and consequently a ceramic bypass **capacitor** of 0.1µF or 0.01µF should be placed between the (red) power wire to the motor and ground. A protection **diode** should also be used. Neither a diode nor a capacitor should be attached to the wire carrying control signals, as it will interfere with the pulse train.

When powering the motor, an AC adapter should only be used with some caution, as its power output may be inadequately smoothed. A voltage regulator is not necessary, but bypass **capacitors** are mandatory. Figure 24-7 shows two hypothetical schematics. The upper section of the figure shows a battery-driven system, possibly using four 1.2V NiMH rechargeable batteries. Since batteries do not generally create voltage spikes, no capacitors are used, but a diode is included to protect the microcontroller from EMF when the servo stops and starts. The lower section of the figure shows the additional precautions that may be necessary when using DC power from an **AC adapter**. The DC-DC **converter**, which derives 6VDC for the motor requires smoothing capacitors (this should be specified in its datasheet), and so does the **voltage regulator**, which delivers regulated 5VDC power to the microcontroller. Once again, the protection diode is included. In both diagrams, the orange wire represents the control wire transmitting pulses to the servo motor.

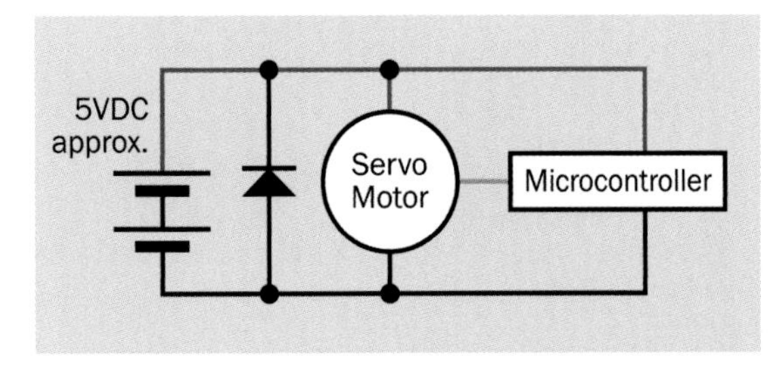

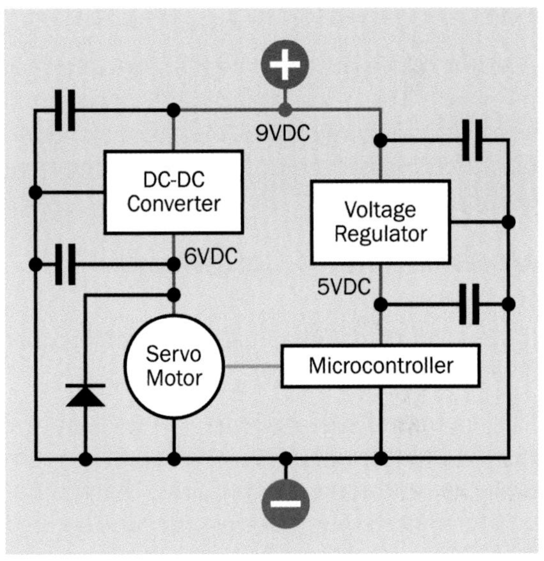

Figure 24-7. *Two possible schematics to run a small servo motor, the upper example using battery power (for example, from four 1.2V NiMH cells) and the lower example using a 9VDC AC adapter. See text for additional explanation.*

Various shaft attachments are available from the same online hobby-electronics suppliers that sell servos. The attachments include discs, single arms, double arms, and four arms in a cross-shaped configuration. A single-arm attachment is often known as a *horn*, and this term may be applied loosely to any kind of attachment. The horn is usually perforated so that other components can be fixed to it by using small screws or nuts and bolts. Figure 24-8 shows a variety of horns.

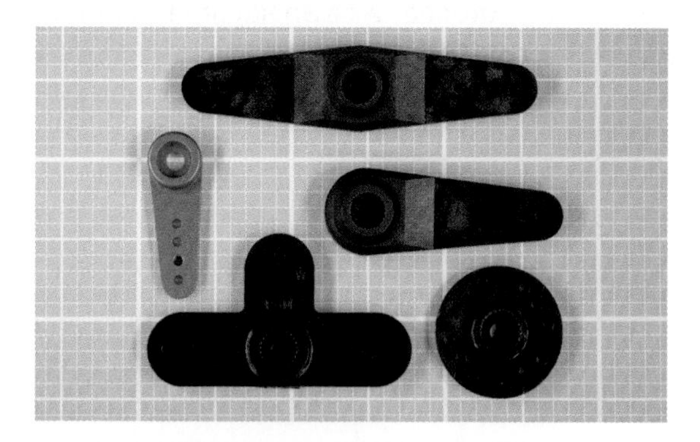

Figure 24-8. *Various shaft attachments, known as horns, are available from motor manufacturers. The blue one is metallic; the others are plastic.*

After the horn is pushed onto the spline (the motor shaft), it is held in place with one central screw. As previously noted, the two major manufacturers of small servos, Futaba and Hitec, have incompatible splines.

Modification for Continuous Rotation

It is possible to modify a small servo motor so that it will rotate continuously.

First the motor case must be opened, and the potentiometer must be centered by using a controller to send some 1.5ms pulses. The potentiometer must then be glued or otherwise secured with its wiper in this precise center position, after which the potentiometer is disconnected from the drive train.

Mechanical stops that would limit the rotation of the motor shaft must be cut away, after which the motor is reassembled. Because the potentiometer has been immobilized, the motor's internal electronics will now "see" the shaft as being in its center position at all times. If the controller sends a pulse instructing the motor to seek a position clockwise or counter-clockwise from center, the motor will rotate in an effort to reach that position. Because the potentiometer will not provide feedback to signal that the motor has achieved its goal, the shaft will continue to rotate indefinitely.

In this mode, the primary distinguishing characteristic of the servo has been disabled, in that it can no longer turn to a specific angle. Also, stopping the servo may be problematic, as it must receive a command that precisely matches the fixed position of the potentiometer. Since the potentiometer may have moved fractionally during the process in which it was immobilized, some trial and error may be needed to determine the pulse width that corresponds with the potentiometer position.

The purpose of modifying a servo for continuous rotation is to take advantage of its high torque, small size, light weight, and the ease of controlling it with a microcontroller.

In response to the interest shown by hobbyists in modifying servos for continuous rotation, some manufacturers now market servos with continuous rotation as a built-in feature. Typically they include a trimmer potentiometer to calibrate the motor, to establish its center-off position.

What Can Go Wrong

Incorrect Wiring

The manufacturer's datasheet should be checked to confirm the color coding of the wires. While a simple DC motor can be reversed by inverting the polarity of its power supply, this is totally inappropriate for a servo motor.

Shaft/Horn Mismatch

Attachments for the spline of one brand of motor may not fit the spline of another brand, and cannot be forced to fit.

Unrealistically Rapid Software Commands

Microcontroller software that positions a servo must allow sufficient time for the servo to respond before the software specifies a new position. It may be necessary to insert delay loops or other wait times in the software.

Jitter

A servo arm that twitches unpredictably usually indicates that the pulse train is being corrupted by external electrical noise. The control wire to the servo should be as short as possible, and should not run closely adjacent to conductors carrying AC or high frequency current switching, or control wires for other servo motors.

Motor Overload

A servo capable of delivering 2 lbs of force 1 inch from its shaft can easily generate enough torque, when it stalls, to break itself free from its mounts, or bend or break any arm or linkage attached to its shaft. Ideally, a relatively "weak link" should be included so that if breakage occurs, it will be predictable and will be relatively easy and cheap to repair.

Unrealistic Duty Cycle

Small servos are designed for intermittent use. Constant cycling will cause wear and tear, especially if the motor has a brushed commutator or metal reduction gears.

Electrical Noise

Brushed motors are always a source of electrical interference, and any servo will also tend to create a voltage dip or surge when it starts and stops. A protection diode may be insufficient to isolate sensitive microcontrollers and other integrated circuit chips. To minimize problems, the servo can be driven by a source of positive voltage that is separate from the regulated power supply used by the chips, and larger filter capacitors may be added to the voltage supply of the microcontroller. A common ground between the motor and the chips is unfortunately unavoidable.

stepper motor

Also often referred to as a *stepping motor*, and sometimes known as a *step motor*. It is a type of *induction motor* but merits its own entry in this encyclopedia as it has acquired significant and unique importance in electronics equipment where precise positioning of a moving part is needed and digital control is available.

OTHER RELATED COMPONENTS

- **DC motor** (See Chapter 22)
- **AC motor** (See Chapter 23)
- **servo motor** (See Chapter 24)

What It Does

A stepper motor rotates its drive shaft in precise steps in response to a timed sequence of pulses (usually one step per pulse). The pulses are delivered to a series of coils or *windings* in the *stator*, which is the stationary section of the motor, usually forming a ring around the *rotor*, which is the part of the motor that rotates. Steps may also be referred to as *phases*, and a motor that rotates in small steps may be referred to as having a high *phase count*.

A stepper motor theoretically draws power for its stator coils at a constant level that does not vary with speed. Consequently the torque tends to decrease as the speed increases, and conversely, it is greatest when the motor is stationary or locked.

The motor requires a suitable control system to provide the sequence of pulses. The control system may consist of a small dedicated circuit, or a microcontroller or computer with the addition of suitable driver transistors capable of handling the necessary current. The torque curve of a motor can be extended by using a controller that increases the voltage as the speed of the control pulses increases.

Because the behavior of the motor is controlled by external electronics, and its interior is usually symmetrical, a stepper motor can be driven backward and forward with equal torque, and can also be held in a stationary position, although the stator coils will continue to consume power in this mode.

How It Works

The stator has multiple poles made from soft iron or other magnetic material. Each pole is either energized by its own coil, or more commonly, several poles share a single, large coil. In all types of stepper motor, sets of stator poles are magnetized sequentially to turn the rotor and can remain energized in one configuration to hold the rotor stationary.

The rotor may contain one or more permanent magnets, which interact with the magnetic fields generated in the stator. Note that this is different from a *squirrel-cage* **AC motor** in which a "cage" is embedded in the rotor and interacts with a rotating magnetic field, but does not consist of permanent magnets.

Three small stepper motors are shown in Figure 25-1. Clockwise from the top-left, they are four-wire, five-wire, and six-wire types (this distinction is explained in the following section). The motor at top-left has a threaded shaft that can engage with a collar, so that as the motor shaft rotates counter-clockwise and clockwise, the collar will be moved down and up.

Figure 25-1. *Three small stepper motors.*

Reluctance Stepper Motors

The simplest form of stepper motor uses a rotor that does not contain permanent magnets. It relies on the principle of *variable reluctance*, reluctance being the magnetic equivalent of electrical *resistance*. The rotor will tend to align its protruding parts with the exterior source(s) of the magnetic field, as this will reduce the reluctance in the system. Additional information about variable reluctance is included in "Reluctance Motor" on page 191 in the section of this encyclopedia dealing with the **AC motor**.

A variable reluctance motor requires an external controller that simply energizes the stator coils sequentially. This is shown in Figure 25-2, where six poles (energized in pairs) are arrayed symmetrically around a rotor with four protrusions, usually referred to as *teeth*. Six stator poles and four teeth are the minimum numbers for reliable performance of a reluctance stepper motor.

In the diagram, the core of each pole is tinted green when it is magnetized, and is gray when it is not magnetized. In each section of this diagram, the stator coils are shown when they have just been energized, and the rotor has not yet had time to respond. External switching to energize the coils has been omitted for simplicity. In a real motor, the rotor would have numerous ridges, and the clearance between them and the stator would be extremely narrow to maximize the magnetic effect.

In a 6-pole reluctance motor where the rotor has four teeth, each time the controller energizes a new pair of poles, the rotor turns by 30 degrees counter-clockwise. This is known as the *step angle*, and means that the motor makes 12 steps in each full 360-degree rotation of its shaft. This configuration is very similar to that of a 3-phase AC *induction motor*, as shown in Figure 23-11 in the **AC motor** section of this encyclopedia. However, the AC motor is designed to be plugged into a power source with a constant frequency, and is intended to run smoothly and continuously, not in discrete steps.

Generally, reluctance motors tend to be larger than those with magnetized rotors, and often require feedback from a sensor that monitors shaft angle and provides this information to control electronics. This is known as a *closed loop* system. Most smaller stepper motors operate in an *open loop* system, where positional feedback is considered unnecessary if the number of pulses to the motor is counted as a means of tracking its position.

Permanent Magnet Stepper Motors

More commonly, the rotor of a stepper motor contains permanent magnets, which require the controller to be capable of reversing the magnetic field created by each of the stator coils, so that they alternately attract and repel the rotor magnets.

In a *bipolar* motor, the magnetic field generated by a coil is reversed simply by reversing the current through it. This is shown diagrammatically in Figure 25-3. In a *unipolar* motor, the magnetic field is reversed by applying positive voltage to the center tap of a coil, and grounding one end or the other. This is shown diagrammatically in Figure 25-4.

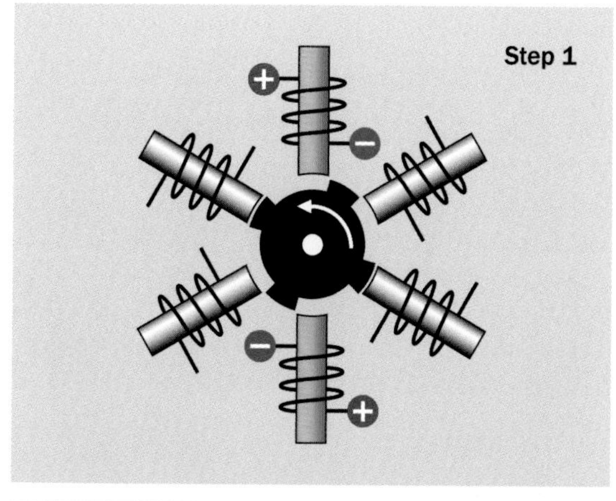

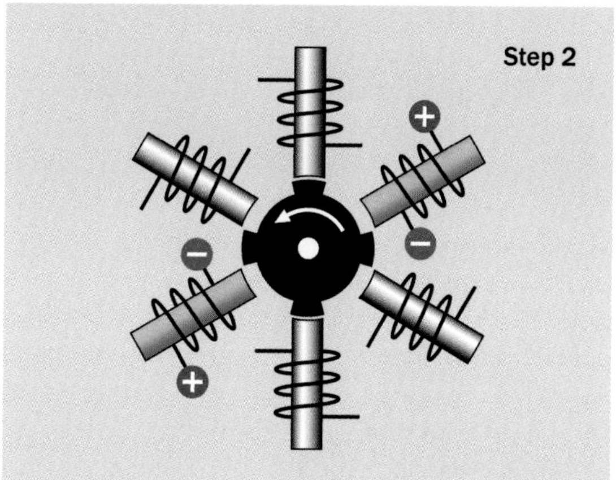

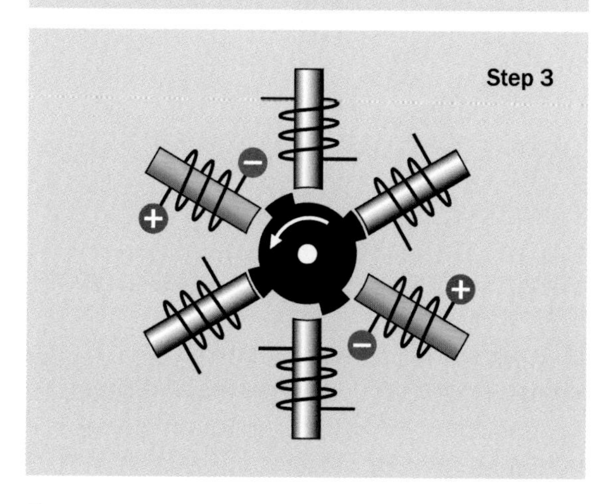

Figure 25-2. *In a variable reluctance stepper motor, the rotor moves to minimize magnetic reluctance each time the next pair of coils is energized. At each step, the coils have been energized a moment before the rotor has had time to respond.*

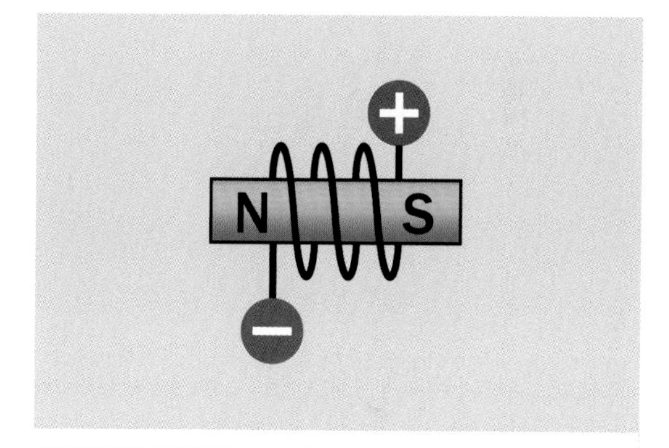

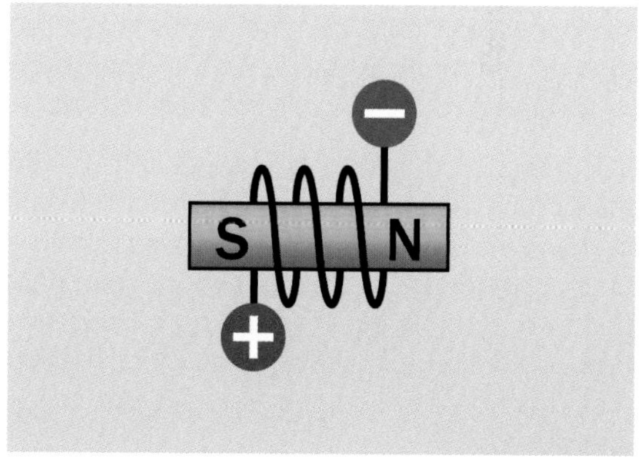

Figure 25-3. *In a bipolar motor, the magnetic field generated by each stator coil is reversed simply by reversing the current through the coil.*

Either type of motor is often designed with an upper and lower deck surrounding a single rotor, as suggested in Figure 25-5. A large single coil, or center-tapped coil, induces a magnetic field in multiple poles in the top deck, out of phase by one step with a second set of poles, energized by their own coil, in the bottom deck. (All three mo-

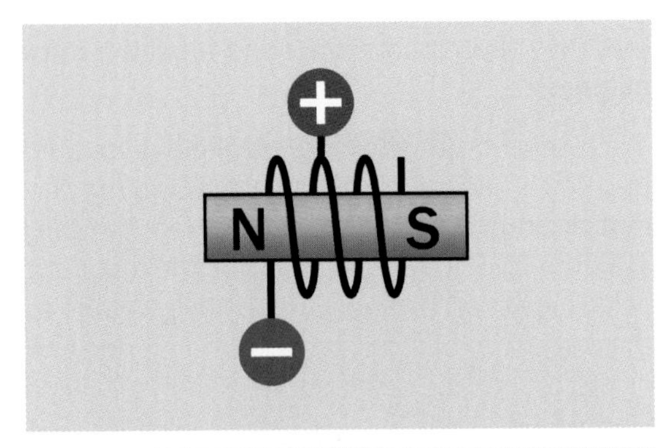

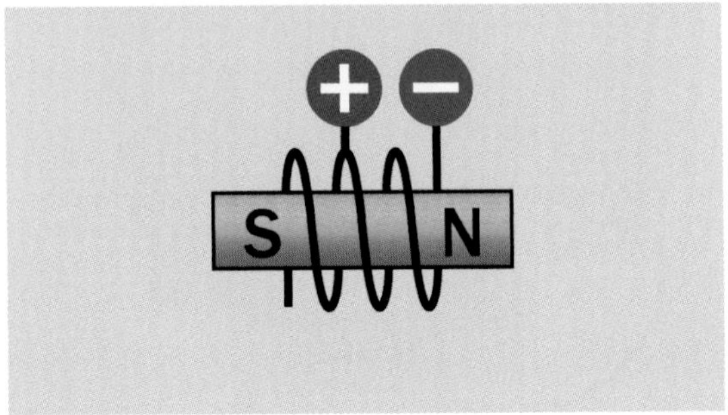

Figure 25-4. *The magnetic field of this coil is reversed by applying positive voltage constantly to a center tap and grounding one end of the coil or the other.*

Figure 25-5. *A simplified rendering of the common "two deck" type of motor. See text for details.*

tors shown in Figure 25-1 are of this type.) The rotor of the motor is tall enough to span both decks, and is rotated by each of them in turn.

In Figure 25-6, the decks of a two-deck four-wire motor have been split apart. The rotor remains in the left-hand section. It is enclosed within a black cylinder that is a permanent magnet divided into multiple poles. In the right-hand section, a coil is visible surrounding metal "teeth" that function as stator poles when the coil is energized.

In Figure 25-7, the same motor has been further disassembled. The coil was secured with a length of tape around its periphery, which has been removed to make the coil visible. The remaining half of the motor, at top-right, contains a second, concealed but identical coil with its own set of poles, one step out of phase with those in the first deck.

Figure 25-6. *A two-deck stepper motor split open to reveal its rotor (left) and one of the stators (right) encircled by a coil.*

Because the field effects in a two-deck stepper motor are difficult to visualize, the remaining diagrams show simplified configurations with a minimum number of stator poles, each with its own coil.

Bipolar Stepper Motors

The most basic way to reverse the current in a coil is by using an *H-bridge* configuration of switches, as shown in Figure 25-8, where the green arrow

Figure 25-7. *The stepper motor from the previous figure, further disassembled.*

indicates the direction of the magnetic field. In actual applications, the switches are solid-state. Integrated circuits are available containing all the necessary components to control a bipolar stepper motor.

Four sequential steps of a bipolar motor are shown in Figure 25-9, Figure 25-10, Figure 25-11, and Figure 25-12. The H-bridge control electronics for each coil are omitted for clarity. As before, energized coils are shown with the pole inside the coil tinted green, while non-energized coils are gray, and the rotor is shown before it has had time to respond to the magnetic field in each step.

Unipolar Motors

The control electronics for a unipolar motor can be simpler than those for a bipolar motor, as off-the-shelf switching transistors can ground one end of the coil or the other. The classic five-wire unipolar stepper motor, often sold to hobbyists and used in robotics projects and similar applications, can be driven by nothing more elaborate than a set of *555 timer* chips. However, this type of motor is less powerful for its size and weight because only half of each coil is energized at a time.

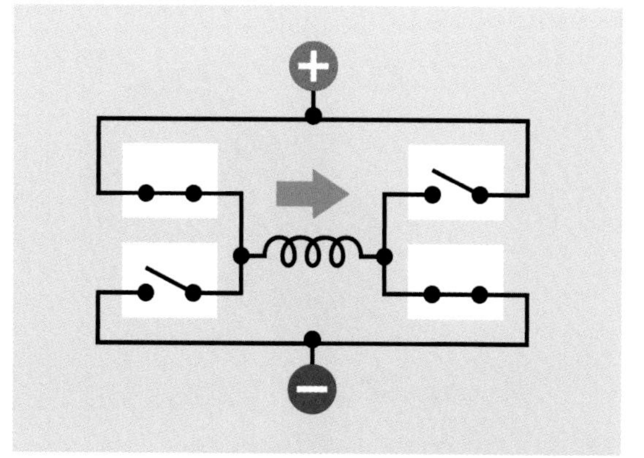

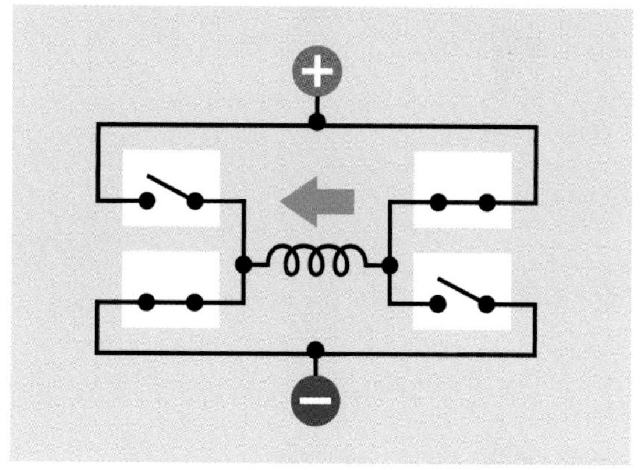

Figure 25-8. *The simplest and most basic way to reverse the current through a coil is via an H-bridge circuit. In practice, the switches are replaced by solid-state components.*

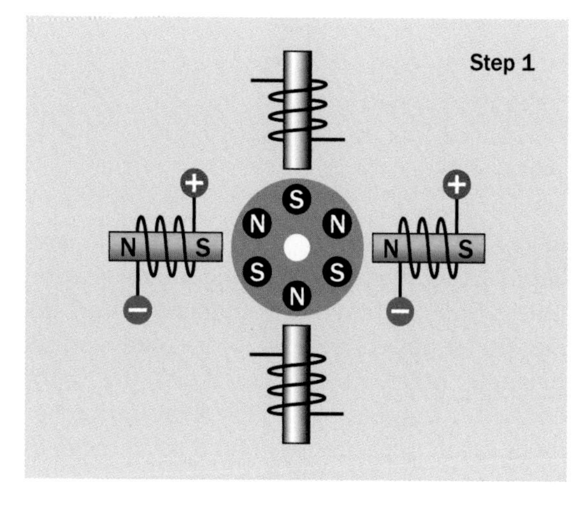

Figure 25-9. *A bipolar stepper motor depicted a moment before the rotor has had time to make its first step in response to magnetic fields created by the stator coils.*

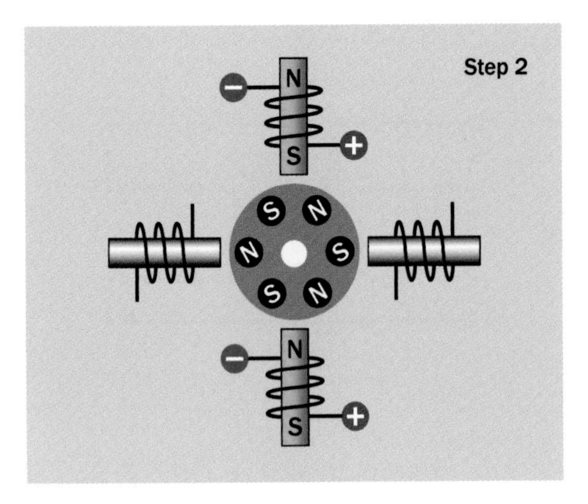

Figure 25-10. *The bipolar stepper motor from the previous figure is shown with its rotor having advanced by one step, and coil polarity changed to induce it to make a second step.*

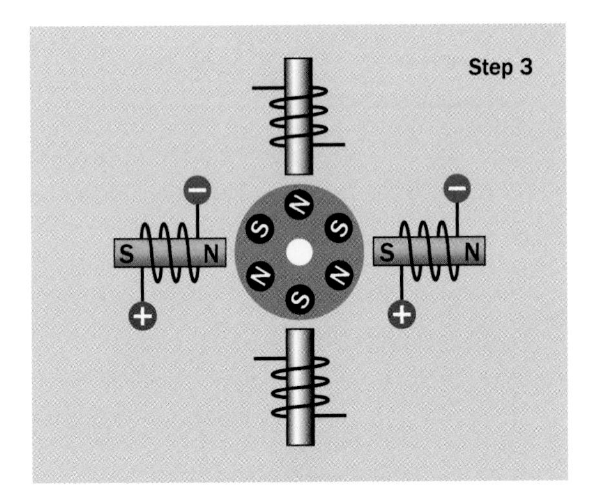

Figure 25-11. *The bipolar stepper motor after taking its second step, immediately before making its third step.*

In Figure 25-13, Figure 25-14, Figure 25-15, and Figure 25-16, the simplest configuration of a unipolar system is shown in diagrammatic form using four stator coils and a rotor containing six magnetic poles. Each figure shows the stator coils when they have just been energized, a moment before the rotor has had time to move in response to them. Coils that are energized are shown with the metal cores tinted green. Wires that are not conducting current are shown in

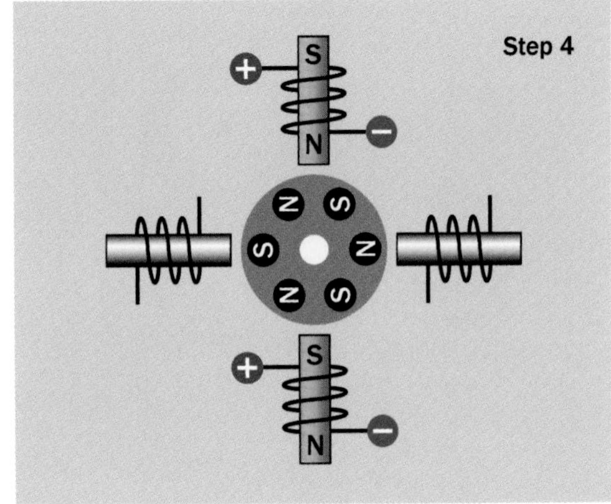

Figure 25-12. *The bipolar stepper motor after taking its third step. When the rotor responds to the new pattern of magnetic fields, its orientation will be functionally identical with that shown in the first step.*

gray. The open and closed positions of switches a, b, c, and d suggest the path that current is taking along the wires that are colored black.

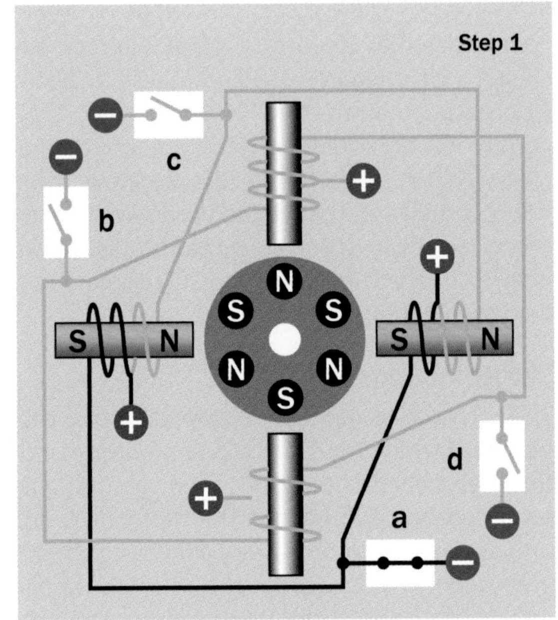

Figure 25-13. *The coils of this unipolar stepper motor are shown an instant after they have been energized, before the rotor has had time to respond by making its first step.*

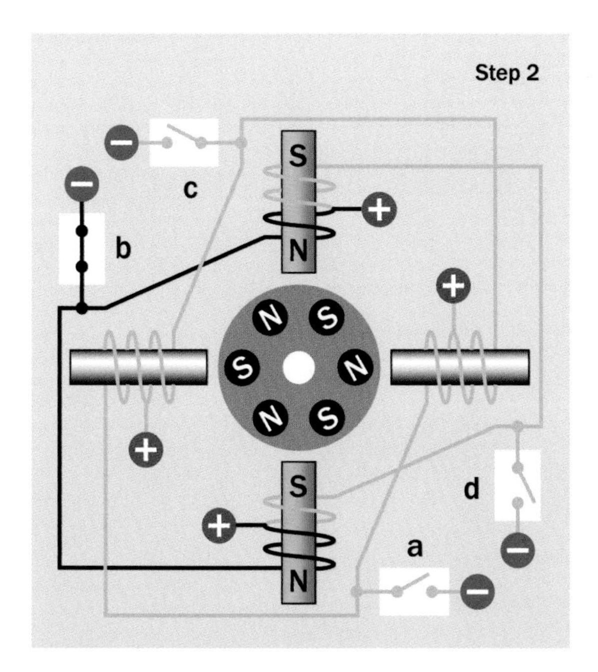

Figure 25-14. *The same motor from the previous figure is shown with coils energized to induce the rotor to make its second step.*

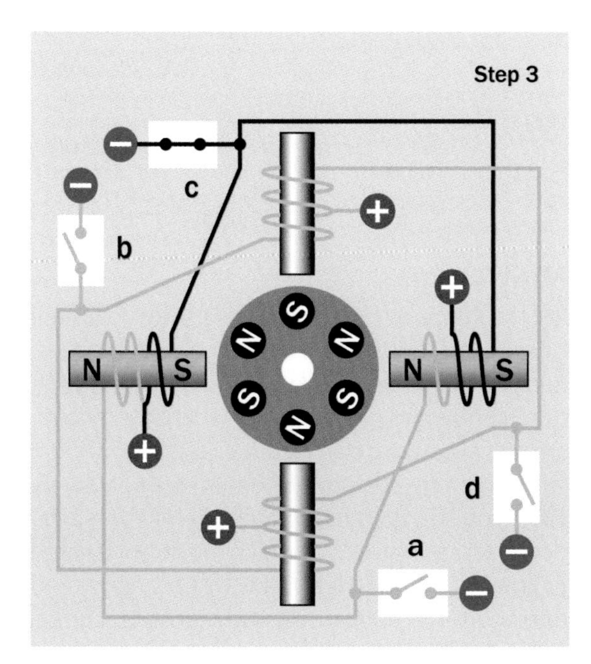

Figure 25-15. *The same motor from the previous figure is shown with coils energized to induce the rotor to make its third step.*

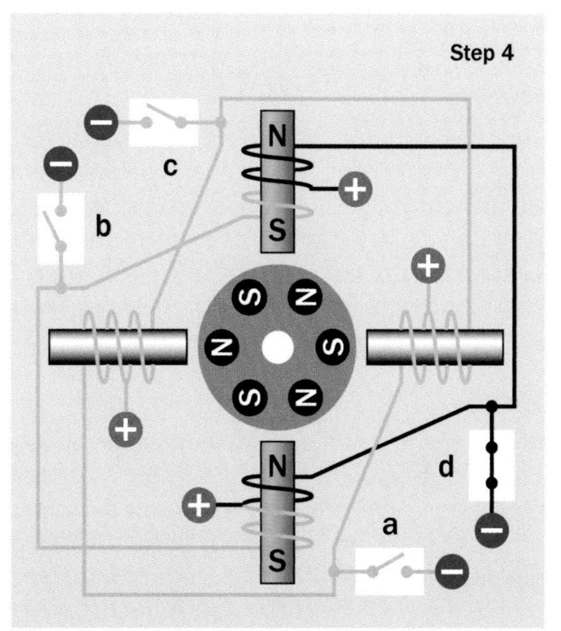

Figure 25-16. *When the rotor makes its fourth step, it will be back in an orientation that is functionally identical with the first figure in this series.*

Note that coils on opposite sides of the motor are energized simultaneously, while the other pair of coils is de-energized. Adjusting the controller so that it overlaps the "on" cycles of the coils can generate more torque, while consuming more power.

A motor containing more stator poles can advance in smaller steps, if the poles are separately energized. However, if the coils have individual windings, this will increase the cost of the motor.

Variants

In addition to bipolar and unipolar variants, previously described, three others are available.

High Phase Count

This term describes any type of stepper motor in which additional poles reduce the step size. The advantages of a high phase count include smoother running at high speed and greater precision when selecting a desired motor position. The additional coils also enable higher power density, but naturally tend to add to the cost of the motor.

Hybrid

This type of motor uses a toothed rotor that provides variable reluctance while also containing permanent magnets. It has become relatively common, as the addition of teeth to the rotor enables greater precision and efficiency. From a control point of view, the motor behaves like a regular permanent-magnet stepper motor.

Bifilar

In this type of motor, also sometimes known as a *universal* stepper motor, two coils are wound in parallel for each stator pole. If there are two poles or sets of poles, and both ends of each winding are accessible via wires that are run out of the motor, there will be eight wires in total. Consequently this type is often referred to as an *8-wire motor*.

The advantage of this scheme is that it allows three possible configurations for the internal coils. By shorting together the wires selectively, the motor can be made to function either in unipolar or bipolar mode.

In Figure 25-17, the upper pair of simplified diagrams depicts one end of one coil connected to the beginning of the other, while positive voltage is applied at the midpoint, as in a unipolar motor. The magnetic polarity of the coil is determined by grounding either end of the coil. The section of each coil that is not conducting current is shown in gray.

The center pair of diagrams shows the adjacent ends of the coils tied together, so that they are now energized in parallel, with the magnetic polarity being determined by the polarity of the voltage, as in a bipolar motor.

The coils may also be connected in series, as shown in the lower pair of diagrams. This will provide greater torque at low speed and lower torque at high speed, while enabling higher-voltage, lower-current operation.

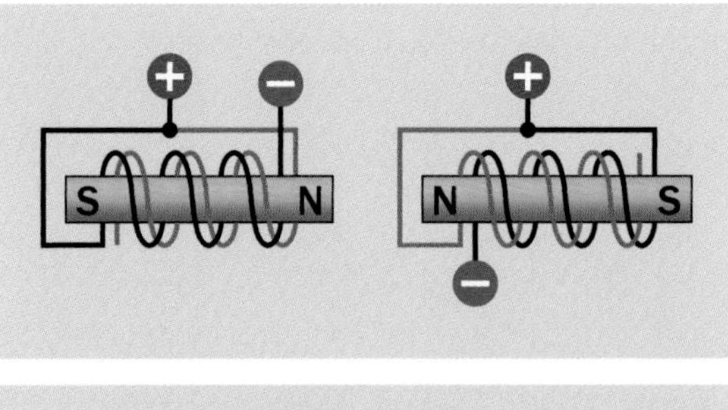

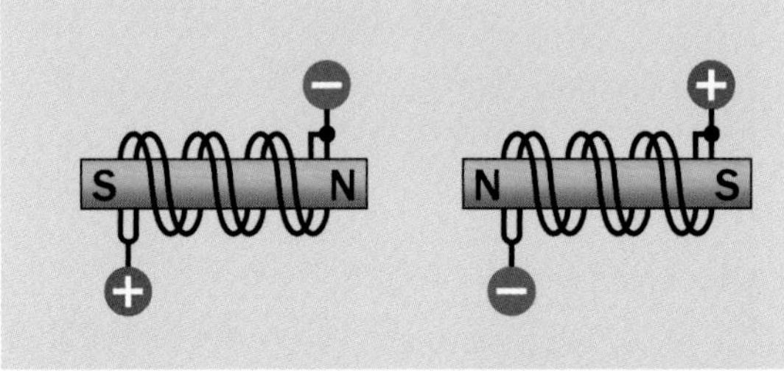

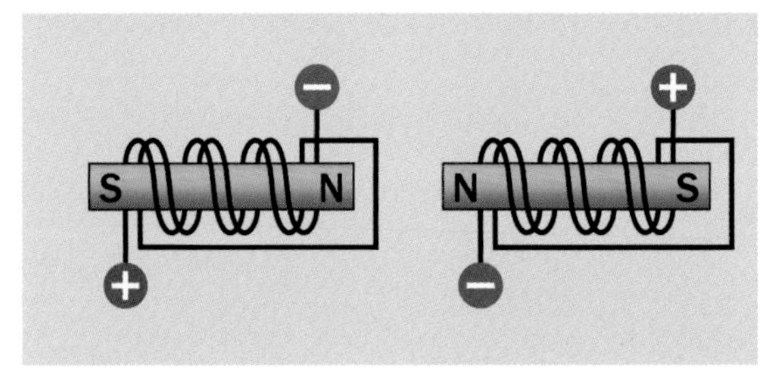

Figure 25-17. *In a bifilar motor, two coils are wound in parallel around each stator pole and can be connected with a center tap to emulate a unipolar motor (upper diagrams), or can be energized in parallel (middle diagrams) or series (lower diagrams) to emulate a bipolar motor.*

Multiphase

In a multiphase motor, multiple stator coils are usually connected in series, with a center tap applied between each pair. A possible configuration is shown in Figure 25-18, where the two diagrams show two consecutive steps in rotation, although the step angle could be halved by changing the voltage polarity in only one location at a time. The way in which the motor is wired enables only one stator coil to be unpowered during any step, because its two ends are at equal potential. Therefore this type of motor is capable of high torque in a relatively small format.

In some multiphase motors, additional wires allow access to both ends of each coil, and the coils are not connected internally. This allows control of the motor to be customized.

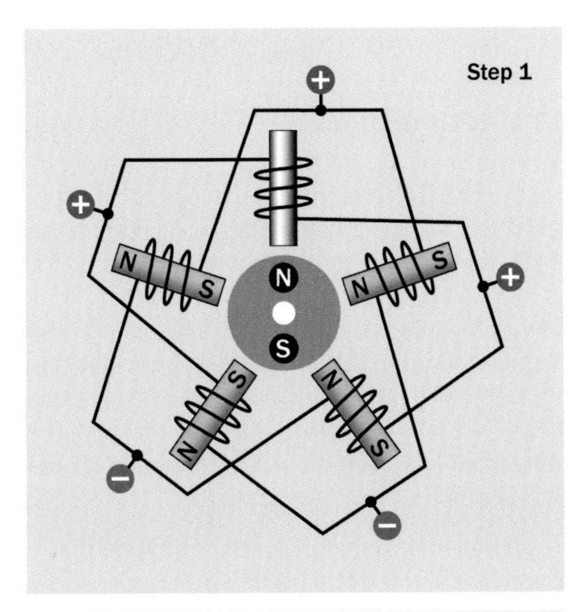

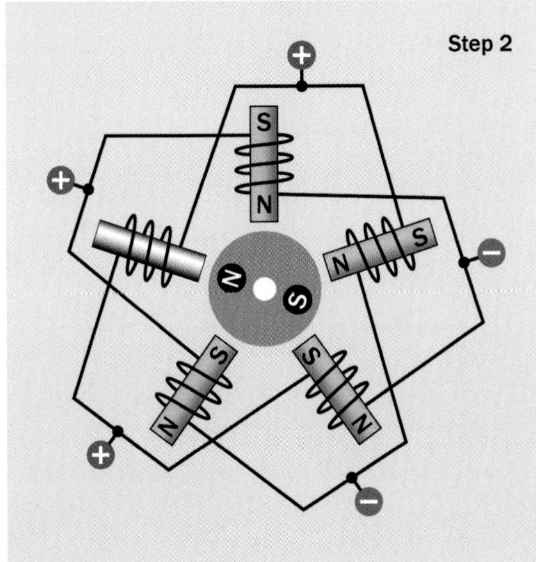

Figure 25-18. *A multiphase stepper motor. By applying voltage in the pattern shown, only one coil is not energized during each step. This enables high torque compared with the size of the motor.*

Microstepping

An appropriately designed stepper motor can be induced to make very small, intermediate steps if the control voltage is modulated to intermediate levels. Step angles as low as 0.007 degrees are claimed by some manufacturers. However, a motor running in this mode is less able to generate torque.

The simplest form of microstepping is half-stepping. To achieve this in a unipolar motor, each coil passes through an "off" state before its magnetic polarity is reversed.

Sensing and Feedback

So long as the series of pulses to the motor allows the rotor ample time to respond, no feedback mechanism from the rotor is necessary to confirm its position, and an open-loop system is sufficient. If sudden acceleration, deceleration, load fluctuations, and/or rotation reversal will occur, or if high speeds are involved, a *closed loop system*, in which a sensor provides positional feedback, may be necessary.

Voltage Control

Rapid stepping of a motor requires rapid creation and collapse of magnetic fields in the stator windings. Therefore, self-inductance of the windings can limit the motor speed. One way to overcome this is to use a higher voltage. A more sophisticated solution is to use a controller that provides a high initial voltage, which is reduced or briefly interrupted when a sensor indicates that coil current has increased sufficiently to overcome the self-inductance of the windings and has reached its imposed limit. This type of controller may be referred to as a *chopper drive* as the voltage is "chopped," usually by power transistors. It is a form of *pulse width modulation*.

Values

The *step angle* of a stepper motor is the angular rotation of its shaft, in degrees, for each full step. This will be determined by the physical construction of the motor. The coarsest step angle is 90 degrees, while sophisticated motors may be capable of 1.8 degrees (without microstepping).

The maximum torque that a motor can deliver is discussed in "Values" on page 178 in the **DC motor** entry of this encyclopedia.

Motor weight and size, shaft length, and shaft diameter are the principal passive values of a stepper motor, which should be checked before it is selected for use.

How to Use it

Stepper motors are used to control the seek action in disk drives, the print-head movement and paper advance in computer printers, and the scanning motion in document scanners and copiers.

Industrial and laboratory applications include the adjustment of optical devices (modern telescopes are often oriented with stepper motors), and valve control in fluid systems.

A stepper motor may be used to power a *linear actuator*, usually via a screw thread (properly known as a lead screw) or worm gear. For more on linear actuators, see "Linear Actuator" on page 178. While the stepper motor will enable greater accuracy than a traditional DC motor, the gearing inevitably will introduce some imprecision.

Advantages of stepper motors include:

- Precise positioning, typically within 3 percent to 5 percent per step. The percentage step error does not accumulate as the motor rotates
- Able to run at a wide range of speeds, including very slow speeds without reduction gearing
- Trouble-free start, stop, and reverse action
- Cheap controller hardware where open-loop applications are acceptable
- High reliability, since no brushes or commutator are involved

Disadvantages include:

- Noise and vibration
- Resonance at low speeds
- Progressive loss of torque at high speeds

Protection Diodes

While a small stepper motor may be driven directly from power transistors, *darlington pairs*, or even *555 timers*, larger motors will create *back-EMF* when the magnetic field of each stator coil is induced or forward EMF when the field is allowed to collapse, and bipolar motors will also induce voltage spikes when the current reverses. In a unipolar motor, while only one-half of the coil is actually energized via its center tap, the other half will have an induced voltage, as the coil acts like a *linear transformer*.

A simplified schematic illustrating diode placement for a bipolar motor is shown in Figure 25-19.

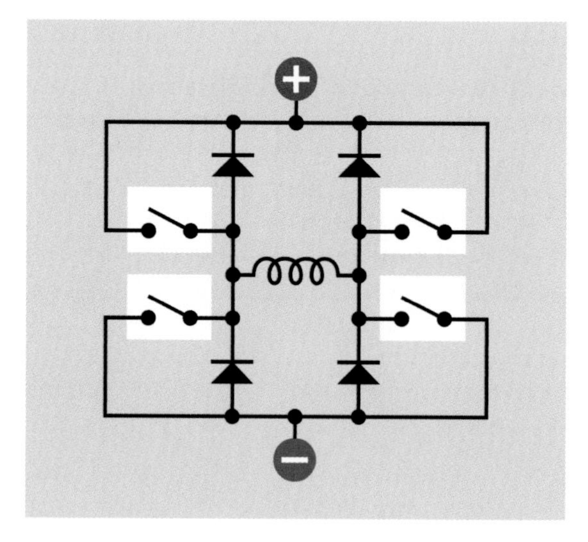

Figure 25-19. *The H-bridge circuit must be augmented with protection diodes to guard against back-EMF created by fluctuating current in the stator coil.*

Integrated circuit chips are available taht incorporate protection diodes, in addition to the necessary power transistors. Stepper motors may also have protection diodes built in. Consult the manufacturer's datasheet for details before attaching a motor to a power source.

Positional Control

The built-in control electronics of a **servo motor** typically turn the shaft to a precisely known position in response to pulse-width modulation from an exterior source such as a **microcontroller**, whereas the angle of rotation of a stepper motor in an open-loop system must be calculated by counting the number of steps from an initial, home position. This limitation of a stepper motor can be overcome by using a closed-loop system, but that will require monitoring the motor, adding complexity to the external controller. The choice between stepper and servo motors should be evaluated on a case-by-case basis.

What Can Go Wrong

General problems affecting all types of motors are listed in "Heat effects" on page 182. Issues relating more specifically to stepper motors are listed in the following sections.

Incorrect Wiring

Because a stepper motor is driven via multiple conductors, there is a significant risk of wiring errors, especially since many motors are not identified with part numbers. The first challenge, then, may be to determine what type of motor it is. When the motor is disconnected from any power, and the shaft is rotated with finger and thumb, a magnetized-rotor motor will not spin as freely as a reluctance motor, because the magnets in the rotor will provide intermittent turning resistance.

If a unipolar motor is relatively small and is fitted with five wires, almost certainly the motor contains two coils, each with a center tap, and their function can be determined by applying positive voltage to the red wire and grounding each of the other wires in turn. Attaching a small piece of tape to the motor shaft will assist in viewing its orientation.

A multimeter set to measure ohms can also be useful in deducing the internal coil connections of the motor, since the end-to-end resistance of a coil should be approximately twice the resistance between the center tap and either end of the coil.

A multiphase motor may have five wires, but in this case, the resistance between any two nonadjacent wires will be 1.5 times the resistance between any two adjacent wires.

Step Loss

In an open-loop system, if the motor skips or misses pulses from the controller, the controller no longer has an accurate assessment of the shaft angle. This is known as *step loss*. Since this can be caused by sudden changes in control frequency, the frequency should be increased (or decreased) gradually. This is known as *ramping* the motor speed. Stepper motors cannot respond instantly to changes in speed, because of inertia in the rotor or in the device that the motor is driving.

Where the motor turns one or more steps beyond its commanded stopping point, this is known as *overshoot*.

Step loss may also occur if the motor continues turning after power has been interrupted (either intentionally or because of an external fault). In an open-loop system, the controller should be designed to reset the motor position when power is initiated.

Excessive Torque

When the motor is stationary and not powered, *detent torque* is the maximum turning force that can be applied without causing the shaft to turn. When the motor is stationary and the controller does deliver power to it, *holding torque* is the maximum turning force that can be applied without causing the shaft to turn, and *pull-in torque* is the maximum torque which the motor can apply to overcome resistance and reach full speed. When the motor is running, *pull-out torque* is the maximum torque the motor can deliver without suffering step loss (pulling it out of sync with its controller). Some or all of these values should be specified on the motor's datasheet. Exceeding any of them will result in step loss.

Hysteresis

When a controller directs a stepper motor to seek a specified position, the term *hysteresis* is often used to mean the total error between the actual position it reaches when turning clockwise, and the actual position it reaches when turning counter-clockwise. This difference may occur because a stepper motor tends to stop a fraction short of its intended position, especially under significant load. Any design that requires precision should be tested under real-world conditions to assess the hysteresis of the motor.

Resonance

A motor has a natural resonant frequency. If it is stepped near that frequency, vibration will tend to be amplified, which can cause positional errors, gear wear (if gears are attached), bearing wear, noise, and other issues. A good datasheet should specify the resonant frequency of the motor, and the motor should run above that frequency if possible. The problem can be addressed by rubber motor mounts or by using a resilient component, such as a drive belt, in conjunction with the drive shaft. *Damping* the vibration may be attempted by adding weight to the motor mount.

Note that if the motor has any significant weight attached directly to its shaft, this will lower its resonant frequency, and should be taken into account.

Resonance may also cause *step loss* (see preceding sections).

Hunting

In a closed-loop system, a sensor on the motor reports its rotational position to the controller, and if necessary, the controller responds by adjusting the position of the motor. Like any feedback system, this entails some lag time, and at certain speeds the motor may start *hunting* or *oscillating* as the controller over-corrects and must then correct its correction. Some closed-loop controllers avoid this issue by running mostly in open-loop mode, using correction only when the motor experiences conditions (such as sudden speed changes), which are likely to cause step loss.

Saturation

While it may be tempting to increase the torque from a stepper motor by upping the voltage (which will increase the current through the stator coils), in practice motors are usually designed so that the cores of the coils will be close to saturation at the rated voltage. Therefore, increasing the voltage may achieve very little increase in power, while causing a significant increase in heat.

Rotor Demagnetization

The permanent magnets in a rotor can be partially demagnetized by excessive heat. Demagnetization can also occur if the magnets are exposed to high-frequency alternating current when the rotor is stationary. Therefore, attempting to run a stepper motor at high speed when the rotor is stalled can cause irrevocable loss of performance.

26 diode

The term **diode** almost always means a semiconductor device, properly known as a *PN junction diode*, although the full term is not often used. It was formerly known as a *crystal diode*. Before that, **diode** usually meant a type of *vacuum tube*, which is now rarely used outside of high-wattage RF transmitters and some high-end audio equipment.

OTHER RELATED COMPONENTS

- **rectifier** (See "Rectification" on page 221)
- **unijunction transistor** (See Chapter 27)
- **LED** (light-emitting diode) (Volume 2)

What It Does

A diode is a two-terminal device that allows current to flow in one direction, known as the *forward direction,* when the *anode* of the diode has a higher positive potential than the *cathode*. In this state, the diode is said to be *forward biased*. If the polarity of the voltage is reversed, the diode is now *reverse biased*, and it will attempt to block current flow, within its rated limits.

Diodes are often used as *rectifiers* to convert alternating current into direct current. They may also be used to suppress voltage spikes or protect components that would be vulnerable to reversed voltage, and they have specialized applications in high-frequency circuits.

A *Zener diode* can regulate voltage, a *varactor diode* can control a high-frequency oscillator, and *tunnel diodes*, *Gunn diodes*, and *PIN diodes* have high-frequency applications appropriate to their rapid switching capability. An **LED** (*light-emitting diode*) is a highly efficient light source, which is discussed in Volume 2 of this encyclopedia. A **photosensitive diode** will adjust its ability to pass current depending on the light that falls upon it, and is included as a *sensor* in Volume 3.

See Figure 26-1 for schematic symbols representing a generic diode.

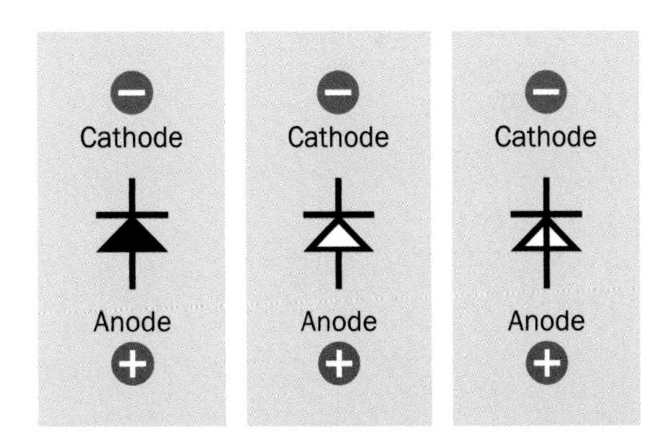

Figure 26-1. *Commonly used schematic symbols for a generic diode. All the symbols are functionally identical. The direction of the arrow formed by the triangle indicates the direction of conventional current (from positive to negative) when the diode is forward-biased.*

The basic diode symbol is modified in various ways to represent variants, as shown in Figure 26-2.

At top
: Each symbol in the group of six indicates a Zener diode. All are functionally identical.

Bottom-left
: Tunnel diode.

Bottom-center
: Schottky diode.

Bottom-right
: Varactor.

A triangle with an open center does not indicate any different function from a triangle with a solid center. The direction of the arrow always indicates the direction of conventional current, from positive to negative, when the diode is forward-biased, although the functionality of Zener diodes and varactors depends on them being reverse-biased, and thus they are used with current flowing opposite to the arrow symbol. The bent line used in the Zener symbol can be thought of as an opened letter Z, while the curled line used in the Schottky diode symbol can be thought of as a letter S, although these lines are sometimes drawn flipped left-to-right.

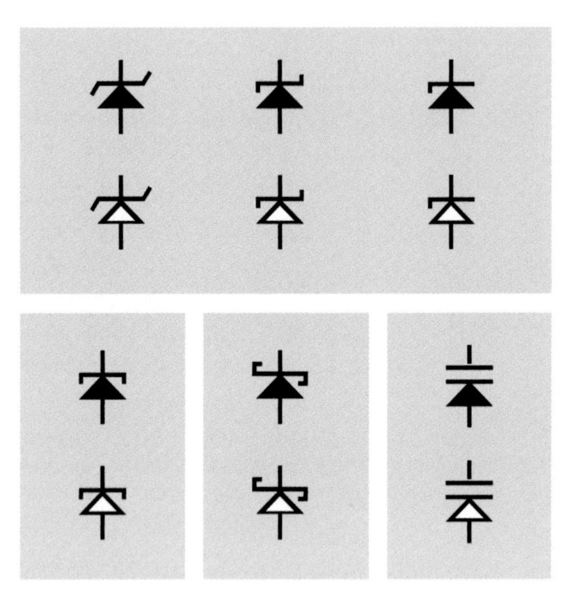

Figure 26-2. *Commonly used schematic symbols for specialized types of diodes. See text for details.*

A range of rectifier and signal diodes is shown in Figure 26-3. (Top: Rectifier diode rated 7.5A at 35VDC. Second from top: Rectifier diode rated 5A at 35VDC. Center: Rectifier diode rated 3A at 35VDC. Second from bottom: 1N4001 Rectifier diode rated 1A at 35VDC. Bottom: 1N4148 signal switching diode rated at 300mA.) All values are for forward continuous current and RMS voltage. Each cylindrical diode is marked with a silver stripe (a black stripe on the 1N4148) to identify its *cathode*, or the end of the diode that should be "more negative" when the component is forward biased. Peak current can greatly exceed continuous current without damaging the component. Datasheets will provide additional information.

Figure 26-3. *Diodes ranging in continuous forward-current capability from 7.5A (top) to 300mA (bottom). See text for additional details.*

How It Works

A PN diode is a two-layer *semiconductor*, usually fabricated from silicon, sometimes from germanium, and rarely from other materials. The layers are doped with impurities to adjust their electrical characteristics (this concept is explained in more detail in Chapter 28). The *N layer* (on the negative, cathode side) has a surplus of electrons, creating a net negative charge. The *P layer* (on the positive, anode side) has a deficit of electrons, creating a net positive charge. The deficit of electrons can also be thought of as a surplus of "positive charges," or more accurately, a surplus of *electron holes*, which can be considered as spaces that electrons can fill.

When the negative side of an external voltage source is connected with the cathode of a diode, and the positive side is connected with the anode, the diode is forward-biased, and electrons and electron holes are forced by mutual repulsion toward the junction between the n and p layers (see Figure 26-4). In a silicon diode, if the potential difference is greater than approximately 0.6 volts, this is known as the *junction threshold voltage*, and the charges start to pass through the junction. The threshold is only about 0.2 volts in a germanium diode, while in a Schottky diode it is about 0.4 volts.

If the negative side of an external voltage source is connected with the anode of a diode and positive side is connected with the cathode, the diode is now reverse-biased, and electrons and electron holes are attracted away from the junction between the n and p layers. The junction is now a *depletion region*, which blocks current.

Like any electronic component, a diode is not 100% efficient. When it is forward-biased and is passing current, it imposes a small voltage drop of around 0.7V for a silicon-based diode (Schottky diodes can impose a drop of as little as 0.2V, germanium diodes 0.3V, and some LEDs between 1.4V and 4V). This energy is dissipated as heat. When the diode is reverse-biased, it is still not 100% efficient, this time in its task of blocking current. The very small amount of current that manages to get through is known as *leakage*. This is almost always less than 1mA and may be just a few µA, depending on the type of diode.

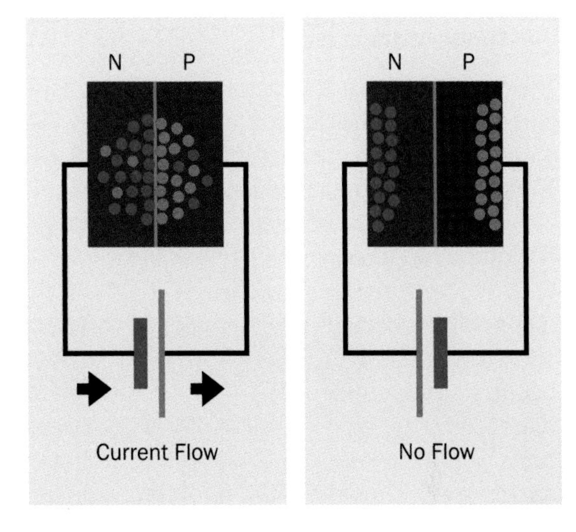

Figure 26-4. *Inside a PN junction diode. Left: in forward-biased mode, voltage from a battery (bottom, with plates colored for clarity) forces charges in the N and P layers toward the central junction of the diode. Current begins to flow. Right: in reverse-biased mode, charges in the N and P layers are attracted away from the central junction, which becomes a depletion region, unable to pass significant current.*

The performance of a theoretical generic PN diode is illustrated in Figure 26-5. The right-hand side of the graph shows that if a diode is forward-biased with a gradually increasing potential, no current passes until the diode reaches its junction threshold voltage, after which the current rises very steeply, as the *dynamic resistance* of the diode diminishes to near zero. The left-hand side of the graph shows that when the diode is reverse-biased with a gradually increasing potential, initially a very small amount of current passes as leakage (the graph exaggerates this for clarity). Eventually, if the potential is high enough, the diode reaches its intrinsic *breakdown voltage*, and once again its effective resistance diminishes to near zero. At either end of the curve, the diode will be easily and permanently damaged by excessive current. With the exception of Zener diodes and varactors, reverse bias on a diode should not be allowed to reach the breakdown voltage level.

The graph in Figure 26-5 does not have a consistent scale on its Y axis, and in many diodes the magnitude of the (reverse-biased) breakdown voltage will be as much as 100 times the magnitude of the (forward-biased) threshold voltage. The graph has been simplified for clarity.

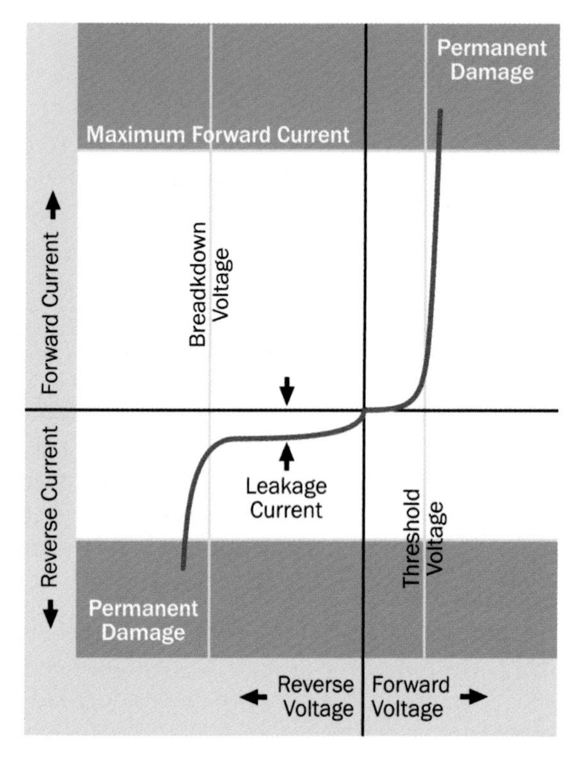

Figure 26-5. *As the forward voltage across a diode reaches the junction threshold, the diode begins passing current. If the voltage across the diode is reversed, initially a small amount of current leakage occurs. Excessive forward or reverse voltage will create sufficient current to destroy the component.*

Variants

Packaging

Some diodes have no information at all printed on them, while others may have a part number. Any additional information is rare. No convention exists for indicating the electrical characteristics of the component by colors or abbreviations. If one terminal is marked in any way, almost certainly it is the cathode. One way to remember the meaning of a stripe on the cathode end of a rectifier diode or signal diode is by thinking of it as resembling the line in the diode schematic symbol.

Signal Diodes

Also known as *switching diodes* and *high-speed diodes*, their small size provides a low junction capacitance, enabling fast response times. They are not designed to withstand high currents. Signal diodes traditionally were packaged with axial leads for through-hole installation (like traditional-style resistors). Although this format still exists, signal diodes are now more commonly available in surface-mount formats.

Rectifier Diodes

Physically larger than signal diodes, and capable of handling higher currents. Their higher junction capacitance makes them unsuitable for fast switching. Rectifier diodes often have axial leads, although different package formats are used where higher currents are involved, and may include a heat sink, or may have provision for being attached to a heat sink.

There are no generally agreed maximum or minimum ratings to distinguish signal diodes from rectifier diodes.

Zener Diode

A Zener diode generally behaves very similarly to a signal or rectifier diode, except that its breakdown voltage is lower.

The Zener is intended to be reverse-biased; that is, conventional current is applied through it "in the wrong direction" compared with conventional diodes. As the current increases, the *dynamic resistance* of the Zener diode decreases. This relationship is shown in Figure 26-6, where the two colored curves represent the performance of different possible Zener diodes. (The curves are adapted from a manufacturer's datasheet.) This behavior allows the Zener to be used in simple voltage-regulator circuits, as it can allow a reverse current to flow at a voltage limited by the diode's breakdown voltage. Other applications for Zener diodes are described in "DC Voltage Regulation and Noise Suppression" on page 224. A typical Zener diode is shown in Figure 26-7.

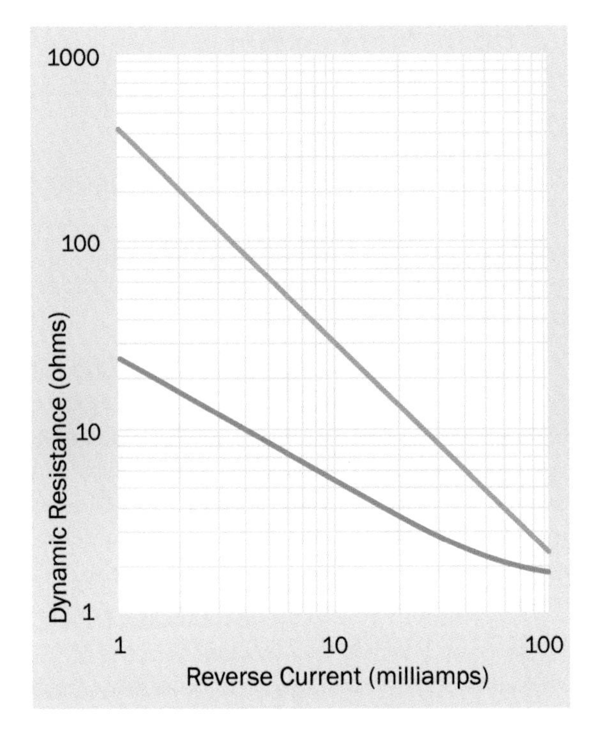

Figure 26-6. *A manufacturer's datasheet may include graphs of this kind, showing the variation in dynamic resistance of two reverse-biased Zener diodes in response to changes in current.*

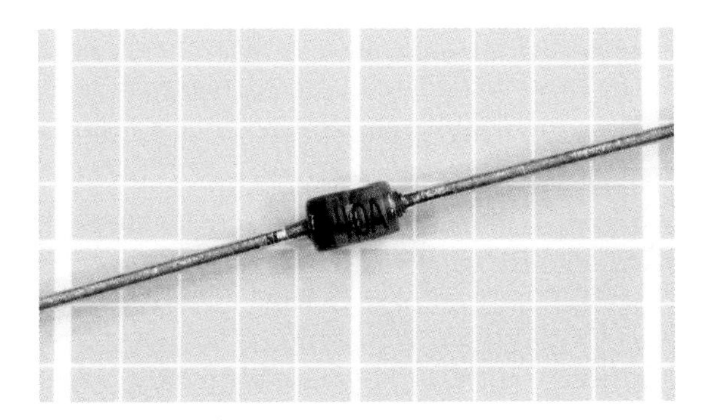

Figure 26-7. *A 1N4740 Zener diode.*

Transient Voltage Suppressor (TVS)

A form of Zener diode designed to protect sensitive devices from transient voltage spikes by clamping them—in other words, diverting the energy to ground. A TVS can absorb as much as 30,000 volts from a lightning strike or static discharge. Typically the Zener diode is incorporated in a network of other diodes in a surface-mount integrated circuit chip.

Zener diodes can also be used in circuits to handle electrostatic discharge (ESD), which can occur when a person unknowingly accumulates an electrostatic potential and then grounds it by touching an electronic device.

Schottky Diode

This type has a low junction capacitance, enabling faster switching than comparable generic silicon diodes. It also imposes a lower forward voltage drop, which can be desirable in low-voltage applications, and allows less power dissipation when a diode is necessary to control current flow. The Schottky diode is fabricated with a semiconductor-to-metal junction, and tends to be slightly more expensive than generic silicon diodes with similar voltage and current specifications.

Varactor Diode

Also known as a *varicap*, this type of diode has variable capacitance controlled by reverse voltage. While other diodes may exhibit this same phenomenon, the varactor is specifically designed to exploit it at very high frequencies. The voltage expands or contracts the depletion region in the junction between the P and N regions, which can be thought of as analogous to moving the plates of a capacitor nearer together or farther apart.

Because the capacitance of a varactor has a low maximum of about 100pF, its uses are limited. It is used extensively in RF applications where its voltage-controlled variable capacitance provides a uniquely useful way to control the frequency of an oscillator circuit. In almost all radio, cellular, and wireless receivers, a varactor controls a phase-locked loop oscillator. In ham radio receivers, it can be used to adjust the tuning of a filter that tracks an incoming radio frequency.

A varactor is always reverse-biased below its breakdown voltage, so that there is no direct conduction. The voltage that controls a varactor must be absolutely free from random fluctuations that would affect its resonant frequency.

Tunnel Diode, Gunn Diode, PIN Diode

Mostly used in very high frequency or microwave applications, where ordinary diodes are unacceptable because they have insufficiently high switching speeds.

Diode Array

Two or more diodes may be encapsulated in a single DIP or (more commonly) surface-mount integrated circuit chip. The internal configuration and the pinouts of the chip will vary from one device to another. Diode arrays may be used for termination of data lines to reduce reflection noise.

Bridge Rectifier

Although this is a diode array, it is commonly indexed in parts catalogues under the term *bridge rectifier*. Numerous through-hole versions are available with ratings as high as 25A, some designed for single-phase input while others process three-phase AC. Screw-terminal components can rectify more than 1,000 volts at 1,000 amps. The package does not usually include any provision for smoothing or filtering the output. See "Rectification" on page 221 for more information on the behavior of a bridge rectifier.

Values

A manufacturer's datasheet for a typical generic diode should define the following values, using abbreviations that may include those in the following list.

- Maximum sustained forward current: I_f or I_o or I_{Omax}
- Forward voltage (the voltage drop imposed by the diode): V_f
- Peak inverse DC voltage (may be referred to as maximum blocking voltage or breakdown voltage): P_{iv} or V_{dc} or V_{br}
- Maximum reverse current (also referred to as leakage): I_r

Datasheets may include additional values when the diode is used with alternating current, and will also include information on peak forward surge current and acceptable operating temperatures.

A typical signal diode is the 1N4148 (included at the bottom of Figure 26-3), which is limited to about 300mA forward current while imposing a voltage drop of about 1V. The component can tolerate a 75V peak inverse voltage. These values may vary slightly among different manufacturers.

Rectifier diodes in the 1N4001/1N4002/1N4003 series have a maximum forward current of 1A and will impose a voltage drop of slightly more than 1V. They can withstand 50V to 1,000V of inverse voltage, depending on the component. Here again, the values may vary slightly among different manufacturers.

Zener diodes have a different specification, as they are used with reverse bias as voltage-regulating devices rather than rectification devices. Manufacturers' data sheets are likely to contain the following terminology:

- Zener voltage (the potential at which the diode begins to allow reverse current flow when it is reverse-biased, similar to breakdown voltage): V_z
- Zener impedance or dynamic resistance (the effective resistance of the diode, specified when it is reverse-biased at the Zener voltage): Z_z

- Maximum or admissible Zener current (or reverse current): I_z or I_{zm}
- Maximum or total power dissipation: P_d or P_{tot}

Zener voltage may be defined within a minimum and maximum range, or as a simple maximum value.

Limits on forward current are often not specified, as the component is not intended to be forward-biased.

How to Use it

Rectification

A *rectifier diode*, as its name implies, is commonly used to rectify alternating current—that is, to turn AC into DC. A half-wave rectifier uses a single diode to block one-half of the AC sinewave. The basic circuit for a half-wave rectifier is shown in Figure 26-8. At top, the diode allows current to circulate counter-clockwise through the load. At bottom, the diode blocks current that attempts to circulate clockwise. Although the output has "gaps" between the pulses, it is usable for simple tasks such as lighting an LED, and with the addition of a smoothing capacitor, can power the coil of a DC relay.

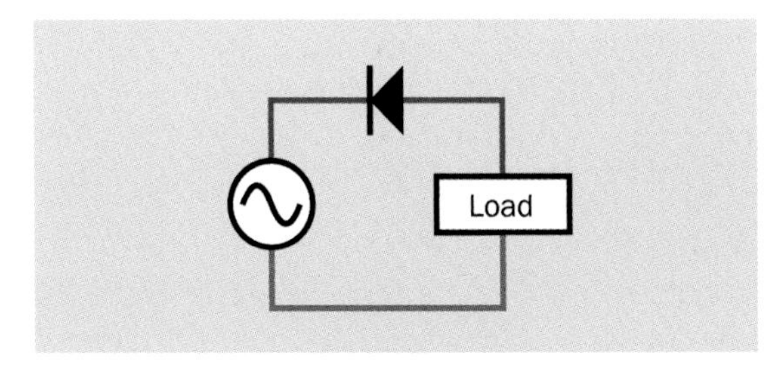

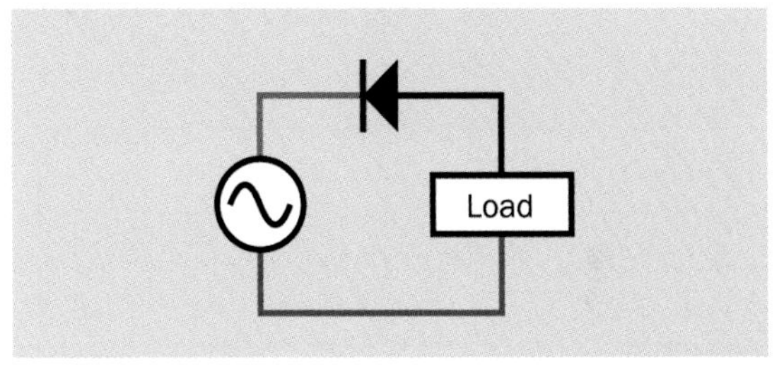

Figure 26-8. *A half-wave rectifier. In this configuration the diode allows AC current to circulate counter-clockwise but blocks it clockwise.*

A full-wave bridge rectifier employs four diodes to provide a more efficient output, usually filtered and smoothed with appropriate **capacitors**. The basic circuit is shown in Figure 26-9. A comparison of input and output waveforms for half-wave and full-wave rectifiers appears in Figure 26-10.

Discrete components are seldom used for this purpose, as off-the-shelf bridge rectifiers are available in a single integrated package. Rectifier diodes as discrete components are more likely to be used to suppress *back-EMF* pulses, as described below.

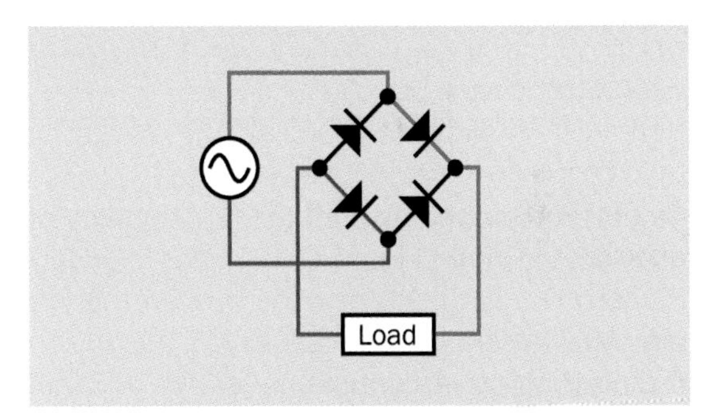

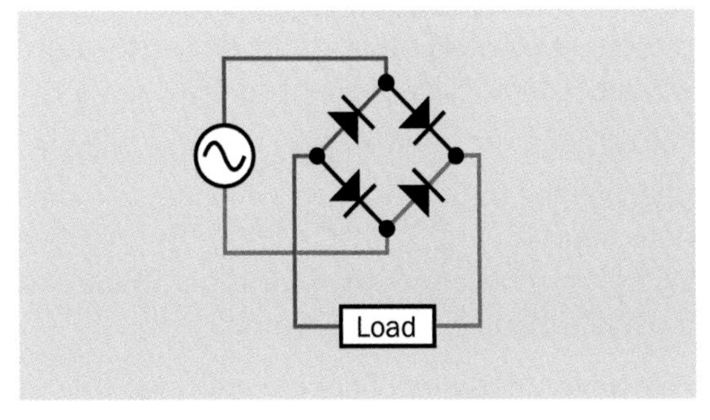

Figure 26-9. *The basic circuit commonly used to form a bridge rectifier, with color added to indicate polarity. Wires shown in black are not passing current because diodes are blocking it. Note that the polarity at the load remains constant.*

An old but widely used design for a full-wave bridge rectifier is shown in Figure 26-11. This unit measured approximately 2″ × 2″ × 1.5″ and was divided into four sections (as indicated by the solder terminals on the right-hand side), each

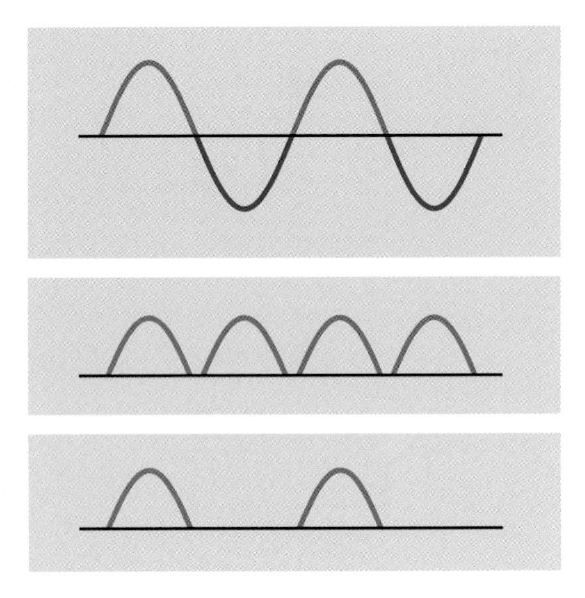

Figure 26-10. *Top: The voltage-amplitude sinewave of an alternating current source that fluctuates between positive voltage (shown red) and negative voltage (shown blue) relative to a neutral (black) baseline. Center: AC current converted by a full-wave rectifier. Because the diodes do not conduct below their threshold voltage, small gaps appear between pulses. Bottom: Output from a half-wave rectifier.*

Figure 26-11. *Prior to the perfection of chip fabrication in the late 1960s, it was common to find selenium rectifiers of this type, measuring about 2" square.*

section corresponding with the functionality of one modern diode. Figure 26-12 shows relatively modern rectifier packages, the one on the left rated at 20A continuous at 800V RMS, the one on the right rated 4A continuous at 200V RMS. In Figure 26-13, the one on the left is rated 4A continuous at 50V RMS, whereas the one on the right is rated 1.5A at 200V RMS.

DC output from rectifier packages is usually supplied via the outermost pins, while the two pins near the center receive AC current. The positive DC pin may be longer than the other three, and is usually marked with a + symbol.

Full-wave bridge rectifiers are also available in surface-mount format. The one in Figure 26-14 is rated for half an amp continuous current.

Figure 26-12. *Full-wave bridge rectifiers are commonly available in packages such as these. See text for details.*

Back-EMF Suppression

A relay coil, motor, or other device with significant inductance will typically create a spike of voltage when it is turned on or off. This *EMF* can be shunted through a rectifier diode to safeguard other components in the circuit. A diode in this configuration may be referred to as a *protection diode*, a *clamp diode*, or *transient suppressor*. See Figure 26-15.

Figure 26-13. *Smaller full-wave bridge rectifiers capable of 1.5A to 4A continuous current.*

Figure 26-14. *This surface-mount component contains four diodes forming a full-wave bridge rectifier circuit, and can pass 0.5A continuous current. It measures approximately 0.2" square.*

Voltage Selection

A diode is sensitive to the relative voltage between its anode and cathode terminals. In other words, if the cathode is at 9V relative to the ground in the circuit, and the anode is at 12V, the 3V difference will easily exceed the threshold voltage, and the diode will pass current. (Actual tolerable values will depend on the forward voltage capability of the diode.) If the voltages are reversed, the diode will block the current.

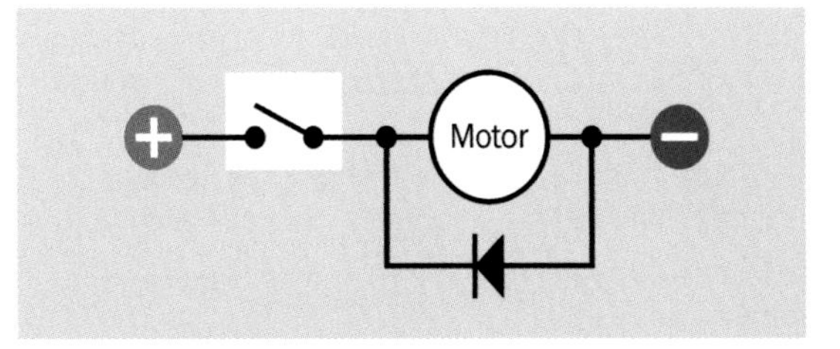

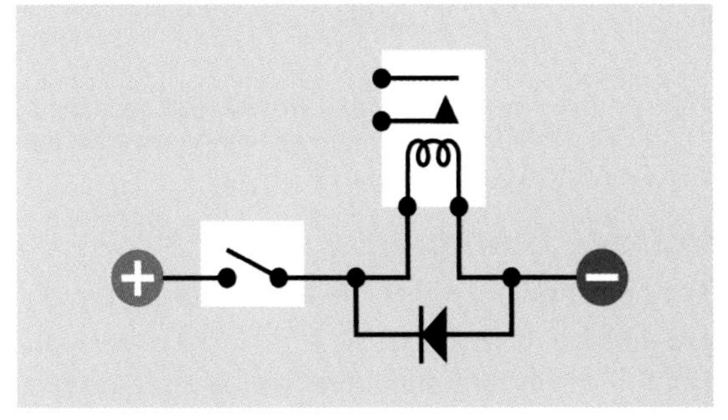

Figure 26-15. *A rectifier diode is very often placed across a motor (top), relay (bottom), or other device with significant inductance that creates a spike of reverse voltage when switched on or off. The surge is shunted through the diode, protecting other components in the circuit.*

This attribute can be used to make a device choose automatically between an AC adapter and a 9V battery. The schematic is shown in Figure 26-16. When an AC adapter that delivers 12VDC is plugged into a wall outlet, the adapter competes with the battery to provide power to a voltage regulator. The battery delivers 9VDC through the lower diode to the cathode side of the upper diode, but the AC adapter trumps it with 12VDC through the upper diode. Consequently, the battery ceases to power the circuit until the AC adapter is unplugged, at which point the battery takes over, and the upper diode now prevents the battery from trying to pass any current back through the AC adapter.

The voltage regulator in this schematic accepts either 12VDC or 9VDC and converts it to 5VDC. (In the case of 12VDC, the regulator will waste more power, which will be dissipated as waste heat.)

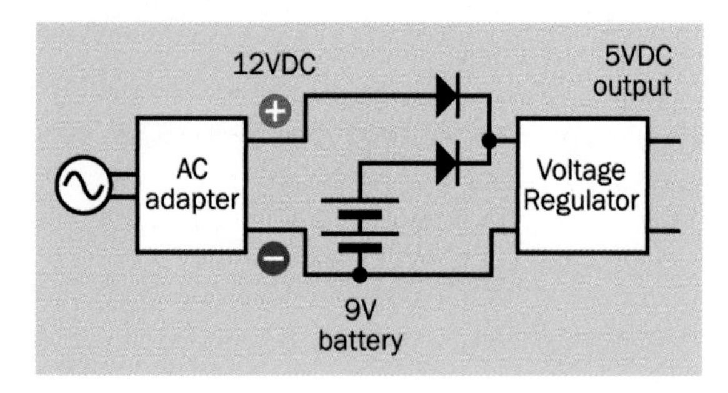

Figure 26-16. *Two diodes with their cathodes tied together will choose automatically between an AC adapter that delivers 12VDC and an internal 9V battery.*

Voltage Clamping

A diode can be used to *clamp* a voltage to a desired value. If an input to a 5V CMOS semiconductor or similarly sensitive device must be prevented from rising out of range, the anode of a diode can be connected to the input and the cathode to a 5V voltage source. If the input rises much above 5.6V, the potential difference exceeds the diode's junction threshold, and the diode diverts the excess energy. See Figure 26-17.

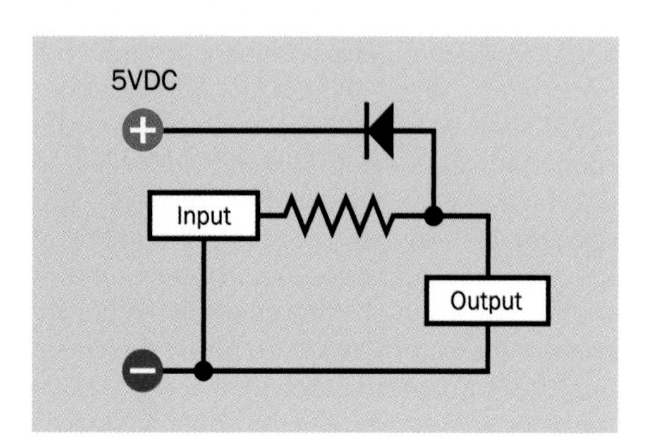

Figure 26-17. *A clamping diode can limit output voltage—in this example, to about 5.6V. If the input rises above that value relative to the common ground, the potential difference across the diode feeds the excess voltage back through it to the 5V source.*

Logic Gate

A signal diode is less than ideal as a logic gate, because it imposes a typical 0.6V voltage reduction, which can be significant in a 5V circuit and is probably unacceptable in a 3.3V circuit. Still, it can be useful on the output side—for example, if two or more outputs from a logic chip or microcontroller are intended to drive, or share, another device such as a single **LED**, as shown in Figure 26-18. In this role, the diodes wired in parallel behave similarly to an OR gate, while preventing either output from the chip from feeding current back into the other output.

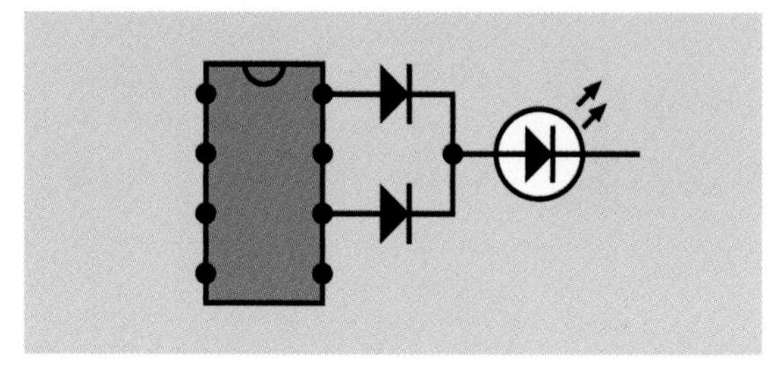

Figure 26-18. *Two or more outputs from a logic chip or microcontroller may be coupled with diodes to power another device, such as an LED, while protecting the chip from backflow of current. The diodes form a logical OR gate.*

DC Voltage Regulation and Noise Suppression

As previously noted, the dynamic resistance of a reverse-biased Zener diode will diminish as the current increases. This relationship begins at the point where breakdown in the diode begins—at its *Zener voltage*--and is approximately linear over a limited range.

The unique behavior of the Zener makes it usable as a very simple voltage controller when placed in series with a resistor as shown in Figure 26-19. It is helpful to imagine the diode and the resistor as forming a kind of voltage divider, with power being taken out at point A in the schematic. If a supply fluctuation increases the input voltage, this will tend to increase the current flowing through the Zener, and its dynamic resistance will diminish accordingly. A lower resistance in its position in the voltage divider will reduce the output voltage at point A, thus tending to compensate for the surge in input voltage.

Conversely, if the load in the circuit increases, and tends to pull down the input voltage, the current

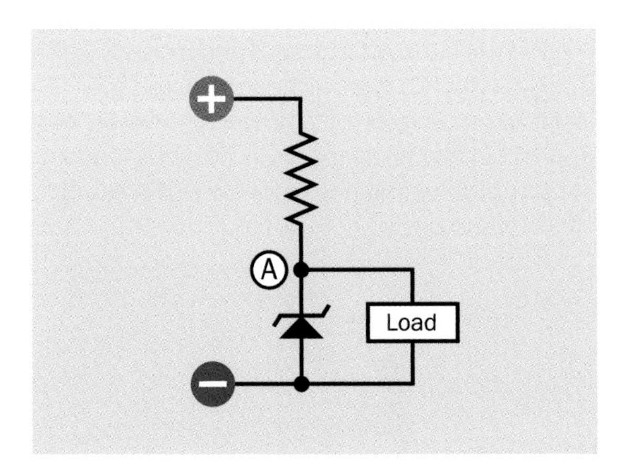

Figure 26-19. *A simplified, basic circuit illustrating the ability of a Zener diode to compensate for variations in the power supply or load in a circuit, creating an approximately constant voltage at point A.*

flowing through the Zener will diminish, and the voltage at point A will tend to increase, once again compensating for the fluctuation in the circuit.

As the series resistor would be a source of heat, a transistor could be added to drive the load, as shown in Figure 26-20.

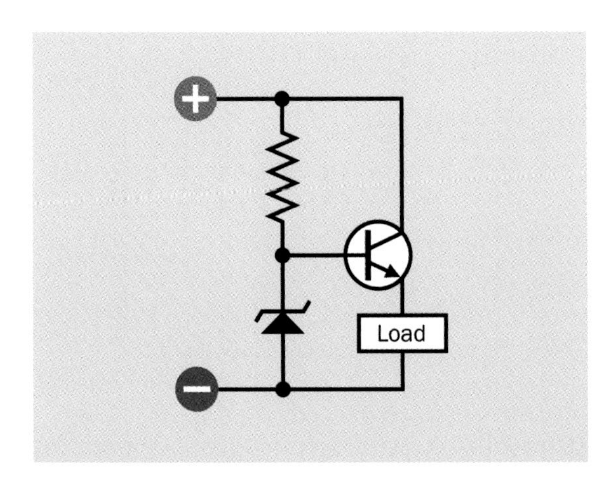

Figure 26-20. *A transistor could be added to the circuit in the previous figure to reduce power waste through the resistor.*

A manufacturer's datasheet may provide guidance regarding the dynamic resistance of a Zener diode in response to current, as previously shown in Figure 26-6. In practice, a packaged **voltage regulator** such as the LM7805 would most likely be used instead of discrete components, since it includes self-calibrating features, requires no series resistor, and is relatively unaffected by temperature. However, the LM7805 contains its own Zener diode, and the principle of operation is still the same.

AC Voltage Control and Signal Clipping

A more practical Zener application would be to limit AC voltage and/or impose clipping on an AC sinewave, using two diodes wired in series with opposed polarities. The basic schematic is shown in Figure 26-21, while clipping of the AC sinewave is illustrated in Figure 26-22. In this application, when one diode is reverse-biased, the other is forward-biased. A forward-biased Zener diode works like any other diode: it allows current to pass relatively freely, so long as the voltage exceeds its threshold. When the AC current reverses, the Zeners trade their functions, so that the first one merely passes current while the second one limits the voltage. Thus, the diodes divert peak voltage away from the load. The Zener voltage of each diode would be chosen to be a small margin above the AC voltage for voltage control, and below the AC voltage for signal clipping.

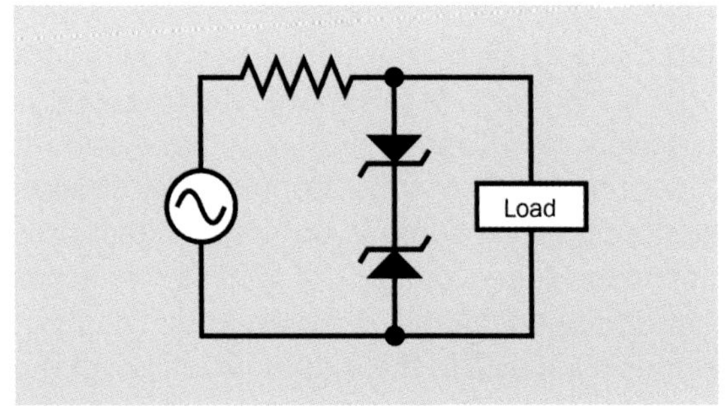

Figure 26-21. *Two Zener diodes placed in series, with opposite polarities, can clip or limit the voltage sinewave of an AC signal.*

Voltage Sensing

A Zener diode can be used to sense a small shift in voltage and provide a switched output in response.

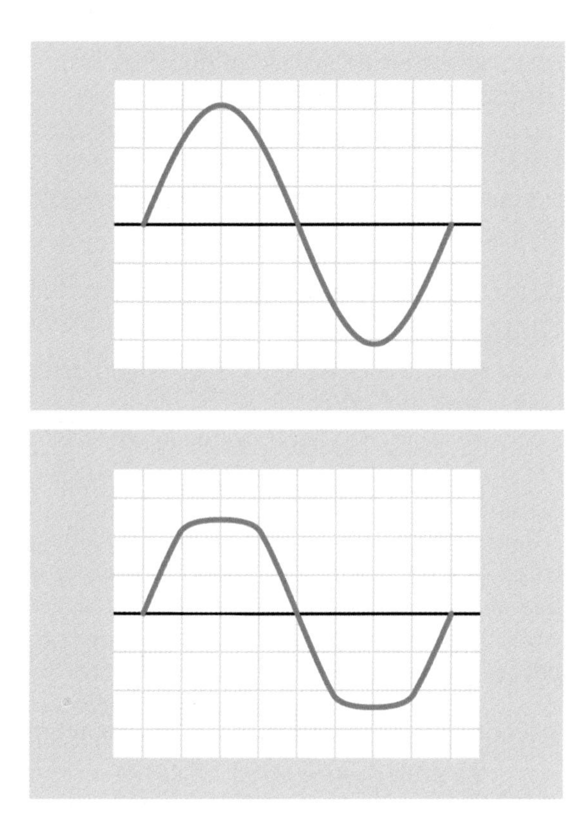

Figure 26-22. *AC input showing a pure sinewave (left) and a clipped version (right) created by Zener diodes wired in series, as in the previous figure.*

In Figure 26-23, the upper schematic shows a Zener diode preventing voltage from reaching the emitter of a PNP transistor while the divided input signal is below the Zener (breakdown) voltage of the diode. In this mode, the transistor is relatively non-conductive, very little current flows through it, and the output is now at near-zero voltage. As soon as the input signal rises above the Zener voltage, the transistor switches on and power is supplied to the output. The input is thus replicated in the output, as shown in the upper portion of Figure 26-24.

In Figure 26-23, the lower schematic shows a Zener diode preventing voltage from reaching the base of an NPN transistor while the input signal is below the Zener (breakdown) voltage of the diode. In this mode, the transistor is relatively non-conductive, and power is supplied to the output. As soon as the input signal rises above the Zener voltage, the transistor is activated, diverting the current to ground and bypassing the output, which is now at near-zero voltage. The input is thus inverted, as shown in the lower portion of Figure 26-24 (provided there is enough current to drive the transistor into saturation).

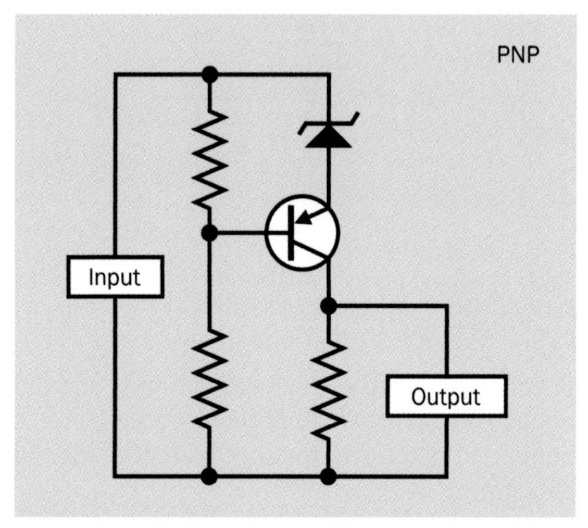

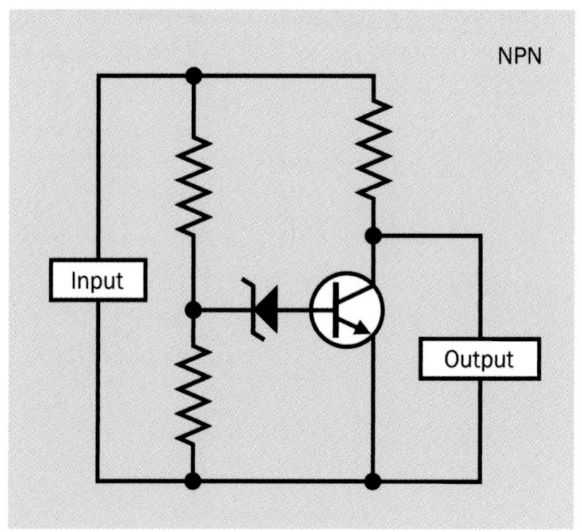

Figure 26-23. *A Zener diode can be used in conjunction with a PNP transistor. See text for details.*

What Can Go Wrong

Overload

If maximum forward current is exceeded, the heat generated is likely to destroy the diode. If the diode is reverse-biased beyond its peak in-

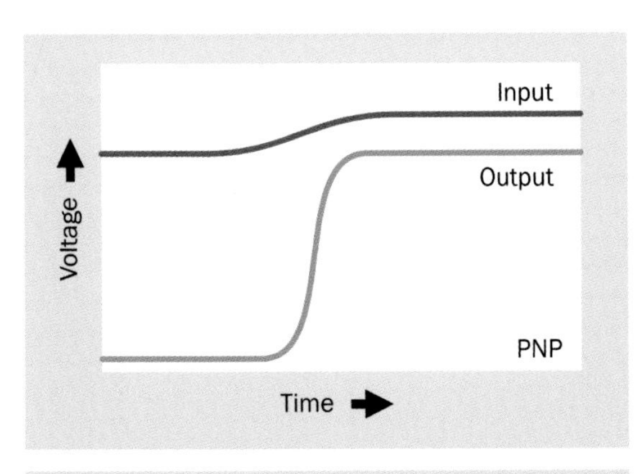

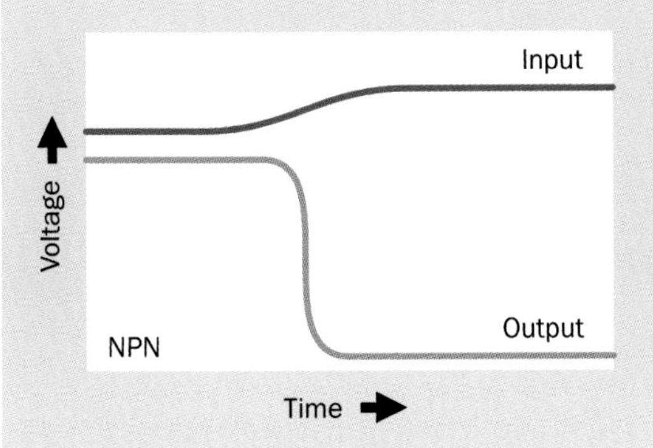

Figure 26-24. *Theoretical output from the transistors in the two previous schematics.*

verse voltage limit, the current will overwhelm the diode's ability to block it, and an *avalanche breakdown* will occur, once again probably destroying the component. The graph in Figure 26-5 illustrates the performance range of a hypothetical generic diode.

Reversed Polarity

Zener diodes look almost identical to other types, and all diodes share the same convention of marking the cathode for identification. Yet Zeners must be reverse-biased while others are forward-biased. This creates a significant risk of installing a diode "the wrong way around," with potentially destructive or at least confusing results, especially when used in a power supply. The very low resistance of a diode to forward current makes it especially vulnerable to burnout if installed incorrectly.

Wrong Type of Diode

If a Zener diode is used accidentally where a signal or rectifier diode is appropriate, the circuit will malfunction, as the Zener will probably have a much lower breakdown voltage, and therefore will not block reverse current. Conversely, if a signal or rectifier diode is used where the circuit calls for a Zener diode, reverse voltage will be clamped (or regulated at the diode's forward voltage value). Since diodes are often poorly marked, a sensible precaution is to store Zener diodes separately from all other types.

unijunction transistor

27

The unijunction transistor (UJT) and programmable unijunction transistor (PUT) are different internally, but are sufficiently similar in function to be combined in this entry.

OTHER RELATED COMPONENTS

- **diode** (See Chapter 26)
- **bipolar transistor** (See Chapter 28)
- **field-effect transistor** (See Chapter 29)

What It Does

Despite their names, the unijunction transistor (UJT) and programmable unijunction transistor (PUT) are not current-amplification devices like bipolar transistors. They are switching components that are more similar to diodes than to transistors.

The UJT can be used to build low- to mid-frequency oscillator circuits, while the PUT provides similar capability with the addition of more sophisticated control, and is capable of functioning at lower currents. The UJT declined in popularity during the 1980s after introduction of components such as the *555 timer*, which offered more flexibility and a more stable output frequency, eventually at a competitive price. UJTs are now uncommon, but PUTs are still available in quantity as through-hole discrete components. Whereas an integrated circuit such as a 555 timer generates a square wave, unijunction transistors in oscillator circuits generate a series of voltage spikes.

The PUT is often used to trigger a **thyristor** (described in Volume 2) and has applications in low-power circuits, where it can draw as little as a few microamps.

Schematic symbols for the two components are shown in Figure 27-1 and Figure 27-2. Although the symbol for the UJT is very similar to the symbol for a field-effect transistor (FET), its behavior is quite different. The bent arrow identifies the UJT, while a straight arrow identifies the FET. This difference is of significant importance.

The schematic symbol for a PUT indicates its function, as it resembles a diode with the addition of a gate connection.

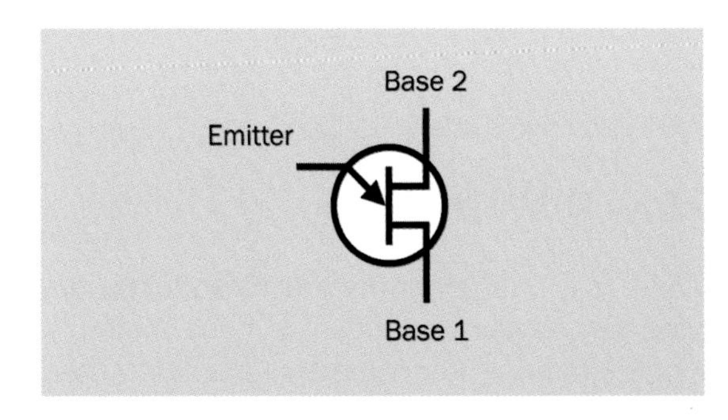

Figure 27-1. *Schematic symbol for a unijunction transistor (UJT). Note the bent arrow. The symbol for a field-effect transistor looks similar, but has a straight arrow. The functionality of the two components is very different.*

In Figure 27-3, the transistors at left and center are old-original unijunction transistors, while the one at right is a programmable unijunction transistor. (Left: Maximum 300mW, 35V interbase voltage. Center: 450mW, 35V interbase voltage. Right: 300mW, 40V gate-cathode forward voltage, 40V anode-cathode voltage.)

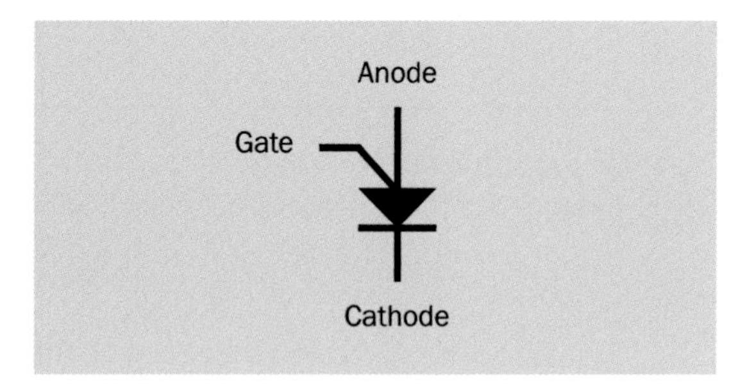

Figure 27-2. *Schematic symbol for a programmable unijunction transistor (PUT). The symbol accurately suggests the similarity in function to a diode, with the addition of a gate to adjust the threshold voltage.*

Figure 27-3. *The unijunction transistors at left and center are becoming obsolete; the one at the right is a programmable unijunction transistor (PUT), still readily available and widely used as a thyristor trigger.*

How It Works

The UJT is a three-terminal semiconductor device, but contains only two sections sharing a single junction—hence its name. Leads attached to opposite ends of a single channel of N-type semiconductor are referred to as base 1 and base 2, with base 2 requiring a slightly higher potential than base 1. A smaller P-type insert, midway between base 1 and base 2, is known as the emitter.

The diagram in Figure 27-4 gives an approximate idea of internal function.

When no voltage is applied to the emitter, a relatively high resistance (usually more than 5K) prevents significant current flow from base 2 to base 1. When the positive potential at the emitter increases to a *triggering voltage* (similar to the *junction threshold voltage* of a forward-biased diode), the internal resistance of the UJT drops very rapidly, allowing current to enter the component via both the emitter and base 2, exiting at base 1. (The term "current" refers, here, to conventional current; electron flow is opposite.) Current flowing from base 2 to base 1 is significantly greater than current flowing from the emitter to base 1.

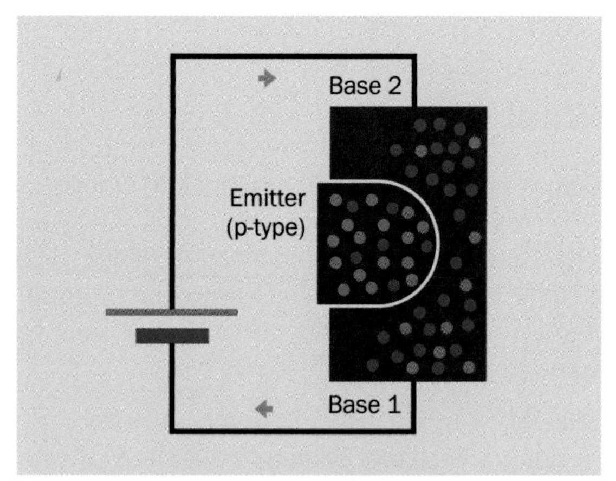

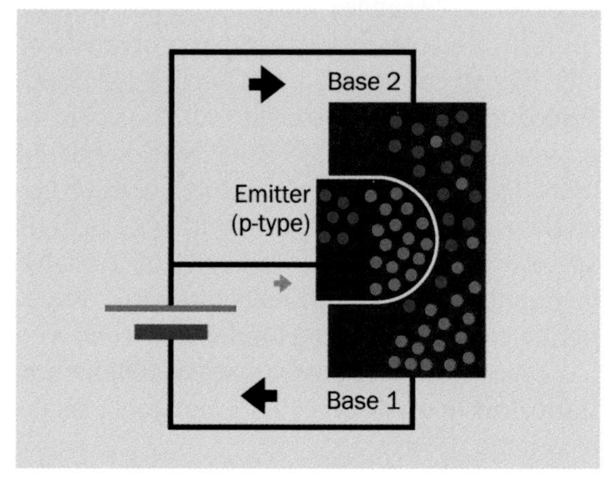

Figure 27-4. *Internal workings of a unijunction transistor.*

The graph in Figure 27-5 outlines the behavior of a UJT. As the voltage applied to the emitter increases, current flowing into the component from the emitter increases slightly, until the triggering voltage is reached. The component's internal resistance now drops rapidly. This pulls down the voltage at the emitter, while the current continues to increase significantly. Because of the drop in resistance, this is referred to as a *negative resistance* region. The resistance actually cannot fall below zero, but its change is negative. After emitter voltage drops to a minimum known as the *valley voltage*, the current continues to increase with a small increase in voltage. On datasheets, the peak current is often referred to as I_P while valley current is I_V.

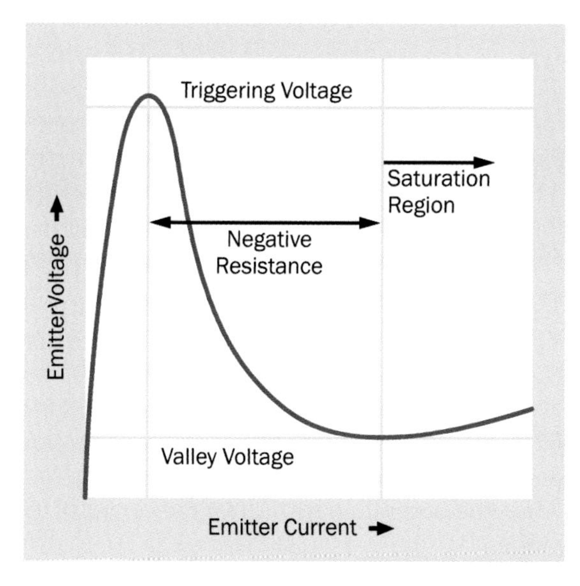

Figure 27-5. *Response curve of a unijunction transistor (UJT). When positive potential at the emitter reaches the triggering voltage, internal resistance drops radically and the component goes through a phase known as "negative resistance" as current increases.*

Figure 27-6 shows a test circuit to demonstrate the function of a UJT, with a volt meter indicating its status. A typical supply voltage would range from 9VDC to 20VDC.

A PUT behaves similarly in many ways to a UJT but is internally quite different, consisting of four semiconducting layers and functioning similarly to a thyristor.

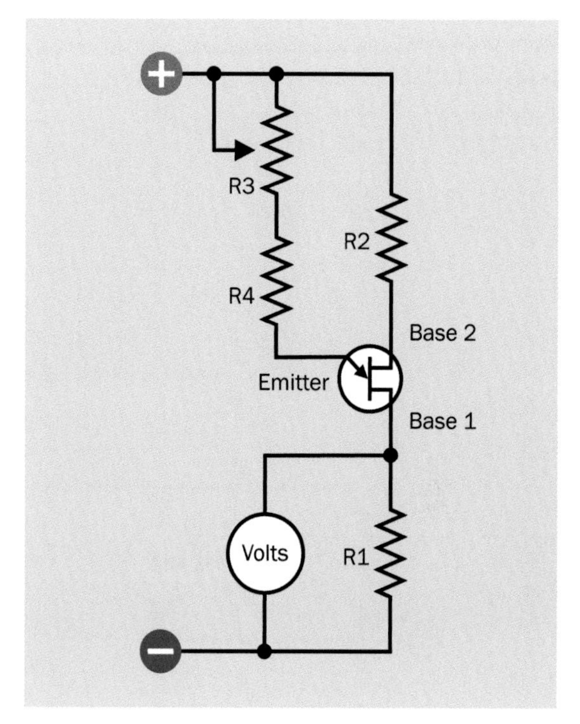

Figure 27-6. *A test circuit for a unijunction transistor (UJT) using a volt meter to show its response as a potentiometer increases the voltage applied to its emitter.*

The PUT is triggered by increasing the voltage on the anode. Figure 27-7 shows a test circuit for a PUT. This component is triggered when the voltage at its anode exceeds a threshold level, while the gate sets the threshold where this occurs. When the PUT is triggered, its internal resistance drops, and current flows from anode to cathode, with a smaller amount of current entering through the gate. This behavior is almost identical to that of a forward-biased diode, except that the threshold level can be controlled, or "programmed," according to the value of the positive potential applied at the gate, with R1 and R2 establishing that potential by functioning as a voltage divider.

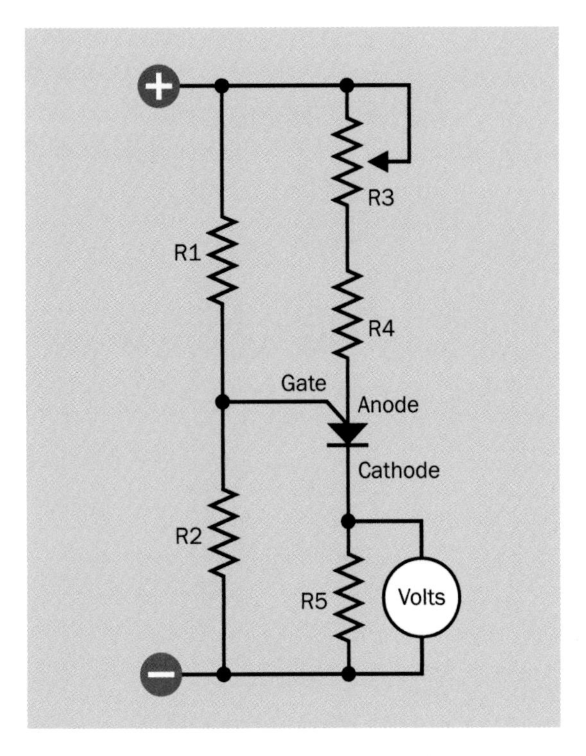

Figure 27-7. *A test circuit for a programmable unijunction transistor (PUT) using a volt meter to show its response as a potentiometer increases the voltage applied to its anode.*

The voltage output of a PUT follows a curve that is very similar to that shown in Figure 27-5, although current and voltage would be measured at the cathode.

Variants

PUTs and UJTs are not made as surface-mount components.

UJTs are usually packaged in black plastic, although older variants were manufactured in cans. PUTs are almost all packaged in black plastic. With the leads pointing downward and the flat side facing toward the viewer, the lead functions of a PUT are usually anode, gate, and cathode, reading from left to right.

Values

The triggering voltage of a UJT can be calculated from the values of R1 and R2 in Figure 27-7 and the voltage at base 1. The term R_{bb} is often used to represent the sum of R1 + R2, with V_{bb} representing the total voltage across the two resistors (this is the same as the supply voltage in Figure 27-6). V_t, the triggering voltage, is given by:

$$V_t = V_{bb} * (R1 / R_{bb})$$

The term ($R1/R_{bb}$) is known as the *standoff ratio*, often represented by the Greek letter ϱ.

Typically the standoff ratio in a UJT is at least 0.7, as R1 is chosen to be larger than R2. Typical values for R1 and R2 could be 180Ω and 100Ω, respectively. If R4 is 50K and a 100K linear potentiometer is used for R3, the PUT should be triggered when the potentiometer is near the center of its range. The emitter saturation voltage is typically from 2V to 4V.

If using a PUT, typical values in the test circuit could be supply voltage ranging from 9VDC to 20VDC, with resistances 28K for R1 and 16K for R2, 20Ω for R5, 280K for R4, and a 500K linear potentiometer for R3. The PUT should be triggered when the potentiometer is near the center of its range.

Sustained forward current from anode to cathode is usually a maximum of 150mA, while from gate to cathode the maximum is usually 50mA. Power dissipation should not exceed 300mW. These values should be lower at temperatures above 25 degrees Centigrade.

Depending on the PUT being used, power consumption can be radically decreased by upping the resistor values by a multiple of 100, while supply voltage can be decreased to 5V. The cathode output from the PUT would then be connected with the base of an NPN transistor for amplification.

How to Use it

Figure 27-8 shows a simple oscillator circuit built around a UJT, Figure 27-9 shows a comparable circuit for a PUT. Initially the supply voltage charges the capacitor, until the potential at the emitter of the UJT or the gate of the PUT reaches the threshold voltage, at which point the capacitor discharges through the emitter and the cycle repeats. Resistor values would be similar to those used in the test circuits previously described, while a capacitor value of 2.2μF would provide a visible pulse of the LED. Smaller capacitor values would enable faster oscillation. In the PUT circuit, adjusting the values of R1 and R2 would allow fine control of triggering the semiconductor.

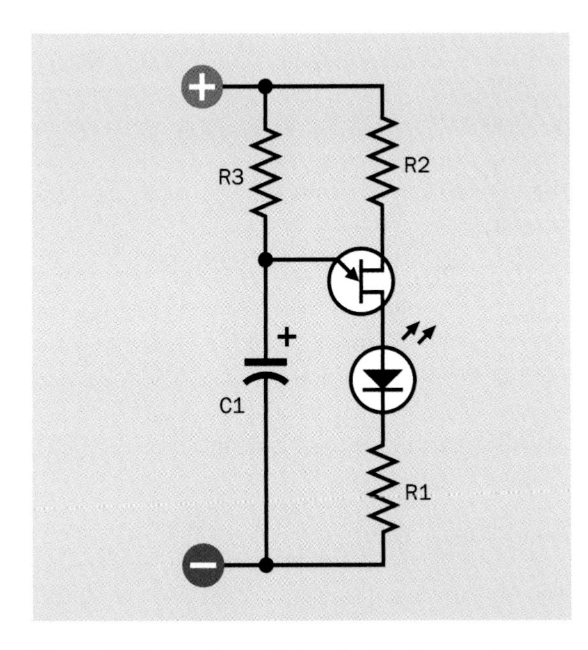

Figure 27-8. *A basic oscillator circuit using a unijunction transistor (UJT). As the capacitor accumulates charge, the voltage on the emitter increases until it triggers the UJT, at which point the capacitor discharges through the emitter.*

Probably the most common use for a PUT at this time is to trigger a **thyristor**.

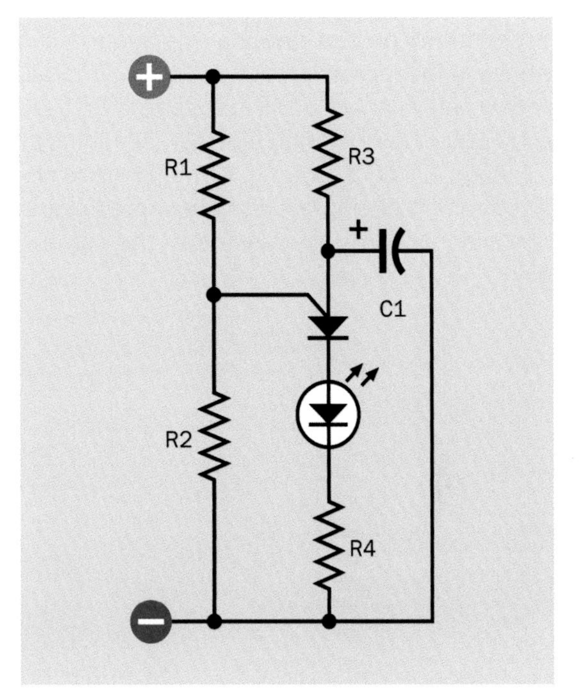

Figure 27-9. *A basic oscillator circuit using a programmable unijunction transistor (PUT). As the capacitor accumulates charge, the voltage on the anode increases until it triggers the PUT, at which point the capacitor discharges through the anode. The gate voltage is preset by R1 and R2 to adjust the triggering voltage.*

What Can Go Wrong

Name Confusion

A programmable unijunction transistor (PUT) is sometimes referred to simply as a "unijunction transistor" (UJT). Bearing in mind the totally different modes of operation of UJT and PUT, the PUT should always be identified by its acronym or by its full name. A circuit will not function if a UJT is substituted for a PUT, or a PUT is substituted for a UJT.

Incorrect Bias

Neither the UJT nor the PUT is designed to operate with reverse bias. In the UJT, a small forward bias should be applied from base 2 to base 1 (that is, base 2 should be at a higher potential relative to base 1) regardless of the voltage on the emitter. The emitter voltage may vary from 0 volts upward. The PUT must be forward biased between its anode and cathode (the anode must have a higher potential relative to the cathode), with an intermediate positive voltage at the gate established by resistors R1 and R2 functioning as a voltage divider (see Figure 27-7). Failure to observe correct biasing will result in unpredictable behavior and possible damage to the component.

Overload

Like any semiconductor, the UJT and the PUT must be protected from excessive current, which can burn out the component. Never connect either of these components directly across a power source without appropriate resistances to limit current flow. Maximum continuous power dissipation for UJTs and PUTs is usually 300mW.

bipolar transistor

28

The word *transistor*, on its own, is often used to mean *bipolar transistor*, as this was the type that became most widely used in the field of discrete semiconductors. However, *bipolar transistor* is the correct term. It is sometimes referred to as a *bipolar junction transistor* or *BJT*.

OTHER RELATED COMPONENTS

- **unijunction transistor** (See Chapter 27)
- **field-effect transistor** (See Chapter 29)
- **diac** (Volume 2)
- **triac** (Volume 2)
- **relay** (Volume 2)
- **solid-state relay** (Volume 2)

What It Does

A bipolar transistor amplifies fluctuations in current or can be used to switch current on and off. In its amplifying mode, it replaced the *vacuum tubes* that were formerly used in the amplification of audio signals and many other applications. In its switching mode it resembles a **relay**, although in its "off" state the transistor still allows a very small amount of current flow, known as *leakage*.

A bipolar transistor is described as a *discrete* semiconductor device when it is individually packaged, with three leads or contacts. A package containing multiple transistors is an *integrated circuit*. A Darlington pair actually contains two transistors, but is included here as a discrete component because it is packaged similarly and functions like a single transistor. Most integrated circuits will be found in Volume 2 of this encyclopedia.

How It Works

Although the earliest transistors were fabricated from germanium, silicon has become the most commonly used material. Silicon behaves like an insulator, in its pure state at room temperature, but can be "doped" (carefully contaminated) with impurities that introduce a surplus of electrons unbonded from individual atoms. The result is an *N-type semiconductor* that can be induced to allow the movement of electrons through it, if it is *biased* with an external voltage. *Forward bias* means the application of a positive voltage, while *reverse bias* means reversing that voltage.

Other dopants can create a deficit of electrons, which can be thought of as a surplus of "holes" that can be filled by electrons. The result is a *P-type semiconductor*.

A bipolar NPN transistor consists of a thin central P-type layer sandwiched between two thicker N-type layers. The three layers are referred to as *collector*, *base*, and *emitter*, with a wire or contact attached to each of them. When a negative charge is applied to the emitter, electrons are forced by mutual repulsion toward the central base layer. If a forward bias (positive potential) is applied to the base, electrons will tend to be attracted out through the base. However, because the base layer is so thin, the electrons are now close to the collector. If the base voltage increases, the additional energy encourages the electrons to jump into the collector, from which they will make their way to the positive current source, which can be thought of as having an even greater deficit of electrons.

Thus, the emitter of an NPN bipolar transistor emits electrons into the transistor, while the collector collects them from the base and moves them out of the transistor. It is important to remember that since electrons carry a negative charge, the flow of electrons moves from negative to positive. The concept of positive-to-negative current is a fiction that exists only for historical reasons. Nevertheless, the arrow in a transistor schematic symbol points in the direction of conventional (positive-to-negative) current.

In a PNP transistor, a thin N-type layer is sandwiched between two thicker P-type layers, the base is negatively biased relative to the emitter, and the function of an NPN transistor is reversed, as the terms "emitter" and "collector" now refer to the movement of electron-holes rather than electrons. The collector is negative relative to the base, and the resulting positive-to-negative current flow moves from emitter to base to collector. The arrow in the schematic symbol for a PNP transistor still indicates the direction of positive current flow.

Symbols for NPN and PNP transistors are shown in Figure 28-1. The most common symbol for an NPN transistor is shown at top-left, with letters C, B, and E identifying collector, base, and emitter. In some schematics the circle in the symbols is omitted, as at top-right.

A PNP transistor is shown in the center. This is the most common orientation of the symbol, since its collector must be at a lower potential than its emitter, and ground (negative) is usually at the bottom of a schematic. At bottom, the PNP symbol is inverted, allowing the positions of emitter and collector to remain the same as in the symbol for the NPN transistor at the top. Other orientations of transistor symbols are often found, merely to facilitate simpler schematics with fewer conductor crossovers. The direction of the arrow in the symbol (pointing out or pointing in) always differentiates NPN from PNP transistors, respectively, and indicates current flowing from positive to negative.

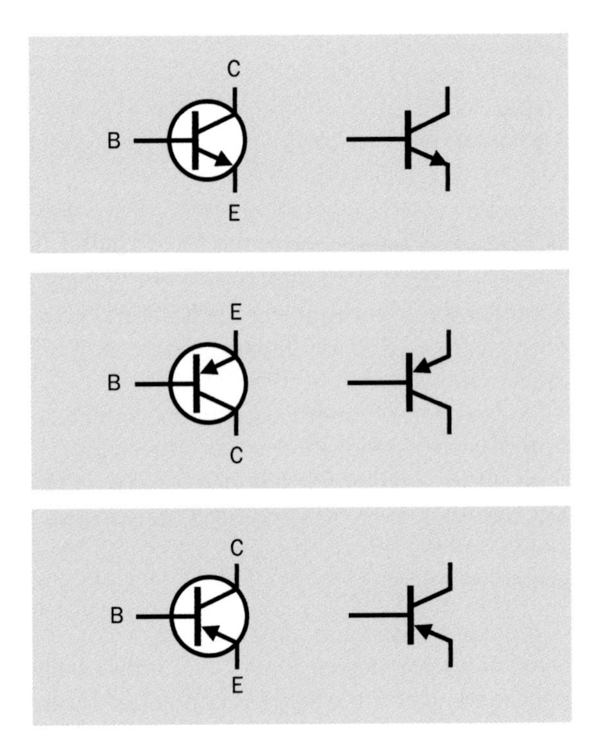

Figure 28-1. *Symbols for an NPN transistor (top) and a PNP transistor (center and bottom). Depending on the schematic in which the symbol appears, it may be rotated or inverted. The circle may be omitted, but the function of the component remains the same.*

NPN transistors are much more commonly used than PNP transistors. The PNP type was more difficult and expensive to manufacture initially, and circuit design evolved around the NPN type. In addition, NPN transistors enable faster switching, because electrons have greater mobility than electron-holes.

To remember the functions of the collector and the emitter in an NPN transistor, you may prefer to think in terms of the collector collecting positive current *into* the transistor, and the emitter emitting positive current *out of* the transistor. To remember that the emitter is always the terminal with an arrow attached to it (both in NPN and PNP schematic symbols), consider that "emitter" and "arrow" both begin with a vowel, while "base" and "collector" begin with consonants. To remember that an NPN transistor symbol has its arrow pointing outward, you can use the mnemonic "N/ever P/ointing i/N."

Current flow for NPN and PNP transistors is illustrated in Figure 28-2. At top-left, an NPN transistor passes no current (other than a small amount of leakage) from its collector to its emitter so long as its base is held at, or near, the potential of its emitter, which in this case is tied to negative or ground. At bottom-left, the purple positive symbol indicates that the base is now being held at a *relatively* positive voltage, at least 0.6 volts higher than the emitter (for a silicon-based transistor). This enables electrons to move from the emitter to the collector, in the direction of the blue arrows, while the red arrows indicate the conventional concept of current flowing from positive to negative. The smaller arrows indicate a smaller flow of current. A resistor is included to protect the transistor from excessive current, and can be thought of as the load in these circuits.

At top-right, a PNP transistor passes no current (other than a small amount of leakage) from its emitter to its collector so long as its base is held at, or near, the potential of the emitter, which in this case is tied to the positive power supply. At bottom-right, the purple negative symbol indicates that the base is now being held at a *relatively* negative voltage, at least 0.6 volts lower than the emitter. This enables electrons and current to flow as shown. Note that current flows into the base in the NPN transistor, but out from the base in the PNP transistor, to enable conductivity. In both diagrams, the resistor that would normally be included to protect the base has been omitted for the sake of simplicity.

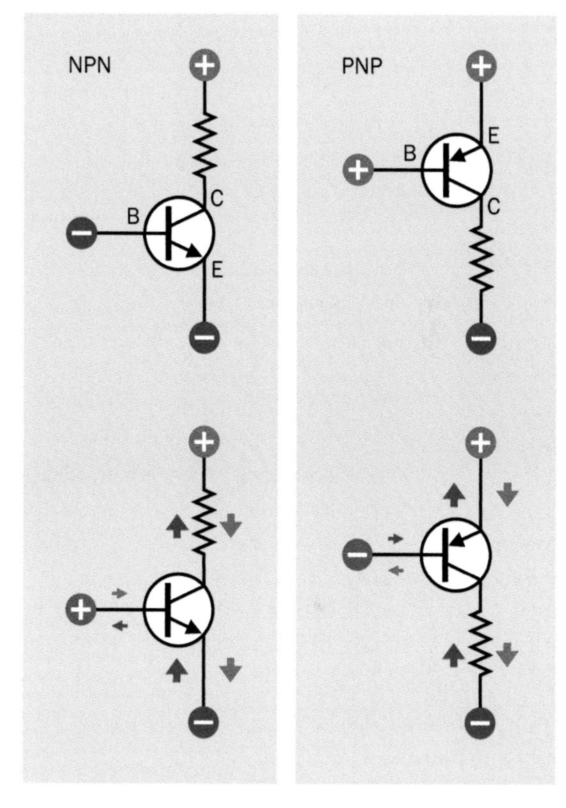

Figure 28-2. *Current flow through NPN and PNP transistors. See text for details.*

An NPN transistor amplifies its base current only so long as the positive potential applied to the collector is greater than the potential applied to the base, and the potential at the base must be greater than the potential at the emitter by at least 0.6 volts. So long as the transistor is biased in this way, and so long as the current values remain within the manufacturer's specified limits, a small fluctuation in current applied to the base will induce a much larger fluctuation in current between the collector and the emitter. This is why a transistor may be described as a *current amplifier*.

A *voltage divider* is often used to control the base potential and ensure that it remains less than the potential on the collector and greater than the potential at the emitter (in an NPN transistor). See Figure 28-3.

See Chapter 10 for additional information about the function of a voltage divider.

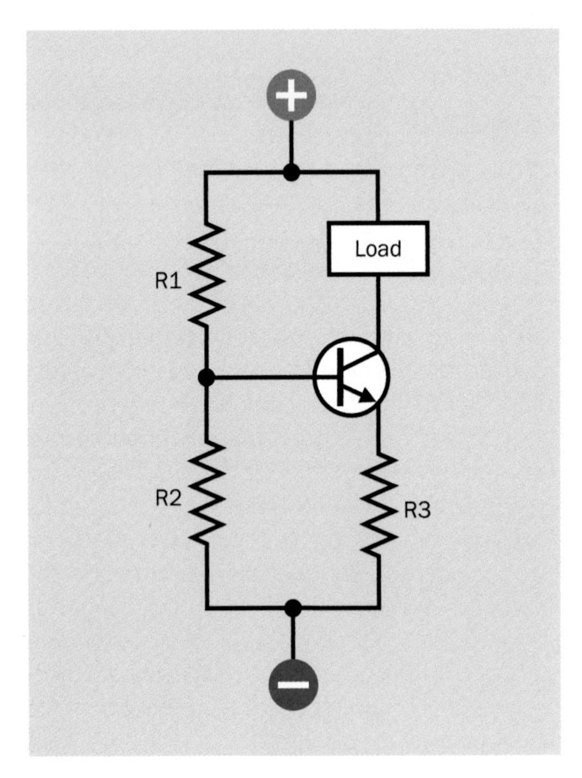

Figure 28-3. *Resistors R1 and R2 establish a voltage divider to apply acceptable bias to the base of an NPN transistor.*

Current Gain

The amplification of current by a transistor is known as its *current gain* or *beta value*, which can be expressed as the ratio of an increase in collector current divided by the increase in base current that enables it. Greek letter β is customarily used to represent this ratio. The formula looks like this:

$$\beta = \Delta I_c \,/\, \Delta I_b$$

where I_c is the collector current and I_b is the base current, and the Δ symbol represents a small change in the value of the variable that follows it.

Current gain is also represented by the term h_{FE}, where E is for the common Emitter, F is for Forward current, and lowercase letter h refers to the transistor as a "hybrid" device.

The beta value will always be greater than 1 and is often around 100, although it will vary from one type of transistor to another. It will also be affected by temperature, voltage applied to the transistor, collector current, and manufacturing inaccuracies. When the transistor is used outside of its design parameters, the formula to determine the beta value no longer directly applies.

There are only two connections at which current can enter an NPN transistor and one connection where it can leave. Therefore, if I_e is the current from the emitter, I_c is the current entering the collector, and I_b is the current entering the base:

$$I_e = I_c + I_b$$

If the potential applied to the base of an NPN transistor diminishes to the point where it is less than 0.6V above the potential at the emitter, the transistor will not conduct, and is in an "off" state, although a very small amount of leakage from collector to emitter will still occur.

When the current flowing into the base of the transistor rises to the point where the transistor cannot amplify the current any further, it becomes *saturated*, at which point its internal impedance has fallen to a minimal value. Theoretically this will allow a large flow of current; in practice, the transistor should be protected by resistors from being damaged by high current resulting from saturation.

Any transistor has maximum values for the collector current, base current, and the potential difference between collector and emitter. These values should be provided in a datasheet. Exceeding them is likely to damage the component.

Terminology

In its *saturated mode*, a transistor's base is saturated with electrons (with no room for more) and the internal impedance between collector and emitter drops as low as it can go.

The *cutoff mode* of an NPN transistor is the state where a low base voltage eliminates all current flow from collector to emitter other than a small amount of leakage.

The *active mode*, or *linear mode*, is the intermediate condition between cutoff and saturated, where the beta value or h_{FE} (ratio of collector current to base current) remains approximately constant. That is, the collector current is almost linearly proportional to the base current. This linear relationship breaks down when the transistor nears its saturation point.

Variants

Small signal transistors are defined as having a maximum collector current of 500 mA and maximum collector power dissipation of 1 watt. They can be used for audio amplification of low-level inputs and for switching of small currents. When determining whether a small-signal transistor can control an inductive load such as a motor or relay coil, bear in mind that the initial current surge will be greater than the rated current draw during sustained operation.

Small switching transistors have some overlap in specification with small signal transistors, but generally have a faster response time, lower beta value, and may be more limited in their tolerance for collector current. Check the manufacturer's datasheet for details.

High frequency transistors are primarily used in video amplifiers and oscillators, are physically small, and have a maximum frequency rating as high as 2,000 MHz.

Power transistors are defined as being capable of handling at least 1 watt, with upper limits that can be as high as 500 watts and 150 amps. They are physically larger than the other types, and may be used in the output stages of audio amplifiers, and in switching **power supplies** (see Chapter 16). Typically they have a much lower current gain than smaller transistors (20 or 30 as opposed to 100 or more).

Sample transistors are shown in Figure 28-4. Top: A 2N3055 NPN power transistor. This type was originally introduced in the late 1960s, and versions are still being manufactured. It is often found in power supplies and in push-pull power amplifiers, and has a total power dissipation rating of 115W. Second row, far left: general purpose switching-amplification PNP power transistor rated for up to 50W power dissipation. Second row, far right: A high-frequency switching transistor for use in lighting ballast, converters, inverters, switching regulators, and motor control systems. It tolerates relatively high voltages (up to 700V collector-emitter peak) and is rated for up to 80W total power dissipation. Second row, center-left and center-right: Two variants of the 2N2222 NPN small signal switching transistor, first introduced in the 1960s, and still very widely used. The metal can is the TO-19 package, capable of slightly higher power dissipation than the cheaper plastic TO-92 package (1.8W vs. 1.5W with a collector temperature no greater than 25 degrees Centigrade).

Packaging

Traditionally, small-signal transistors were packaged in small aluminum "cans" about 1/4" in diameter, and are still sometimes found in this form. More commonly they are embedded in buds of black plastic. Power transistors are packaged either in a rectangular module of black plastic with a metal back, or in a round metal "button." Both of these forms are designed to dissipate heat by being screw-clamped to a *heat sink*.

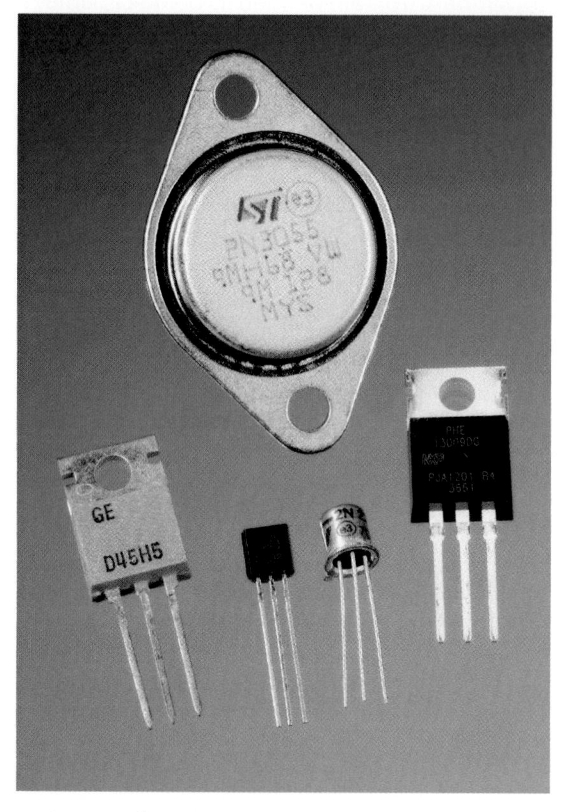

Figure 28-4. *Samples of commonly used transistors. See text for details.*

Connections

Often a transistor package provides no clue as to which lead is the emitter, which lead is the base, and which lead is the collector. Old can-style packaging includes a protruding tab that usually points toward the emitter, but not always. Where power transistors are packaged in a metal enclosure, it is typically connected internally with the collector. In the case of surface-mount transistors, look for a dot or marker that should identify the base of a bipolar transistor or the gate of a field-effect transistor.

A through-hole transistor usually has its part number printed or engraved on its package, although a magnifying glass may be necessary to see this. The component's datasheet may then be checked online. If a datasheet is unavailable, meter-testing will be necessary to confirm the functions of the three transistor leads. Some multimeters include a transistor-test function, which may validate the functionality of a transistor while also displaying its beta value. Otherwise, a meter can be put in diode-testing mode, and an unpowered NPN transistor should behave as if diodes are connected between its leads as shown in Figure 28-5. Where the identities of the transistor's leads are unknown, this test will be sufficient to identify the base, after which the collector and emitter may be determined empirically by testing the transistor in a simple low-voltage circuit such as that shown in Figure 28-6.

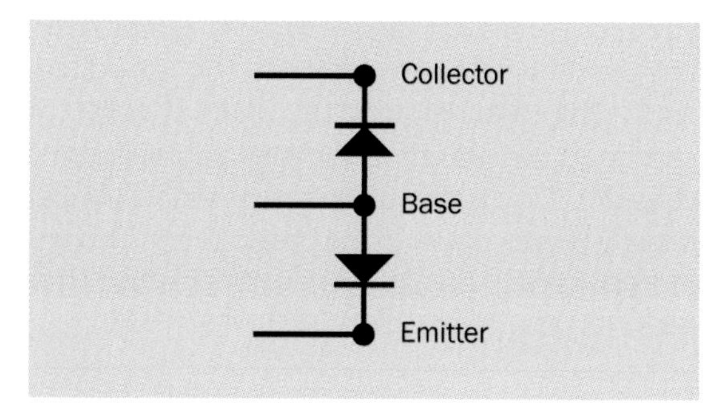

Figure 28-5. *An NPN transistor can behave as if it contains two diodes connected as shown. Where the functions of the leads of the transistor are unknown, the base can be identified by testing for conductivity.*

How to Use it

The following abbreviations and acronyms are common in transistor datasheets. Some or all of the letters following the initial letter are usually, but not always, formatted as subscripts:

h_{FE}
: Forward current gain

β
: Same as hFE

V_{CEO}
: Voltage between collector and emitter (no connection at base)

V_{CBO}
: Voltage between collector and base (no connection at emitter)

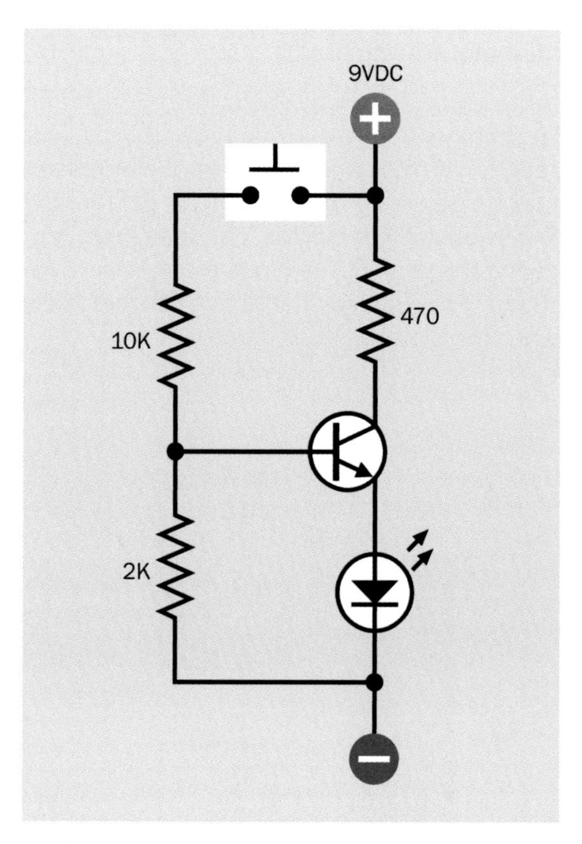

Figure 28-6. *This simple schematic can be used to breadboard-test a transistor empirically, determining its functionality and the identities of its collector and emitter leads.*

V_{EBO}
: Voltage between emitter and base (no connection at collector)

V_{CEsat}
: Saturation voltage between collector and emitter

V_{BEsat}
: Saturation voltage between base and emitter

I_c
: Current measured at collector

I_{CM}
: Maximum peak current at collector

I_{BM}
: Maximum peak current at base

P_{TOT}
: Total maximum power dissipation at room temperature

T_J
: Maximum junction temperature to avoid damage

Often these terms are used to define "absolute maximum values" for a component. If these maximums are exceeded, damage may occur.

A manufacturer's datasheet may include a graph showing the *safe operating area* (SOA) for a transistor. This is more common where power transistors are involved, as heat becomes more of an issue. The graph in Figure 28-7 has been adapted from a datasheet for a silicon diffused power transistor manufactured by Philips. The safe operating area is bounded at the top by a horizontal segment representing the maximum safe current, and at the right by a vertical segment representing the maximum safe voltage. However, the rectangular area enclosed by these lines is reduced by two diagonal segments representing the *total power limit* and the *second breakdown limit*. The latter refers to the tendency of a transistor to develop internal localized "hot spots" that tend to conduct more current, which makes them hotter, and able to conduct better—ultimately melting the silicon and causing a short circuit. The total power limit and the second breakdown limit reduce the safe operating area, which would otherwise be defined purely by maximum safe current and maximum safe voltage.

Uses for discrete transistors began to diminish when integrated circuits became cheaper and started to subsume multi-transistor circuits. For instance, an entire 5-watt audio amplifier, which used to be constructed from multiple components can now be bought on a chip, requiring just a few external capacitors. More powerful audio equipment typically uses integrated circuits to

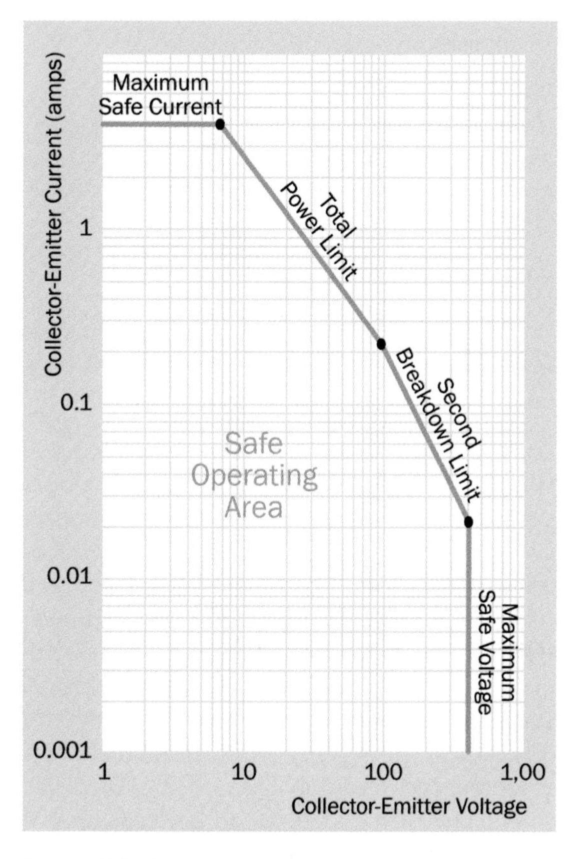

Figure 28-7. *Adapted from a Philips datasheet for a power transistor, this graph defines a safe operating area (SOA) for the component. See text for details.*

process inputs, but will use individual power transistors to handle high-wattage output.

Darlington Pairs

Discrete transistors are useful in situations where current amplification or switching is required at just one location in a circuit. An example would be where one output pin from a **microcontroller** must switch a small motor on and off. The motor may run on the same voltage as the microcontroller, but requires considerably more current than the typical 20mA maximum available from a microcontroller output. A *Darlington pair* of transistors may be used in this application. The overall gain of the pair can be 100,000 or more. See Figure 28-8. If a power source feeding through a potentiometer is substituted for the microcontroller chip, the circuit can function as a motor speed control (assuming that a generic DC motor is being used).

In the application shown here, the microcontroller chip must share a common ground (not shown) with the transistors. The optional resistor may be necessary to prevent leakage from the first transistor (when in its "off" state) from triggering the second. The diode protects the transistors from voltage transients that are likely when the motor stops and starts.

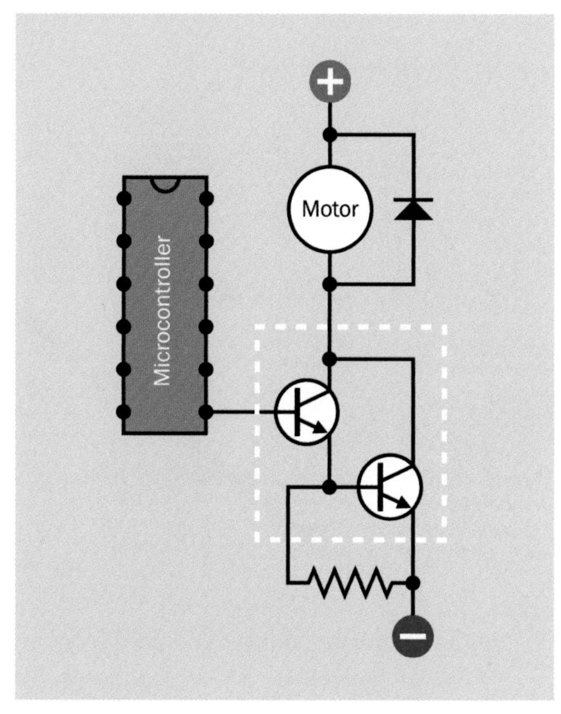

Figure 28-8. *Where the emitter of one NPN transistor is coupled to the base of another, they form a Darlington pair (identified by the dashed rectangle in this schematic). Multiplying the gain of the first transistor by the gain of the second gives the total gain of the pair.*

A Darlington pair can be obtained in a single transistor-like package, and may be represented by the schematic symbol shown in Figure 28-9.

Various through-hole Darlington packages are shown in Figure 28-10.

Seven or eight Darlington pairs can be obtained in a single integrated chip. Each transistor pair in these chips is typically rated at 500mA, but they

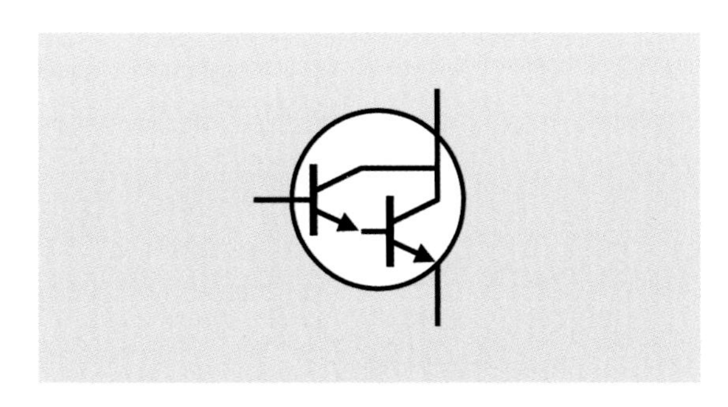

Figure 28-9. *When a Darlington pair is embedded in a single transistor-like package, it may be represented by this schematic symbol. The leads attached to the package can be used as if they are the emitter, base, and collector of a single NPN transistor.*

Figure 28-10. *Various packaging options for Darlington pairs. From left to right: The 2N6426 contains a Darlington pair rated to pass up to 500mA continuous collector current. The 2N6043 is rated for 8A continuous. The ULN2003 and ULN2083 chips contain seven and eight Darlington pairs, respectively.*

can be connected in parallel to allow higher currents. The chip usually includes protection diodes to allow it to drive inductive loads directly.

A typical schematic is shown in Figure 28-11. In this figure, the microcontroller connections are hypothetical and do not correspond with any actual chip. The Darlington chip is a ULN2003 or similar, containing seven transistor pairs, each with an "input" pin on the left and an "output" pin opposite it on the right. Any of pins 1 through 7 down the left side of the chip can be used to control a device connected to a pin on the opposite side.

A high input can be thought of as creating a negative output, although in reality the transistors inside the chip are sinking current via an external device—a motor, in this example. The device can have its own positive supply voltage, shown here as 12VDC, but must share a common ground with the microcontroller, or with any other component which is being used on the input side. The lower-right pin of the chip shares the 12VDC supply because this pin is attached internally to clamp diodes (one for each Darlington pair), which protect against surges caused by inductive loads. For this reason, the motor does not have a clamp diode around it in the schematic.

The Darlington chip does not have a separate pin for connection with positive supply voltage, because the transistors inside it are sinking power from the devices attached to it.

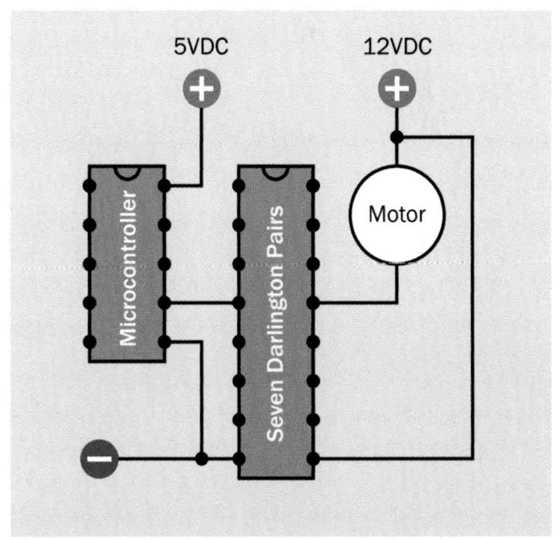

Figure 28-11. *A chip such as the ULN2003 contains seven Darlington pairs. It will sink current from the device it is driving. See text for details.*

A surface-mount Darlington pair is shown in Figure 28-12. This measures just slightly more than 0.1" long but is still rated for up to 500mA collector current or 250mW total power dissipation (at a component temperature no higher than 25 degrees Centigrade).

Figure 28-12. *A surface-mount package for a Darlington pair. Each square in the background grid measures 0.1". See text for additional details.*

Amplifiers

Two basic types of transistor amplifiers are shown in Figure 28-13 and Figure 28-14. The *common-collector* configuration has current gain but no voltage gain. The capacitor on the input side blocks DC current from entering the amplifier circuit, and the two resistors forming a voltage divider on the base of the transistor establish a voltage midpoint (known as the *quiescent point* or *operating point*) from which the signal to be amplified may deviate above and below.

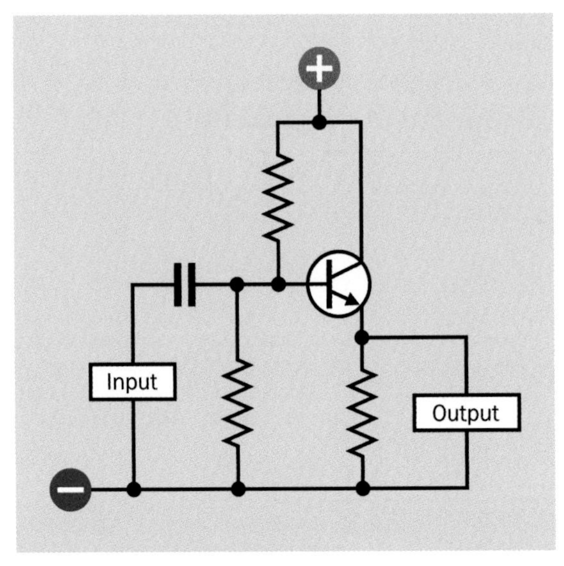

Figure 28-13. *The basic schematic for a common-collector amplifier. See text for details.*

The *common-emitter* amplifier provides voltage gain instead of current gain, but inverts the phase of the input signal. Additional discussion of amplifier design is outside the scope of this encyclopedia.

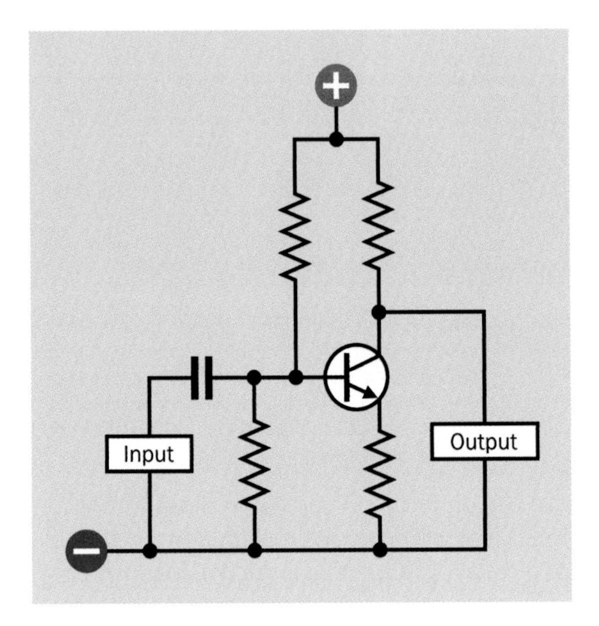

Figure 28-14. *The basic schematic for a common-emitter amplifier. See text for details.*

In switching applications, modern transistors have been developed to handle a lot of current compared with earlier versions, but still have some limitations. Few power transistors can handle more than 50A flowing from collector to emitter, and 1,000V is typically a maximum value. Electromechanical relays continue to exist because they retain some advantages, as shown in the table in Figure 28-15, which compares switching capabilities of transistors, **solid-state relays**, and electromechanical **relays**.

	Transistor	Solid-State Relay	Electro-Mechanical Relay
Long-term reliability	Excellent	Excellent	Limited
NC contacts?	No	Yes	Yes
DT contacts?	No	No	Yes
Ability to switch large current	Limited	Some	Good
Ability to switch AC	No	Yes	Yes
Can be triggered by AC voltage?	No	Yes	Yes
Suitability for miniaturization	Excellent	Poor	Poor
Vulnerable to heat	Yes	Yes	Not very
Vulnerable to corrosion	No	No	Yes
OK at high speed	Excellent	Good	Poor
Price advantage for low-voltage low-current	Yes	No	No
Price advantage for high-voltage high-current	No	No	Yes
Current leakage when "off"	Yes	Yes	No
Trigger circuit isolated from switched circuit	No	Yes	Yes

Figure 28-15. *A comparison of characteristics of switching devices.*

What Can Go Wrong

Wrong Connections on a Bipolar Transistor

Failing to identify a transistor's leads or contacts correctly can obviously be a potential source of damage, but swapping the collector and emitter accidentally will not necessarily destroy the transistor. Because of the inherent symmetry of the device, it will in fact function with collector and emitter connections reversed. Rohm, a large semiconductor manufacturer, has included this scenario in its general information pages and concludes that the primary indicator of transposed connections is that the β value, or h_{FE} , drops to about 1/10th of specification. If you are using a transistor that works but provides much less amplification than you expect, check that the emitter and collector leads are not transposed.

Wrong Connections on a Darlington Pair Chip

While a single-component package for a Darlington pair functions almost indistinguishably from a single transistor, multiple Darlington pairs in a DIP package may create confusion because the component behaves differently from most other chips, such as logic chips.

A frequent error is to ground the output device instead of applying positive power to it. See Figure 28-11 and imagine an erroneous connection of negative power instead of the 12VDC positive power.

Additional confusion may be caused by reading a manufacturer's datasheet for a Darlington pair chip such as the ULN2003. The datasheet depicts the internal function of the chip as if it contains logic inverters. While the chip can be imagined as behaving this way, in fact it contains bipolar transistors that amplify the current applied to the base of each pair. The datasheet also typically will not show the positive connection that should be made to the common-diode pin (usually at bottom-right), to provide protection from surges caused by inductive loads. This pin must be distinguished carefully from the common-ground pin (usually at bottom-left). The positive connection to the common-diode pin is optional; the common-ground connection is mandatory.

Soldering Damage

Like any semiconductor, transistors are vulnerable to heat and can be damaged while soldering, although this seldom happens if a low-wattage iron is used. A copper alligator clip can be applied as a heat sink to each lead before it is soldered.

Excessive Current or Voltage

During use, a transistor will be damaged if it is subjected to current or voltage outside of its rated range. Passing current through a transistor without any series resistance to protect it will almost certainly burn it out, and the same thing can happen if incorrect resistor values are used.

The maximum wattage that a transistor can dissipate will be shown in its datasheet. Suppose, for example, this figure is 200mW, and you are using a 12VDC supply. Ignoring the base current, the maximum collector current will be 200 / 12 = approximately 15mA. If the transistor's emitter is connected to ground, and the load applied to the transistor output has a high impedance, and if we ignore the transresistance, Ohm's Law suggests that a resistor that you place between the collector and the supply voltage should have a resistance of at least 12 / 0.015 = 800 ohms.

When transistors are used in switching applications, it is customary for the base current to be 1/5th of the collector current. In the example discussed here, a 4.7K resistor might be appropriate. A meter should be used to verify actual current and voltage values.

Excessive Leakage

In a Darlington pair, or any other configuration where the output from one transistor is connected with the base of another, leakage from the first transistor while in its "off" state can be amplified by the second transistor. If this is unacceptable, a bypass resistor can be used to divert some of the leakage from the base of the second transistor to ground. Of course the resistor will also steal some of the base current when the first transistor is active, but the resistor value is typically chosen so that it takes no more than 10% of the active current. See Figure 28-8 for an example of a bypass resistor added to a Darlington pair.

field effect transistor

29

The term **field-effect transistor** encompasses a family primarily consisting of the *junction field-effect transistor* (or *JFET*, which is the simplest generic type) and the *metal-oxide semiconductor field-effect transistor* (or *MOSFET*, also sometimes known as an *insulated-gate field-effect transistor*, or *IGFET*). Because the principles of operation overlap considerably, the entire -FET family is grouped in this entry.

OTHER RELATED COMPONENTS

- **bipolar transitor** (See Chapter 28)
- **unijunction transistor** (See Chapter 27)
- **diode** (See Chapter 26)

What It Does

A field-effect transistor creates an electric field to control current flowing through a channel in a semiconductor. MOSFETs of microscopic size form the basis of complementary metal oxide semiconductor (CMOS) *integrated circuit chips*, while large discrete MOSFETs are capable of switching substantial currents, in lamp dimmers, audio amplifiers, and motor controllers. FETs have become indispensable in computer electronics.

A **bipolar transistor** is generally thought of as a *current amplifier* because the current passing through it is controlled by a smaller amount of current passing through the base. By contrast, all FETs are considered to be *voltage amplifiers*, as the control voltage establishes field intensity, which requires little or no current. The negligible leakage through the gate of an FET makes it ideal for use in low-power applications such as portable hand-held devices.

How It Works

This section is divided into two subsections, describing the most widely used FETs: JFETs and MOSFETs.

JFETs

A *junction field-effect transistor* (or *JFET*) is the simplest form of FET. Just as a bipolar transistor can be of NPN or PNP type, a JFET can have an *N-channel* or *P-channel*, depending whether the channel that transmits current through the device is negatively or positively *doped*. A detailed explanation of semiconductor doping will be found in the **bipolar transistor** entry.

Because negative charges have greater mobility, the N-channel type allows faster switching and is more commonly used than the P-channel type. A schematic symbol for it is shown in Figure 29-1 alongside the schematic for an NPN transistor. These symbols suggest the similarity of the devices as amplifiers or switches, but it is important to remember that the FET is a primarily a voltage amplifier while the bipolar transistor is a current amplifier.

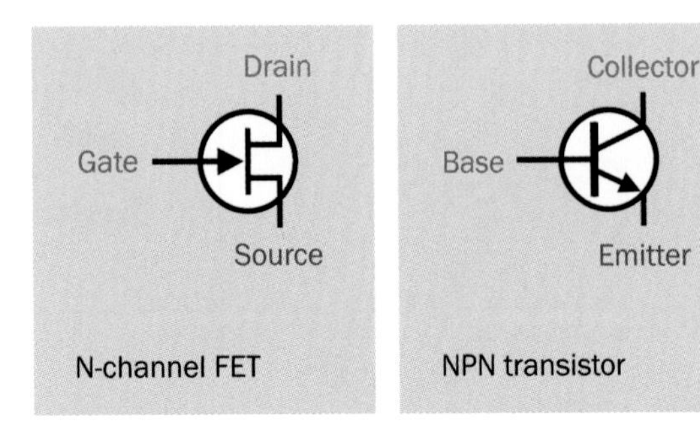

Figure 29-1. *A comparison between schematic symbols for N-channel JFET (left) and NPN bipolar transistor (right) suggests their functional similarity as switches or amplifiers, although their behavior is markedly different.*

Three JFETs are shown in Figure 29-2. The N-channel J112 type is supplied by several manufacturers, the figure showing two samples, one from Fairchild Semiconductor (left) and the other from On Semiconductor (right). Although the full part numbers are different, the specifications are almost identical, including a drain-gate voltage of 35V, a drain-source voltage of 35V, and a gate current of 50mA. The metal-clad 2N4392 in the center has similar values but is three times the price, with a much higher power dissipation of 1.8W, compared with 300mW and 350mW for the other two transistors respectively.

Figure 29-2. *Junction Field Effect Transistors (JFETs). See text for details.*

Schematic symbols for N-channel and P-channel JFETs are shown in Figure 29-3, N-channel being on the left while P-channel is on the right. The upper-left and lower-left symbol variants are both widely used and are functionally identical. The upper-right and lower-right variants likewise mean the same thing. Because the upper variants are symmetrical, an S should be added to clarify which terminal is the source. In practice, the S is often omitted, allowing some ambiguity. While the source and drain of some JFETs are in fact interchangeable, this does not apply to all types.

The circle around each symbol is occasionally omitted when representing discrete components, and is almost always omitted when multiple FETs are shown connected to form an integrated circuit.

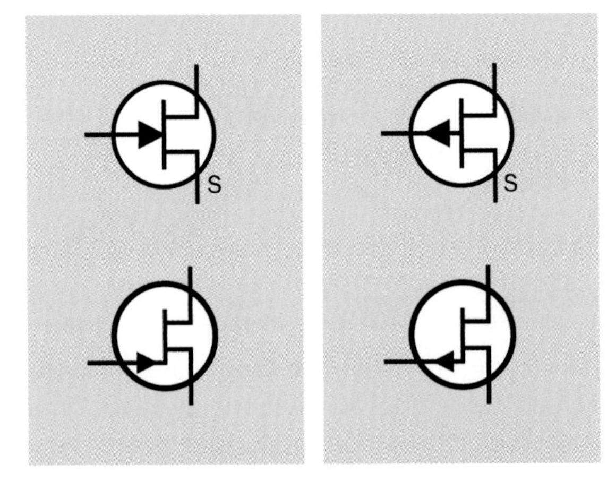

Figure 29-3. *Schematic symbols for junction field-effect transistors (JFETs). Left: N-channel. Right: P-channel. The symbols at top and bottom on each side are functionally identical. Circles may be omitted. Letter S may be omitted from the symmetrical symbol variants, even though this creates some ambiguity.*

The internal function of an N-channel JFET is shown diagrammatically in Figure 29-4. In this component, the source terminal is a source of electrons, flowing relatively freely through the N-doped channel and emerging through the drain. Thus, conventional current flows nonintuitively from the drain terminal, to the source terminal, which will be of lower potential.

The JFET is like a normally-closed switch. It has a low resistance so long as the gate is at the same potential as the source. However, if the potential of the gate is reduced below the potential of the source—that is, the gate acquires a more relatively negative voltage than the source—the current flow is *pinched off* as a result of the field created by the gate. This is suggested by the lower diagram in Figure 29-4.

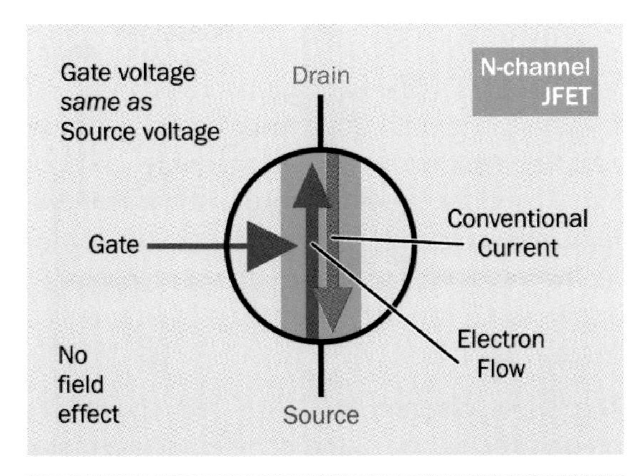

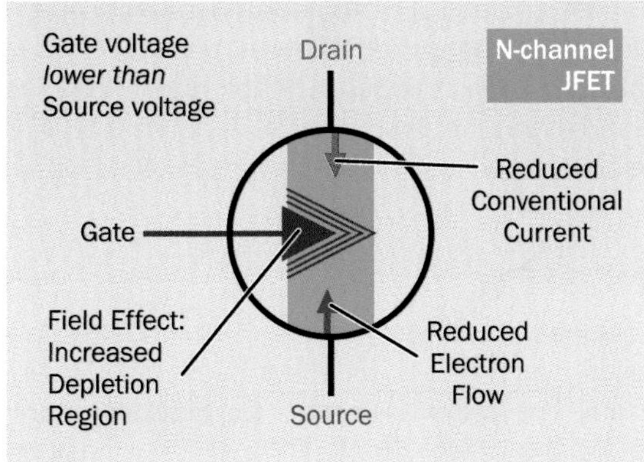

Figure 29-4. *At top, conventional current flows freely from drain to source through the channel of an N-doped JFET. At bottom, the lowered voltage of the gate relative to the source creates a field effect that pinches off the flow of current.*

The situation for a P-channel JFET is reversed, as shown in Figure 29-5. The source is now positive (but is still referred to as the source), while the drain can be grounded. Conventional current now flows freely from source to drain, so long as the gate is at the same positive potential as the source. If the gate voltage rises above the source voltage, the flow is pinched off.

A bipolar transistor tends to block current flow by default, but becomes less resistive when its base is forward-biased. Therefore it can be referred to as an *enhancement device*. By contrast,

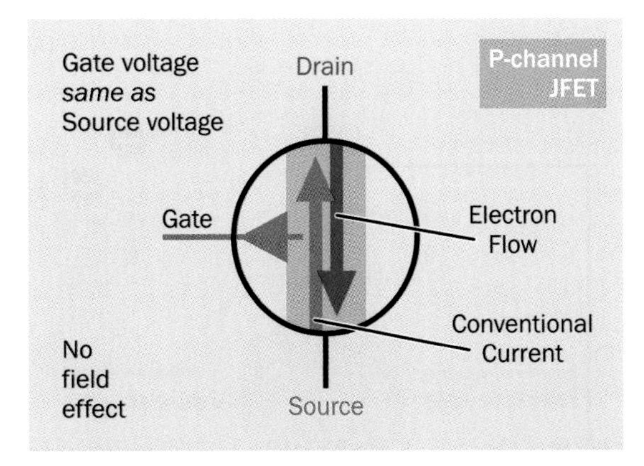

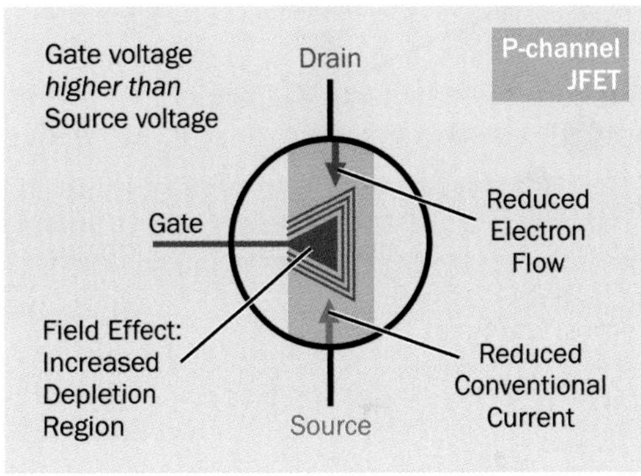

Figure 29-5. *At top, conventional current flows freely from source to drain through the channel of a P-doped JFET. At bottom, the higher voltage of the gate relative to the source creates a field effect that pinches off the flow of current.*

an N-channel JFET allows current to flow by default, and becomes more resistive when its base is reverse-biased, which widens the *depletion layer* at the base junction. Consequently it can be referred to as a *depletion device*.

The primary characteristicts of a junction field-effect transistor relative to an NPN bipolar transistor are summarized in the table in Figure 29-6.

JFET Behavior

The voltage difference between gate and source of a JFET is usually referred to as V_{gs} while the voltage difference between drain and source is referred to as V_{ds}.

	NPN Bipolar Transistor	N-Channel JFET
Type of amplifier	Current	Voltage
Active bias	Positive	Negative
Unbiased state	Nonconductive	Conductive
Biased state	More conductive	Less conductive

Figure 29-6. *This table contrasts the characteristics of an N-channel JFET with those of an NPN bipolar transistor.*

Suppose the gate of an N-channel JFET is connected with the source, so that $V_{gs}=0$. Now if V_{ds} increases, the current flowing through the channel of the JFET also increases, approximately linearly with V_{ds}. In other words, initially the JFET behaves like a low-value resistor in which the voltage across it, divided by the amperage flowing through it, is approximately constant. This phase of the JFET's behavior is known as its *ohmic region*. While the unbiased resistance of the channel in a JFET depends on the component type, it is generally somewhere between 10Ω and 1K.

If V_{ds} increases still further, eventually no additional flow of current occurs. At this point the channel has become *saturated*, and this plateau zone is referred to as the *saturation region*, often abbreviated I_{dss}, meaning "the saturated drain current at zero bias." Although this is a nearly constant value for any particular JFET, it may vary somewhat from one sample of a component to another, as a result of manufacturing variations.

If V_{ds} continues to increase, the component finally enters a *breakdown* state, sometimes referred to by its full formal terminology as *drain-source breakdown*. The current passing through the JFET will now be limited only by capabilities of the external power source. This breakdown state can be destructive to the component, and is comparable to the breakdown state of a typical **diode**.

What if the voltage at the gate is reduced below the voltage at the source—such as V_{gs} becomes negative? In its ohmic region, the component now behaves as if it has a higher resistance, and it will reach its saturation region at a lower current value (although around the same value for V_{ds}). Therefore, by reducing the voltage on the gate relative to the voltage at the source, the effective resistance of the component increases, and in fact it can behave as a *voltage-controlled resistor*.

The upper diagram in Figure 29-7 shows this graphically. Below it, the corresponding graph for a P-channel JFET looks almost identical, except that the current flow is reversed and is pinched off as the gate voltage rises above the source voltage. Also, the breakdown region is reached more quickly with a P-channel JFET than with an N-channel JFET.

MOSFETs

MOSFETs have become one of the most widely used components in electronics, everywhere from computer memory to high-amperage switching power supplies. The name is an acronym for *metal-oxide semiconductor field-effect transistor*. A simplified cross-section of an N-channel MOSFET is shown in Figure 29-8.

Two MOSFETs are shown in Figure 29-9.

Like a JFET, a MOSFET has three terminals, identified as drain, gate, and source, and it functions by creating a field effect that controls current flowing through a channel. (Some MOSFETS have a fourth terminal, described later). However, it has a metal source and drain making contact with each end of the channel (hence the term "metal" in its acronym) and also has a thin layer of silicon dioxide (hence the term "oxide" in its acronym) separating the gate from the channel, thus raising the impedance at the gate to at least 100,000 gigaohms and reducing gate current essentially to zero. The high gate impedance of a

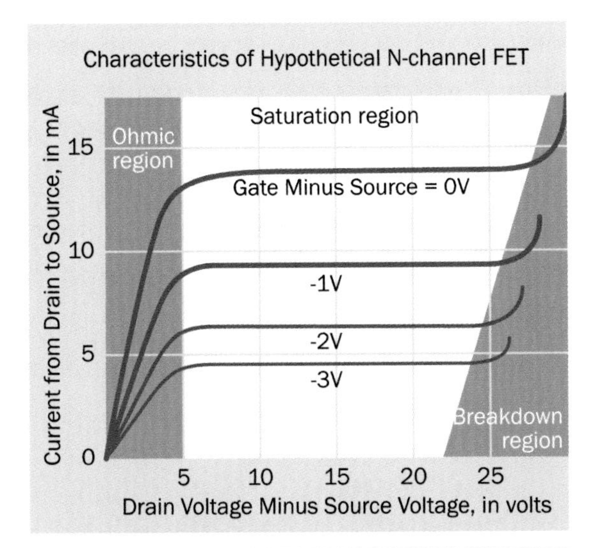

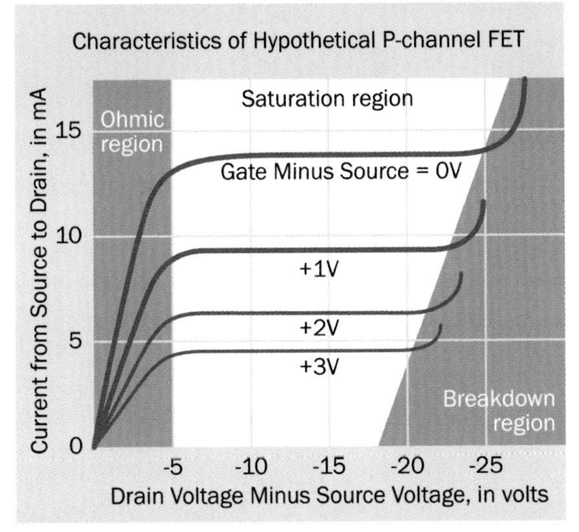

Figure 29-7. *The top graph shows current passing through the channel of an N-channel JFET, depending on gate voltage and source voltage. The lower graph is for a P-channel JFET.*

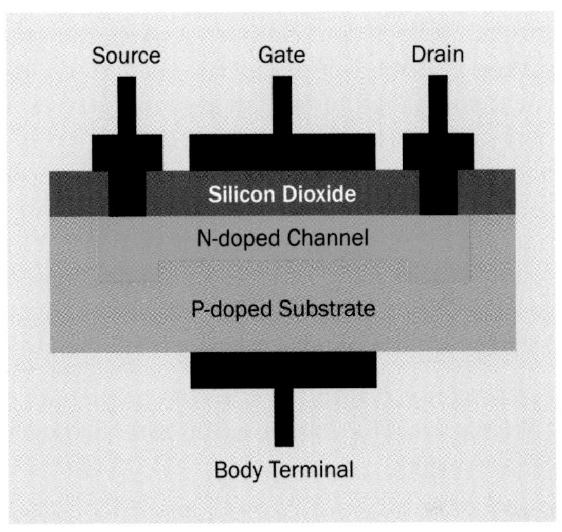

Figure 29-8. *Simplified diagram of an N-channel MOSFET. The thickness of the silicon dioxide layer has been greatly exaggerated for clarity. The black terminals are metallic.*

Figure 29-9. *Two MOSFETs. At left, the TO-220 package claims a drain current of up to 65A continuous, and a drain-to-source breakdown voltage 100V. At right, the smaller package offers a drain current of 175mA continuous, and a drain-to-source breakdown voltage of 300V.*

MOSFET allows it to be connected directly to the output of a digital integrated circuit. The layer of silicon dioxide is a *dielectric*, meaning that a field appled to one side creates an opposite field on the other side. The gate attached to the surface of the layer functions in the same way as one plate of a capacitor.

The silicon dioxide also has the highly desirable property of insulating the gate from the channel, thus preventing unwanted reverse current. In a JFET, which lacks a dielectric layer, if source voltage is allowed to rise more than about 0.6V higher than gate voltage, the direct internal connection between gate and channel allows negative charges to flow freely from source to gate, and as the internal resistance will be very low, the resulting current can be destructive. This is why the JFET must always be reverse-biased.

A MOSFET is freed from these restrictions, and the gate voltage can be higher or lower than the source voltage. This property enables an N-channel MOSFET to be designed not only as a depletion device, but alternatively as an *enhancement device*, which is "normally off" and can be switched on by being forward-biased. The primary difference is the extent to which the channel in the MOSFET is N-doped with charge carriers, and therefore will or will not conduct without some help from the gate bias.

In a depletion device, the channel conducts, but applying negative voltage to the gate can pinch off the current.

In an enhancement device, the channel does not conduct, but applying positive voltage to the gate can make it start to do so.

In either case, a shift of bias from negative to positive encourages channel conduction; the depletion and enhancement versions simply start from different points.

This is clarified in Figure 29-10. The vertical (logarithmic) scale suggests the current being conducted through the channel of the MOSFET, while the green curve describes the behavior of a depletion version of the device. Where this curve crosses the center line representing 0 volts bias, the channel is naturally conductive, like a JFET. Moving left down the curve, as reverse bias is applied (shown on the horizontal axis), the component becomes less conductive until finally its conductivity reaches zero.

Meanwhile on the same graph, the orange curve represents an enhancement MOSFET, which is nonconductive at 0 volts bias. As forward bias increases, the current also increases—similar to a bipolar transistor.

To make things more confusing, a MOSFET, like a JFET, can have a P-doped channel; and once again it can function in depletion or enhancement mode. The behavior of this variant is shown in Figure 29-11. As before, the green curve shows the behavior of a depletion MOSFET, while the orange curve refers to the enhancement version. The horizontal axis now shows the voltage difference between the gate and the drain terminal. The depletion component is naturally conductive at zero bias, until the gate voltage increases above the drain voltage, pinching off the current flow. The enhancement component is not conductive until reverse bias is applied.

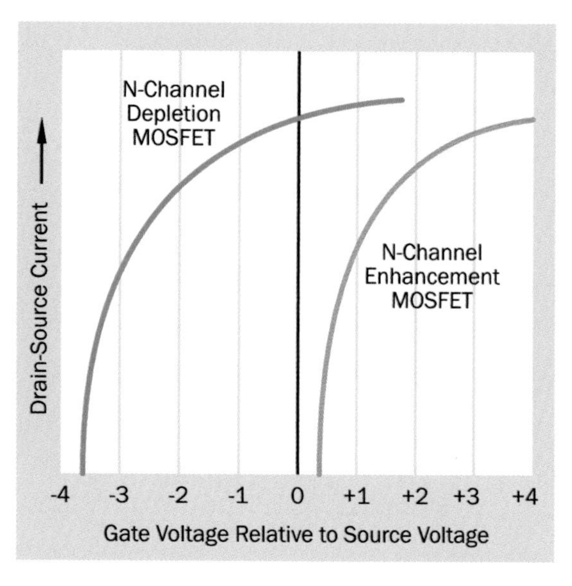

Figure 29-10. *The current conduction of depletion and enhancement N-channel MOSFETs. See text for details. (Influenced by The Art of Electronics by Horowitz and Hill.)*

Figure 29-12 shows schematic symbols that represent depletion MOSFETs. The two symbols on the left are functionally identical, representing N-channel versions, while the two symbols on the right represent P-channel versions. As in the case of JFETs, the letter "S" should be (but often is not) added to the symmetrical versions of the symbols, to clarify which is the source terminal. The left-pointing arrow identifies the components as N-channel, while in the symbols on the right, the right-pointing arrows indicate P-channel MOSFETs. The gap between the two vertical lines in each symbol suggests the silicon dioxide dielectric. The right-hand vertical line represents the channel.

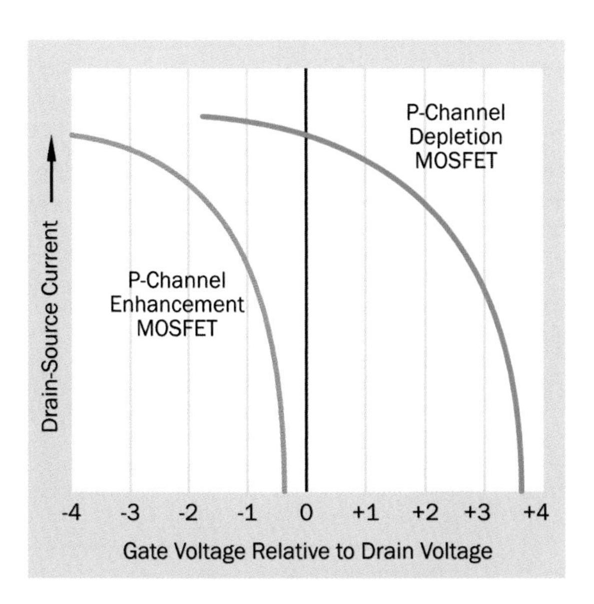

Figure 29-11. *The current conduction of depletion and enhancement P-channel MOSFETs. See text for details.*

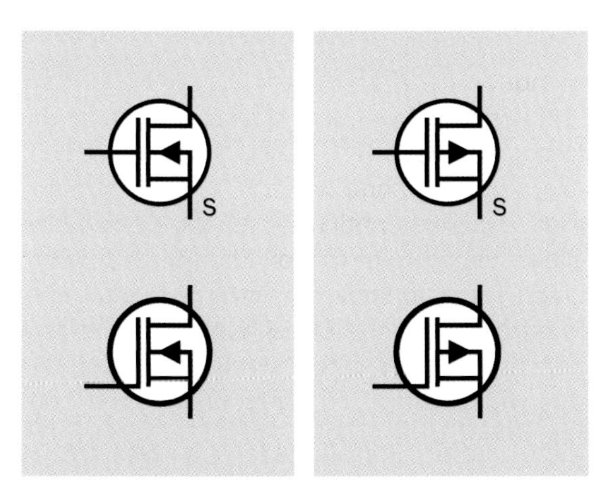

Figure 29-12. *Schematic symbols for depletion MOSFETs. These function similarly to JFETs. The two symbols on the left are functionally identical, and represent N-channel depletion MOSFETs. The two symbols on the right are both widely used to represent P-channel depletion MOSFETs.*

For enhancement MOSFETs, a slightly different symbol uses a broken line between the source and drain (as shown in Figure 29-13) to remind us that these components are "normally off" when zero-biased, instead of "normally on." Here again a left-pointing arrow represents an N-channel MOSFET, while a right-pointing arrow represents a P-channel MOSFET.

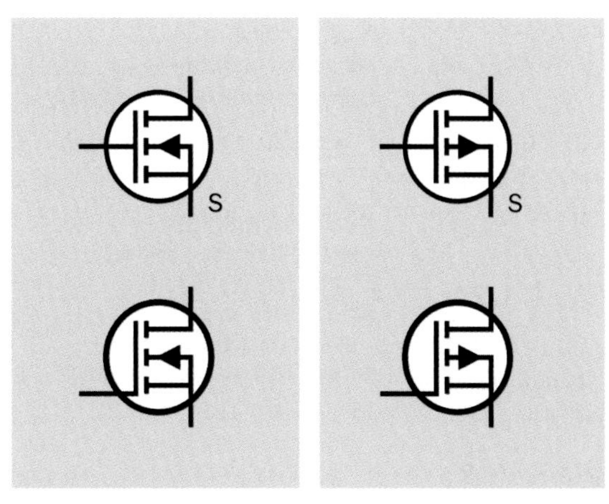

Figure 29-13. *Schematic symbols for enhancement MOSFETs. The two on the left are functionally identical, and represent N-channel enhancement MOSFETs. The two on the right represent P-channel enhancement MOSFETs.*

Because there is so much room for confusion regarding MOSFETs, a summary is presented in Figure 29-14 and Figure 29-15. In these figures, the relevant parts of each schematic symbol are shown disassembled alongside text explaining their meaning. Either of the symbols in Figure 29-14 can be superimposed on either of the symbols in Figure 29-15, to combine their functions. So, for instance, if the upper symbol in Figure 29-14 is superimposed on the lower symbol in Figure 29-15, we get an N-channel MOSFET of the enhancement type.

N-channel MOSFET
Source relatively negative
Drain relatively positive
Gate relatively negative
changes conductivity as gate becomes more negative

P-channel MOSFET
Source relatively positive
Drain relatively negative
Gate relatively positive
changes conductivity as gate becomes more positive

Figure 29-14. *Either of the two symbols can be combined with either of the two symbols in the next figure, to create one of the four symbols for a MOSFET. See text for details.*

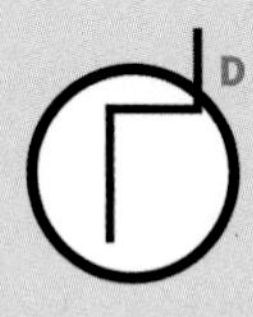

Depletion MOSFET
normally ON
Change of gate voltage (more negative for N-channel, more positive for P-channel) pinches off the current flow.

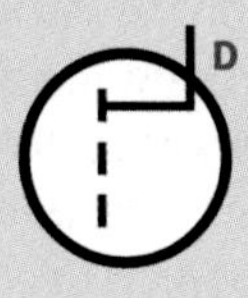

Enhancement MOSFET
Normally OFF
Change of gate voltage (less positive for N-channel, less negative for P-channel) pinches off the current flow.

Figure 29-15. *Either of the two symbols can be combined with either of the two symbols from the previous figure, to create one of the four symbols for a MOSFET. See text for details.*

In an additional attempt to clarify MOSFET behavior, four graphs are provided in Figure 29-16, Figure 29-17, Figure 29-18, and Figure 29-19. Like JFETs, MOSFETs have an initial ohmic region, followed by a saturation region where current flows relatively freely through the device. The gate-to-source voltage will determine how much flow is permitted. However, it is important to pay close attention to the graph scales, which differ for each of the four types of MOSFET.

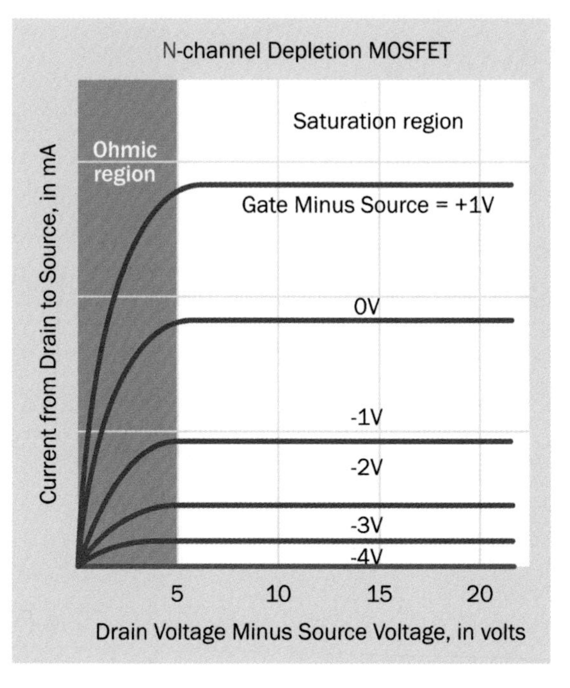

Figure 29-16. *Current flow through a depletion-type, N-channel MOSFET.*

In all of these graphs, a bias voltage exists, which allows zero current to flow (represented by the graph line superimposed on the horizontal axis). In other words, the MOSFET can operate as a switch. The actual voltages where this occurs will vary with the particular component under consideration.

The N-channel, enhancement-type MOSFET is especially useful as a switch because in its normally-off state (with zero bias) it presents a very high resistance to current flow. It requires a relatively low positive voltage at the gate, and effectively no gate current, to begin conducting conventional current from its drain terminal to its source terminal. Thus it can be driven directly by typical 5-volt logic chips.

Depletion-type MOSFETs are now less commonly used than the enhancement-type.

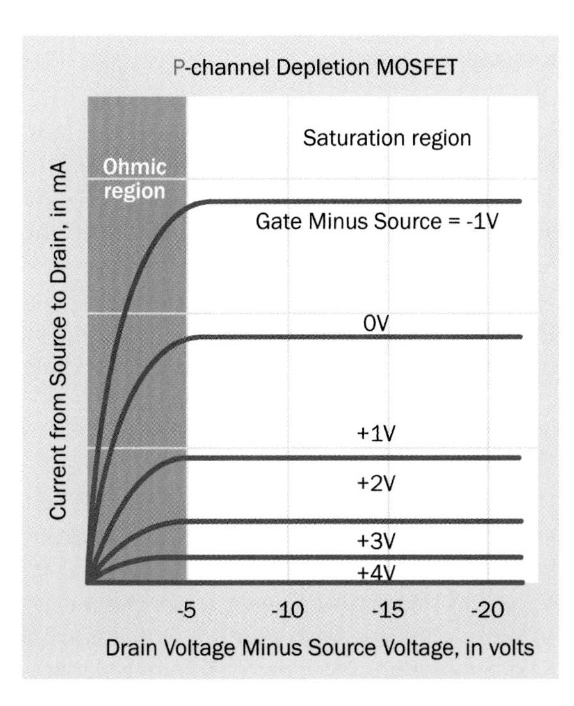

Figure 29-17. *Current flow through a depletion-type, P-channel MOSFET.*

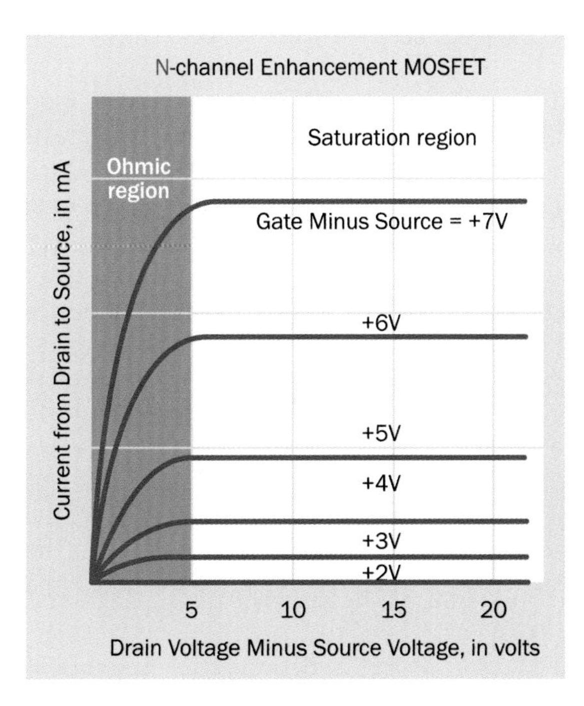

Figure 29-18. *Current flow through an enhancement-type, N-channel MOSFET.*

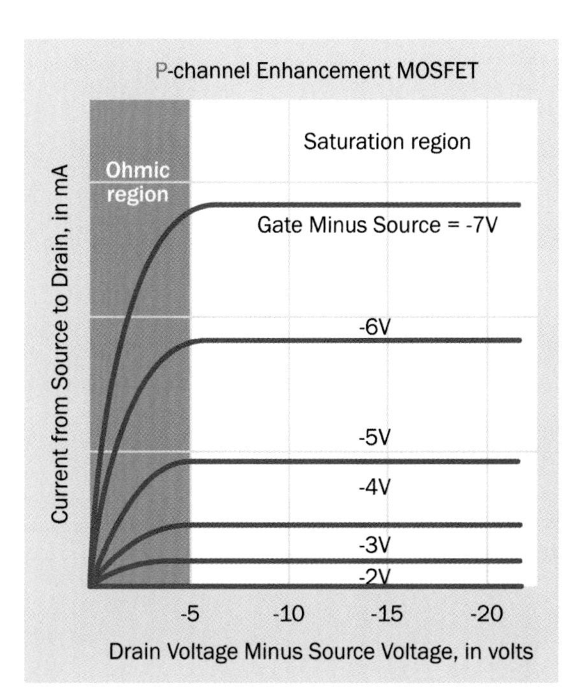

Figure 29-19. *Current flow through an enhancement-type, P-channel MOSFET.*

The Substrate Connection

Up to this point, nothing has been said about a fourth connection available on many MOSFETs, known as the *body terminal*. This is connected to the substrate on which the rest of the component is mounted, and acts as a diode junction with the channel. It is typically shorted to the source terminal, and in fact this is indicated by the schematic symbols that have been used so far. It is possible, however, to use the body terminal to shift the threshold gate voltage of the MOSFET, either by making the body terminal more negative than the source terminal (in an N-channel MOSFET) or more positive (in a P-channel MOSFET). Variants of the MOSFET schematic symbols showing the body terminal are shown in Figure 29-20 (for depletion MOSFETS) and Figure 29-21 (for enhancement MOSFETS).

A detailed discussion of the use of the body terminal to adjust characteristics of the gate is beyond the scope of this encyclopedia.

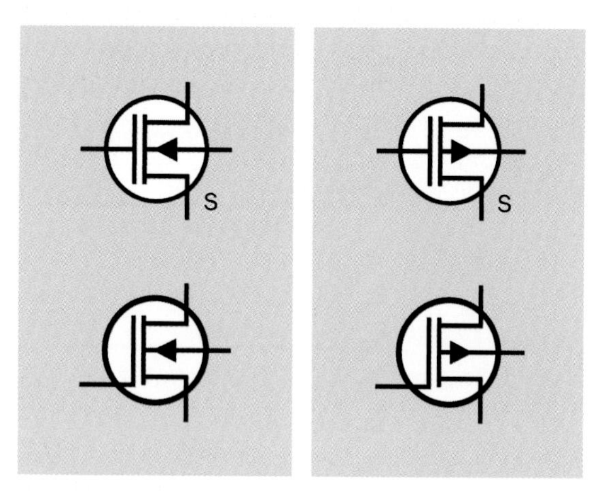

Figure 29-20. *Schematic symbol variants for depletion MOSFETs, showing the body terminal separately accessible instead of being tied to the source terminal.*

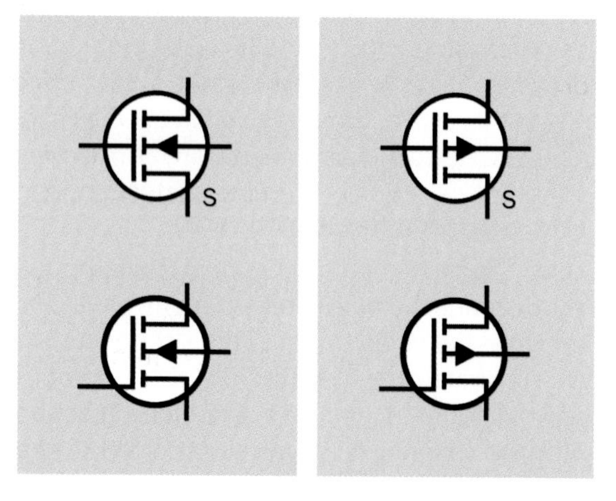

Figure 29-21. *Schematic symbol variants for enhancement MOSFETs, showing the body terminal separately accessible instead of being tied to the source terminal.*

Variants

A few FET variants exist in addition to the two previously discussed.

MESFET

The acronym stands for *MEtal-Semiconductor Field Effect Transistor*. This FET variant is fabricated from gallium arsenide and is used primarily in radio frequency amplification, which is outside the scope of this encyclopedia.

V-Channel MOSFET

Whereas most FET devices are capable of handling only small currents, the *V-channel MOSFET* (which is often abbreviated as a *VMOS FET* and has a V-shaped channel as its name implies) is capable of sustained currents of at least 50A and voltages as high as 1,000V. It is able to pass the high current because its channel resistance is well under 1Ω. These devices, commonly referred to as *power MOSFETs*, are available from all primary semiconductor manufacturers and are commonly used in switching power supplies.

Trench MOS

The TrenchMOS or Trenchgate MOS is a MOSFET variant that encourages current to flow vertically rather than horizontally, and includes other innovations that enable an even lower channel resistance, allowing high currents with minimal heat generation. This device is finding applications in the automobile industry as a replacement for electromechanical **relays**.

Values

The maximum values for JFETs, commonly found listed in datasheets, will specify V_{ds} (the drain-source voltage, meaning the potential difference between drain and source); V_{dg} (the drain-gate voltage, meaning the potential difference between drain and gate); V_{gsr} (the reverse gate-source voltage); gate current; and total device dissipation in mW. Note that the voltage differences are relative, not absolute. Thus a voltage of 50V on the drain and 25V on the source might be acceptable in a component with a V_{ds} of 25V. Similarly, while a JFET's "pinch-off" effect begins as the gate becomes "more negative" than the source, this can be achieved if, for example, the source has a potential of 6V and the gate has a potential of 3V.

JFETs and MOSFETs designed for low-current switching applications have a typical channel resistance of just a few ohms, and a maximum switching speed around 10Mhz.

The datasheet for a MOSFET will typically include values such as gate threshold voltage, which may be abbreviated V_{gs} (or V_{th}) and establishes the relative voltage at which the gate starts to play an active role; and the maximum on-state drain current, which may be abbreviated $I_{d(on)}$ and establishes the maximum limiting current (usually at 25 degree Centigrade) between source and gate.

How to Use it

The combination of a very high gate impedance, very low noise, very low quiescent power consumption in its off state, and very fast switching capability makes the MOSFET suitable for many applications.

P-Channel Disadvantage

P-channel MOSFETs are generally less popular than N-channel MOSFETS because of the higher resistivity of P-type silicon, resulting from its lower carrier mobility, putting it at a relative disadvantage.

Bipolar Substitution

In many instances, an appropriate enhancement-type MOSFET can be substituted for a bipolar transistor with better results (lower noise, faster action, much higher impedance, and probably less power dissipation).

Amplifier Front Ends

While MOSFETs are well-suited for use in the front end of an audio amplifier, chips containing MOSFETs are now available for this specific purpose.

Voltage-Controlled Resistor

A simple voltage-controlled resistor can be built around a JFET or MOSFET, so long as its performance remains limited to the linear or ohmic region.

Compatibility with Digital Devices

A JFET may commonly use power supplies in the range of 25VDC. However, it can accept the high/low output from a 5V digital device to control its gate. A 4.7K pullup resistor is an appropriate value to be used if the FET is to be used in conjunction with a TTL digital chip that may have a voltage swing of only approximately 2.5V between its low and high thresholds.

What Can Go Wrong

Static Electricity

Because the gate of a MOSFET is insulated from the rest of the component, and functions much like a plate of a capacitor, it is especially likely to accumulate static electricity. This static charge may then discharge itself into the body of the component, destroying it. A MOSFET is particularly vulnerable to electrostatic discharge because its oxide layer is so thin. Special care should be taken either when handling the component, or when it is in use. Always touch a grounded object or wear a grounded wrist band when handling MOSFETs, and be sure that any circuit using MOSFETs includes appropriate protection from static and voltage spikes.

A MOSFET should not be inserted or removed while the circuit in which it performs is switched on or contains residual voltage from undischarged capacitors.

Heat

Failure because of overheating is of special concern when using power MOSFETs. A Vishay Application Note ("Current Power Rating of Power Semiconductors") suggests that this kind of component is unlikely to operate at less than 90 degrees Centigrade in real-world conditions, yet the power handling capability listed in a datasheet usually assumes an industry standard of 25 degrees Centigrade.

On the other hand, ratings for continuous power are of little relevance to switching devices that have duty cycles well below 100%. Other factors also play a part, such as the possibility of power surges, the switching frequency, and the integrity of the connection between the component and its heat sink. The heat sink itself creates uncertainty by tending to average the temperature of the component, and of course there is no simple way to know the actual junction temperature, moment by moment, inside a MOSFET.

Bearing in mind the accumulation of unknown factors, power MOSFETs should be chosen on an extremely conservative basis. According to a tutorial in the *EE Times*, actual current switched by a MOSFET should be less than half of its rated current at 25 degrees, while one-fourth to one-third are common. Figure 29-22 shows the real-world recommended maximum drain current at various temperatures. Exceeding this recommendation can create additional heat, which cannot be dissipated, leading to further accumulation of heat, and a thermal runaway condition, causing eventual failure of the component.

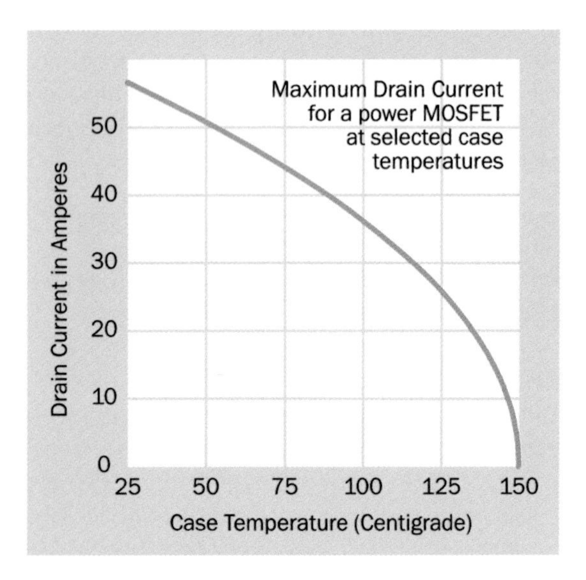

Figure 29-22. *Maximum advised drain current through a power MOSFET, related to case temperature of the component. Derived from EE Times Power MOSFET Tutorial.*

Wrong Bias

As previously noted, applying forward bias to a JFET can result in the junction between the gate and the source starting to behave like a forward-biased diode, when the voltage at the gate is greater than the voltage at the source by approximately 0.6V or more (in an N-channel JFET). The junction will present relatively little resistance, encouraging excessive current and destructive consequences. It is important to design devices that allow user input in such a way that user error can never result in this eventuality.

Schematic Symbols

This section contains a compilation of schematic symbols for components that have been described in this volume. They are sequenced primarily in alphabetical order, as this section is intended for use as an index. However, symbols that have a strong similarity are grouped together; thus **potentiometer** is found adjacent to **resistor**, and all types of transistors are in the same group.

The symbol variants shown in each blue rectangle are functionally identical.

Where a component has mandatory polarity or is commonly used with a certain polarity, a red plus sign (+) has been added for guidance. This sign is not part of the symbol. In the case of polarized capacitors, where a plus sign is normally shown (or should be shown) with the symbol, the plus sign is a part of the symbol and appears in black.

This is not intended to be an exhaustive compilation of symbol variants. Some uncommon ones may not be here. However, the list should be sufficient to enable identification of components in this volume.

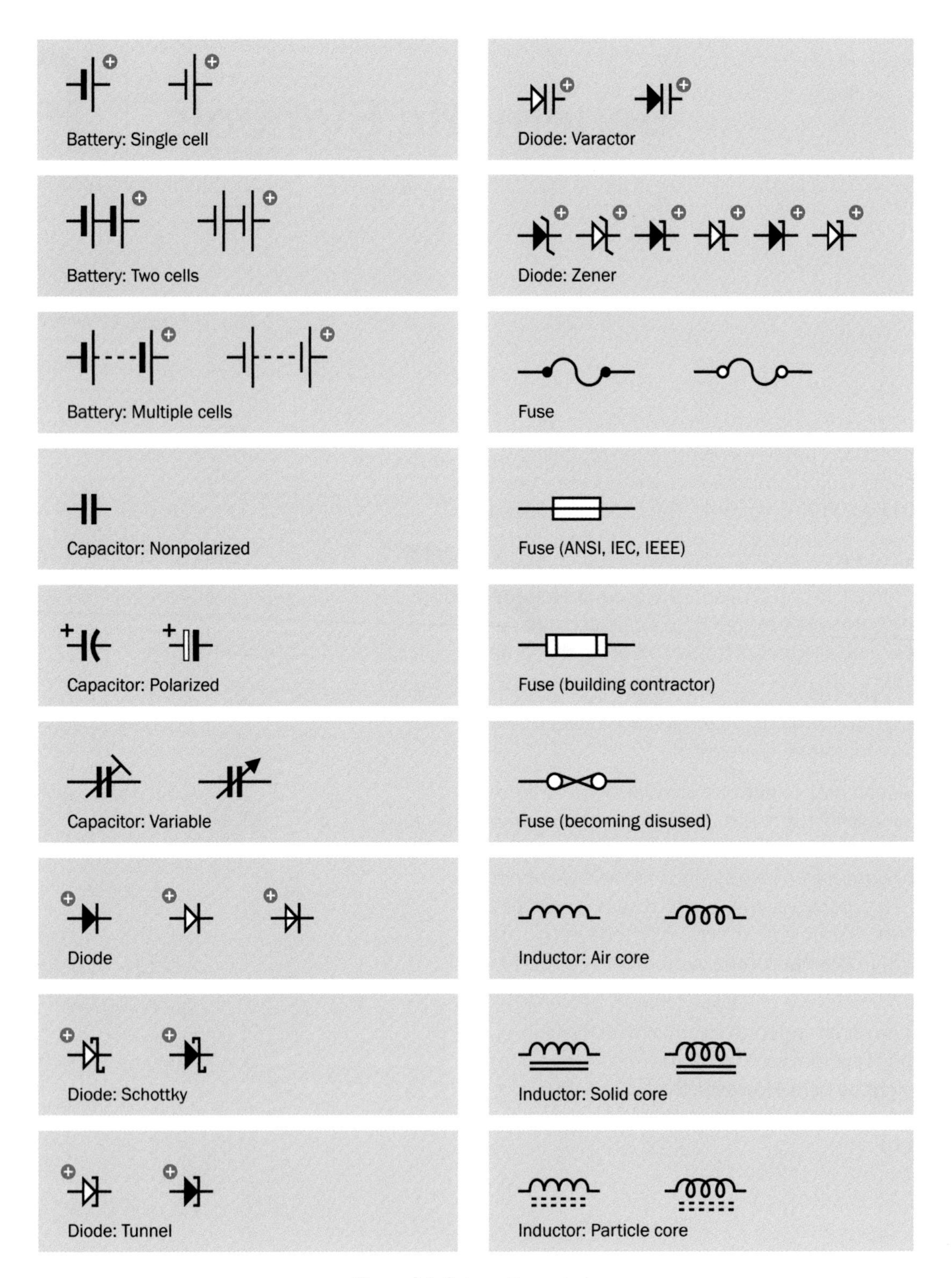

Figure A-1. *Schematic symbols*

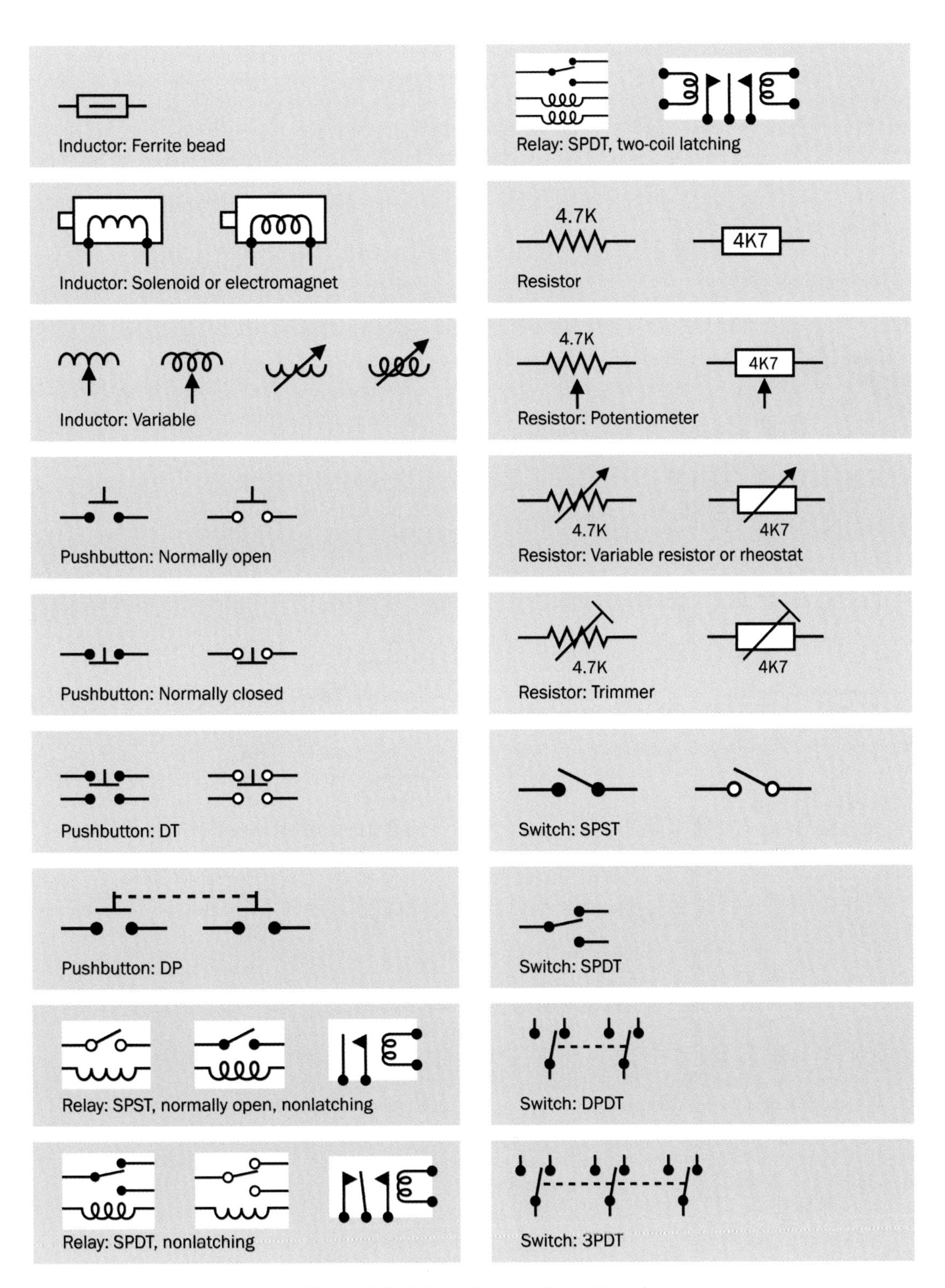

Figure A-2. *Schematic symbols, continued*

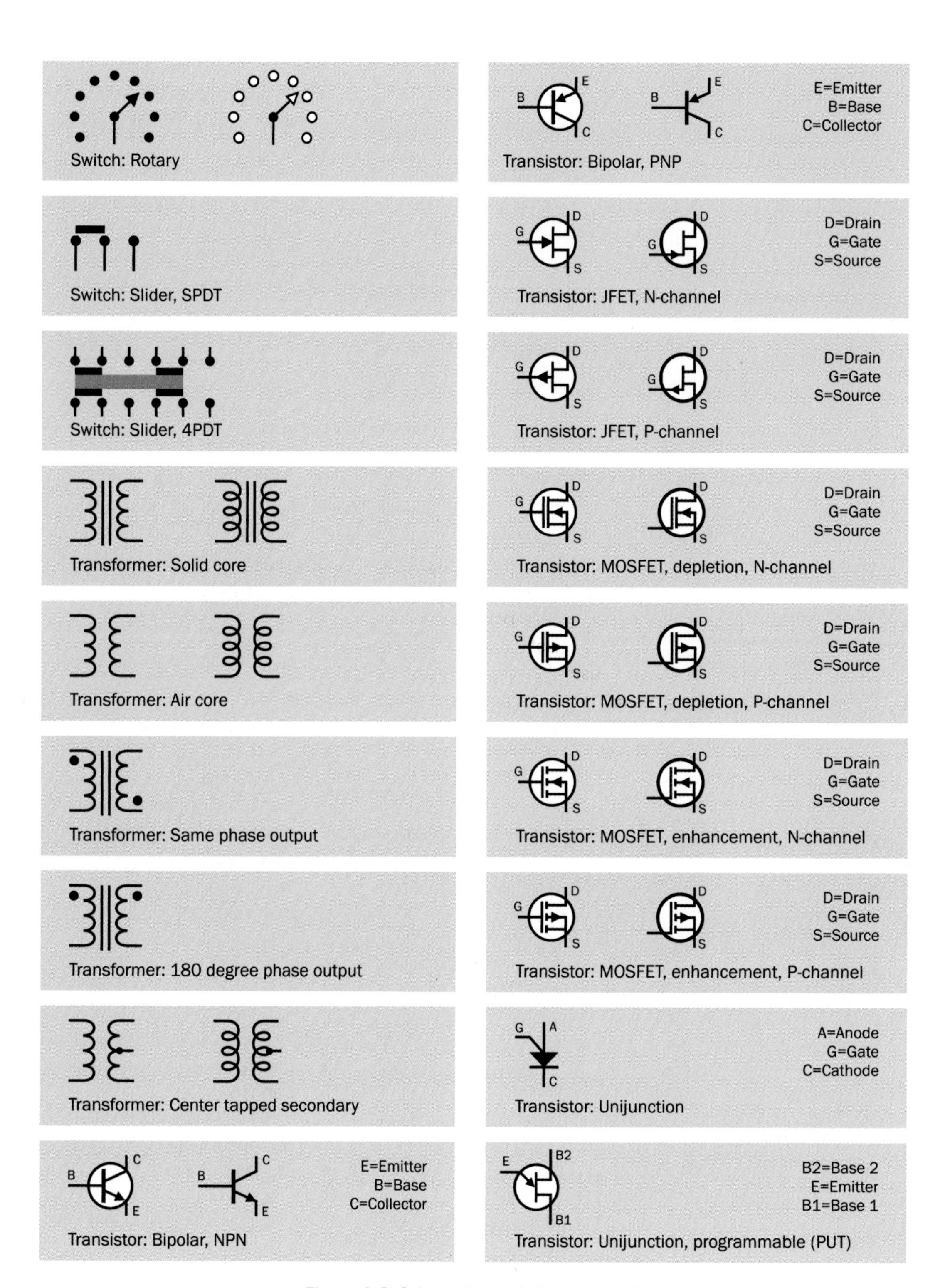

Figure A-3. *Schematic symbols, continued*

Index

Symbols

A

B

We'd like to hear your suggestions for improving our indexes. Send email to index@oreilly.com.

C

D

E

F

G

H

I

J

K

L

M

N

O

P

Q

R

S

T

U

V

W

Z